Advances in Intelligent Systems and Computing

Volume 244

Series Editor

Janusz Kacprzyk, Warsaw, Poland

For further volumes:
http://www.springer.com/series/11156

Van-Nam Huynh · Thierry Denœux
Dang Hung Tran · Anh Cuong Le
Son Bao Pham
Editors

Knowledge and Systems Engineering

Proceedings of the Fifth International Conference KSE 2013, Volume 1

Springer

Editors
Van-Nam Huynh
School of Knowledge Science
Japan Advanced Institute of Science and Technology
Ishikawa
Japan

Thierry Denœux
Universite de Technologie de Compiegne
Compiegne Cedex
France

Dang Hung Tran
Faculty of Information Technology
Hanoi National University of Education
Hanoi
Vietnam

Anh Cuong Le
Faculty of Information Technology
University of Engineering and Technology - VNU Hanoi
Hanoi
Vietnam

Son Bao Pham
Faculty of Information Technology
University of Engineering and Technology - VNU Hanoi
Hanoi
Vietnam

ISSN 2194-5357 ISSN 2194-5365 (electronic)
ISBN 978-3-319-02740-1 ISBN 978-3-319-02741-8 (eBook)
DOI 10.1007/978-3-319-02741-8
Springer Cham Heidelberg New York Dordrecht London

Library of Congress Control Number: 2013950936

Printed on acid-free paper

Springer is part of Springer Science+Business Media (www.springer.com)

Preface

This volume contains papers presented at the Fifth International Conference on Knowledge and Systems Engineering (KSE 2013), which was held in Hanoi, Vietnam, during 17–19 October, 2013. The conference was jointly organized by Hanoi National University of Education and the University of Engineering and Technology, Vietnam National University. The principal aim of KSE Conference is to bring together researchers, academics, practitioners and students in order to not only share research results and practical applications but also to foster collaboration in research and education in Knowledge and Systems Engineering.

This year we received a total of 124 submissions. Each of which was peer reviewed by at least two members of the Program Committee. Finally, 68 papers were chosen for presentation at KSE 2013 and publication in the proceedings. Besides the main track, the conference featured six special sessions focusing on specific topics of interest as well as included one workshop, two tutorials and three invited speeches. The kind cooperation of Yasuo Kudo, Tetsuya Murai, Yasunori Endo, Sadaaki Miyamoto, Akira Shimazu, Minh L. Nguyen, Tzung-Pei Hong, Bay Vo, Bac H. Le, Benjamin Quost, Sébastien Destercke, Marie-Hélène Abel, Claude Moulin, Marie-Christine Ho Ba Tho, Sabine Bensamoun, Tien-Tuan Dao, Lam Thu Bui and Tran Dinh Khang in organizing these special sessions and workshop is highly appreciated.

As a follow-up of the Conference, two special issues of the Journal of *Data & Knowledge Engineering* and *International Journal of Approximate Reasoning* will be organized to publish a small number of extended papers selected from the Conference as well as other relevant contributions received in response to subsequent calls. These journal submissions will go through a fresh round of reviews in accordance with the journals' guidelines.

We would like to express our appreciation to all the members of the Program Committee for their support and cooperation in this publication. We would also like to thank Janusz Kacprzyk (Series Editor) and Thomas Ditzinger (Senior Editor, Engineering/Applied Sciences) for their support and cooperation in this publication.

Last, but not the least, we wish to thank all the authors and participants for their contributions and fruitful discussions that made this conference a success.

Hanoi, Vietnam
October 2013

Van-Nam Huynh
Thierry Denœux
Dang Hung Tran
Anh Cuong Le
Son Bao Pham

Organization

Honorary Chairs

Van Minh Nguyen – Hanoi National University of Education, Vietnam
Ngoc Binh Nguyen – VNU University of Engineering and Technology, Vietnam

General Chairs

Cam Ha Ho – Hanoi National University of Education, Vietnam
Anh Cuong Le – VNU University of Engineering and Technology, Vietnam

Program Chairs

Van-Nam Huynh – Japan Advanced Institute of Science and Technology, Japan
Thierry Denœux – Université de Technologie de Compiègne, France
Dang Hung Tran – Hanoi National University of Education, Vietnam

Program Committee

Akira Shimazu, Japan
Azeddine Beghdadi, France
Son Bao Pham, Vietnam
Benjamin Quost, France
Bernadette Bouchon-Meunier, France
Binh Thanh Huynh, Vietnam
Bay Vo, Vietnam
Cao H, Tru, Vietnam
Churn-Jung Liau, Taiwan
Dinh Dien, Vietnam
Claude Moulin, France
Cuong Nguyen, Vietnam
Dritan Nace, France
Duc Tran, USA
Duc Dung Nguyen, Vietnam
Enrique Herrera-Viedma, Spain
Gabriele Kern-Isberner, Germany
Hiromitsu Hattori, Japan
Hoang Truong, Vietnam
Hung V. Dang, Vietnam
Hung Son Nguyen, Poland
Jean Daniel Zucker, France

Jérôme Lang, France
Jing Liu, China
Jiuyong Li, Australia
Jonathan Lawry, UK
Kenji Satou, Japan
Lam T. Bui, Vietnam
Bac H. Le, Vietnam
Loannis Parissis, France
Marie-Helene Abel, France
Martin Steffen, Norway
Masahiro Inuiguchi, Japan
Michel Riveill, France
Mina Ryoke, Japan
Minh-Dung Phan, Thailand
Mitsuru Ikeda, Japan
Minh L. Nguyen, Japan
Noboru Takagi, Japan
Peter Whigham, New Zealand
Phayung Meesad, Thailand
Quang-Huy Nguyen, France
Quang Uy Nguyen, Ireland
Sabine Bensamoun, France
Sadaaki Miyamoto, Japan
Serge Stinckwich, France
Sébastien Destercke, France
Si Quang Le, UK
Son Doan, USA
Tien-Tuan Dao, France
Tetsuya Murai, Japan
Thanh Binh Nguyen, Vietnam
Thanh Tri Nguyen, Vietnam
Thanh-Thuy Nguyen, Vietnam
The Duy Bui, Vietnam
The Loc Nguyen, Vietnam
Thomas Huynh, USA
Tho Hoan Pham, Vietnam
Thepchai Supnithi, Thailand
The Dung Luong, Vietnam
Tran Dinh Khang, Vietnam
Tsutomu Fujinami, Japan
Tzung-Pei Hong, Taiwan
Vladik Kreinovich, USA
Xiaoshan Li, Macau
Xuan Hoai Nguyen, Vietnam
Xuan-Hieu Phan, Vietnam
Yasuo Kudo, Japan

Contents

Part I
Keynote Addresses

What Ontological Engineering Can Do for Solving Real-World Problems

Riichiro Mizoguchi

Abstract. Ontological engineering works as a theory of content and/or content technology. It provides us with conceptual tools for analyzing problems in a right way by which we mean analysis of underlying background of the problems as well as their essential properties to obtain more general and useful solutions. It also suggests that we should investigate the problems as deeply as possible like philosophers to reveal essential and intrinsic characteristics hidden in the superficial phenomena/appearance. Knowledge is necessarily something about existing entities and their relations, and ontology is an investigation of being, and hence ontology contributes to facilitation of our knowledge about the world in an essential manner.

There exist a lot of problems to be solved in the real world. People tend to solve them immediately after they realize needs to solve them. One of the issues here is that necessary consideration about the nature of those problems is often skipped to get solutions quickly,which sometimes leads to ad-hoc solutions and/or non-optimal solutions. This is why ontological engineering can make a reasonable contribution to improving such situations.

In my talk, after a brief introduction to ontological engineering, I explain technological aspects of ontological engineering referring to my experiences. One of the important conceptual techniques is separation of what and how in procedures/algorithms. Then, I show you a couple of concrete examples of deployment of such conceptual tools in several domains.

Riichiro Mizoguchi
Research Center for Service Science, Japan Advanced Institute of Science and Technology, 1-1 Asahidai, Nomi, Ishikawa, Japan

V.-N. Huynh et al. (eds.), *Knowledge and Systems Engineering, Volume 1*, Advances in Intelligent Systems and Computing 244,
DOI: 10.1007/978-3-319-02741-8_1,

Argumentation for Practical Reasoning

Phan Minh Dung

Abstract. We first present a short introduction illustrating how argumentation could be viewed as an universal mechanism humans use in their practical reasoning where by practical reasoning we mean both commonsense reasoning and reasoning by experts as well as their integration. We then present logic-based argumentation employing implicit or explicit assumptions. Logic alone is not enough for practical reasoning as it can not deal with quantitative uncertainties. We explain how probabilities could be integrated with argumentation to provide an integrated framework for jury-based (or collective multiagent) dispute resolution.

Phan Minh Dung
Department of Computer Science and Information Management,
Asian Institute of Technology, Thailand

V.-N. Huynh et al. (eds.), *Knowledge and Systems Engineering, Volume 1*,
Advances in Intelligent Systems and Computing 244,
DOI: 10.1007/978-3-319-02741-8_2,

Legal Engineering and Its Natural Language Processing

Akira Shimazu and Minh Le Nguyen

Abstract. Our society is regulated by a lot of laws which are related mutually. When we view a society as a system, laws can be viewed as the specifications for the society. Such a system-oriented aspect of laws have not been studied well so far. In the upcoming e-Society, laws have more important roles in order to achieve a trustworthy society and we expect a methodology which treats a system-oriented aspect of laws. Legal Engineering is the new field that studies the methodology and applies information science, software engineering and artificial intelligence to laws in order to support legislation and to implement laws using computers. So far, as studies on Legal Engineering, Shimazu group of JAIST proposed the logical structure model of law paragraphs, the coreference model of law texts, the editing model of law texts and so on, and implemented their models. Tojo group of JAIST verified whether several related ordinances of Toyama prefecture in Japan contains contradictions or not. Ochimizu group of JAIST studied the model for designing a law-implementation system and proposed the accountability model for the law-implementation system. Futatsugi group of JAIST proposed the formal description and the verification method of legal domains. As laws are written in natural language, natural language processing is essential for Legal Engineering. In this talk, after the aim, the approach and the problems of Legal Engineering are introduced, studies on natural language processing for Legal Engineering are introduced.

Akira Shimazu · Minh Le Nguyen
School of Information Science, Japan Advanced Institute of Science and Technology

V.-N. Huynh et al. (eds.), *Knowledge and Systems Engineering, Volume 1*,
Advances in Intelligent Systems and Computing 244,
DOI: 10.1007/978-3-319-02741-8_3,

Part II
KSE 2013 Main Track

A Hierarchical Approach for High-Quality and Fast Image Completion

Thanh Trung Dang, Azeddine Beghdadi, and Mohamed-Chaker Larabi

Abstract. Image inpainting is not only the art of restoring damaged images but also a powerful technique for image editing e.g. removing undesired objects, recomposing images, etc. Recently, it becomes an active research topic in image processing because of its challenging aspect and extensive use in various real-world applications. In this paper, we propose a novel efficient approach for high-quality and fast image restoration by combining a greedy strategy and a global optimization strategy based on a pyramidal representation of the image. The proposed approach is validated on different state-of-the-art images. Moreover, a comparative validation shows that the proposed approach outperforms the literature in addition to a very low complexity.

1 Introduction

Image inpainting, also known as blind image completion, is not only the art of restoring damaged images; but also a powerful technique in many real-world applications, such as image editing (removing undesired objects, restoring scratches), film reproduction (deleting logos, subtitles, and so on), or even creating artistic effects (reorganizing objects, smart resizing of images, blending images). Recently, it becomes an active research topic in image processing because of its challenging aspect and extensive use in various real-world applications. This topic began by skillful and professional artists in museum to manually restore the old painting.

Digital image inpainting tries to mimic this very precise process in an automatic manner on computers. Because the completion is performed blindly without

Thanh Trung Dang · Azeddine Beghdadi
L2TI, Institut Galilée, Université Paris 13, France
e-mail: {dang.thanhtrung,azeddine.beghdadi}@univ-paris13.fr

Mohamed-Chaker Larabi
XLIM, Dept. SIC, Université de Poitiers, France
e-mail: chaker.larabi@univ-poitiers.fr

V.-N. Huynh et al. (eds.), *Knowledge and Systems Engineering, Volume 1*,
Advances in Intelligent Systems and Computing 244,
DOI: 10.1007/978-3-319-02741-8_4, © Springer International Publishing Switzerland 2014

reference to original images, the aim of digital image completion is only restoring the damaged image by maintaining its naturalness, i.e undetectable by viewers. However, this task is extremely difficult in the case of high resolution and structured images. On the one hand, the restored parts should not be visible or perceptually annoying to human viewers when filled; on the other hand, the used algorithm needs to be robust, efficient and requiring minimal user interactions and quick feedbacks.

An image inpainting algorithm often works in two stages. First the missing or damaged regions are identified (inpainting regions or target regions). Second, these regions are filled in the most natural manner possible. Up to now, there is no approach for automatically detecting damaged regions to be restored. For the sake of simplicity, they are usually marked manually using image editing softwares. Several approaches have been proposed in the literature and they may be categorized into two main groups [1]: geometry-oriented methods and texture-oriented methods.

The methods of the first group are designed to restore small or thin regions such as scratches or blotches, overlaid text, subtitles, etc. In this group, the image is modeled as a function of smoothness and the restoration is solved by interpolating the geometric information within the adjacent regions into the target region. Approaches falling in this category show good performance in propagating smooth level lines or gradient but they have the tendency to generate synthesis artifacts or blur effects in the case of large missing regions [2, 3, 4].

Whereas, the objective of the methods in the second group is to recover larger areas where the texture is assumed to be spatially stationary. Texture is modeled through probability distribution of the pixel brightness values. The pixel intensity distribution depends on only its neighborhood. This group could be further subdivided into two subgroups named: greedy strategy [5, 6, 7, 8] and global optimization strategy [10, 11, 12]. Greedy strategies have acceptable computation time and take into account human perception features (priority is designed based on the salient structures considered as important for human perception). However, some problems such as local optimization and patch selection may limit the efficiency of these approaches. In contrast, global optimization strategies often provide better results. But, they are computationally expensive. This is mainly due to the fact that time complexity increases linearly both with the number of source pixels and unknown pixels.

In this study, we propose a novel approach for high-quality and fast image completion by combining both greedy and global optimization strategies based on a pyramidal representation of the image [13]. The use of pyramidal representation is twofold: first it allows accounting for the multi-scale characteristics of the HVS; second it offers a good way to accelerate the completion process. It is worth noticing that a perceptual pyramidal representation [16] would be better but at the expense of increased computational complexity.

The proposal is directed by the observation that the human visual system is more sensitive to salient structures being stable and repetitive at different scales. Also, a hierarchical completion is a suitable solution for preserving high frequency components in a visually plausible way, and thus generates high-quality outputs. Namely, a top-down completion is implemented from top level (the lowest resolution) to the bottom level (the original resolution). A greedy algorithm is applied for the lowest resolution to complete the damaged regions and create a good initialization accounting for the human perception for the next level. At each higher level, a relation map, called shift-map, is interpolated from adjacently lower level and then optimized by a global optimization algorithm, i.e. multi-label graph-cuts [12, 14]. Experimental results highlight a noticeable improvement in both implementation performance and quality of the inpainted image. To affirm the performance of our implementation, the running time is calculated in comparison with some typical inpainting methods. To confirm the quality of our results, the viewer can visually evaluate outputs of inpainting approaches in conjunction with some objective inpainting quality metrics [17, 18].

The rest of the paper is organized as follows. More details of our framework are introduced in section 2. Section 3 is dedicated to experimental results and comparison with the state-of-the-art methods. Finally, this paper ends with some conclusions and future works.

2 Our Proposal

The inpainting problem could be considered as an optimal graph labeling where a shift-map represents the selected label for each unknown pixels and it could be solved by optimizing an energy function using multi-label graph cuts. Because an unknown pixel in the damaged regions could originate from any pixel in the source regions, the global optimization strategies can be computationally infeasible. Moreover, they consider fairly possible label assignments but this does not fit with human perception. In term of inpainting quality, fair assignments may lead to unexpected bias for optimization. In terms of speed, a huge label set requires high computational load.

Our method is designed to overcome these limitations. In order to reduce the memory and computational requirements, a hierarchical approach for optimizing the graph labeling is developed. This hierarchy could provide enough-good results for inpainting problem, even though optimality cannot be guaranteed. In order to take into account human perception, a greedy strategy is applied at the lowest resolution to generate a suitable initialization for the next pyramidal levels. The priority of greedy strategy is designed based on the salient structures considered as one of the most important features for the HVS. An algorithmic description of our framework is given in the Fig. 1.

For details, some notations that are similar to those in paper [7] are adopted. The whole image domain, I, is composed of two disjoint regions: the inpainting region (or target region) Ω, and the source region Φ ($\Phi = I - \Omega$). According to the above idea, a set of images $G_0, G_1, ..., G_N$ with various levels of details is generated using pyramidal operators, where $G_0 = I$ is the input or original image [13]. The inpainting regions are also reduced to the eliminated areas level by level.

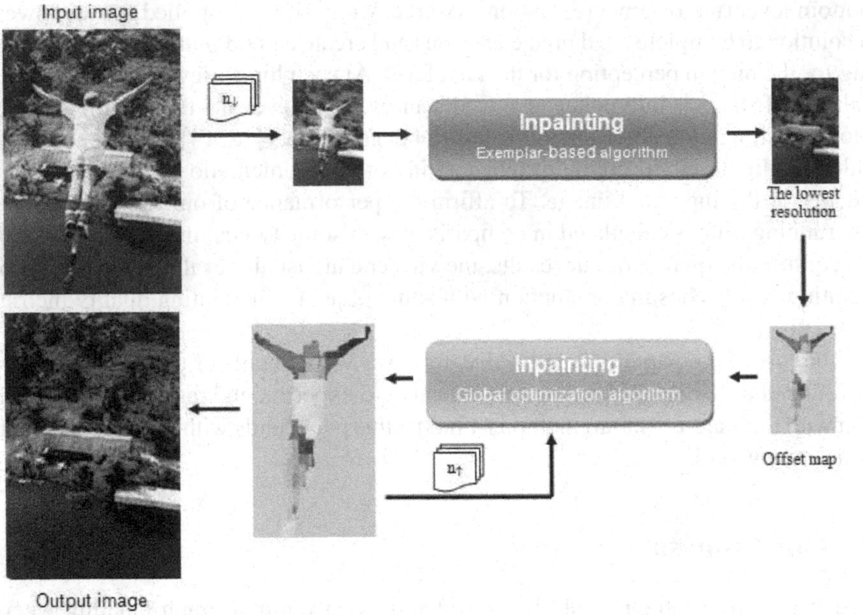

Fig. 1 Our scheme

2.1 Lowest Level Completion

In order to take into account HVS properties, a greedy strategy is applied for the lowest resolution. In our framework, an extension of [7] is developed to complete the reduced inpainting image. The algorithm for a single resolution image repeats the following steps (Fig. 2):

1. *Initialization*: Identify inpaiting boundary, $\delta\Omega$. If there is no pixel on the boundary, the algorithm is terminated.
2. *Priority estimation*: Compute the priority, $P(p)$, for all pixels on boundary, $p \in \delta\Omega$ and select randomly a pixel p with the highest priority.
3. *Patch match*: Find the patch or window Ψ_q that is most similar to Ψ_p thus minimizing mean squared error with existing pixels.
4. *Patch filling*: Fill the missing information in patch Ψ_p by copying the corresponding pixels from patch Ψ_q.
5. *Update*: Update the shift-map, SM_N, defining the relation between filled pixels and their sources and return to the step 1 for next iteration.

In this strategy, a good priority definition is very important because a decision taken based on it could not be changed anymore. Many models for priority have been proposed in the literature [5, 6, 7, 8, 9]. In this work, we used the priority model proposed in [7], namely window-based priority, which is more robust than the

a. Priority estimation b. Patch match c. Inpaint d. Shift definition

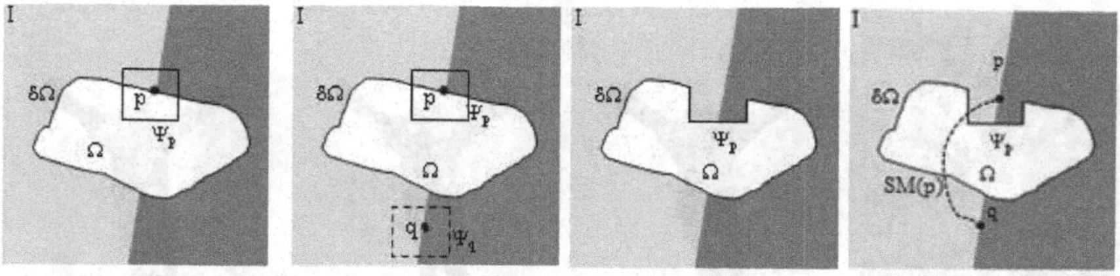

Fig. 2 The greedy strategy

others. After inpainting the image at the lowest resolution, a complete shift-map is generated and used as an initialization for the completion of next levels.

2.2 Higher Level Completion

Since the principle of inpainting is to fill in unknown pixels ($p(x_p, y_p) \subset \Omega$) using the most plausible source pixels ($q(x_q, y_q) \in \Phi$), a relationship between them needs to be defined. This relation can be characterized by a shift-map determining an offset from known pixel to unknown one for each coordinate in the image (Fig. 3b). The shift-map can be formulated by eq. (1). Then the output pixel $O(p)$ is derived from the input pixel $I(p + SM(p))$.

$$SM(p) = \begin{cases} (\triangle x, \triangle y) & p(x,y) \in \Omega \\ (0,0) & \text{otherwise} \end{cases} \qquad (1)$$

The naturalness of the resulting image is one of the most important issue of inpainting. Therefore, the used shift-map has to comply with such a requirement. In [12], authors proposed a solution to evaluate the shift-map by designing an energy function and optimizing it by a graph-cut algorithm. The energy function is defined as follows:

$$EM = \alpha \sum_{p \in \Omega} E_d(SM(p)) + (1 - \alpha) \sum_{(p,q) \in NB} E_s(SM(p), SM(q)) \qquad (2)$$

Where E_d is a data term providing external requirements and E_s is a smoothness term defined over a set of neighboring pixels, NB. α is a user defined weight balancing the two terms fixed to $\alpha = 0.5$ in our case. Once the graph and energy function are given, the shift-map labeling is computed using multi-label graph-cuts algorithm [14, 15].

2.2.1 A. Data Term

The data term E_d is used to include external constraints. Because the unknown pixels are filled thanks to the known ones, the data term assumes that no pixels in the hole are used in the output image. The detail of the data term is given by Eq. (3):

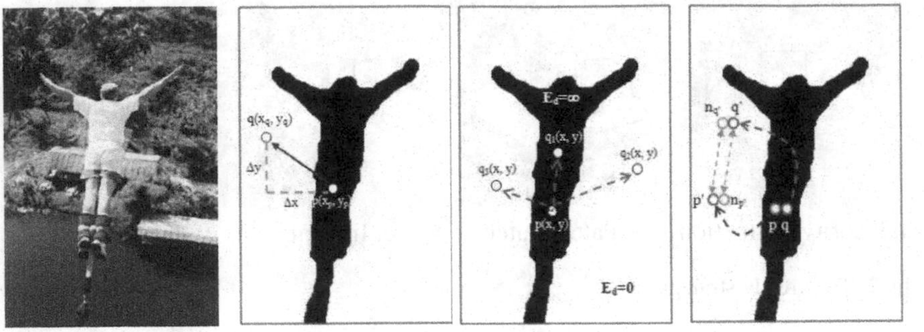

(a) An original image (b) An offset (c) Data term (d) Smoothness term

Fig. 3 Algorithm Operators

$$E_d(SM(p)) = \begin{cases} \infty & (x+\triangle x, y+\triangle y) \in \Omega \\ 0 & \text{otherwise} \end{cases} \tag{3}$$

In some cases, the specific pixels in the input image can be forced to appear or disappear in the output image by setting $E_d = \infty$. For example, saliency map can be used to weight the data term. Therefore, a pixel with a high saliency value should be kept and a pixel with a low saliency value should be removed (Fig. 3c).

2.2.2 B. Smoothness Term

The smoothness term represents discontinuity between two neighbor pixels $p(x_p, y_p)$ and $q(x_q, y_q)$. In paper [12], the authors proposed an effective formula for smoothness term which takes into account both color differences and gradient differences between corresponding spatial neighbors in the output image and in the input image to create good stitching. This treatment is represented as eq. (4) (Fig. 3d):

$$E_s(SM(p), SM(q)) = \begin{cases} 0 & SM(p) = SM(q) \\ \beta\delta M(SM(p)) + \gamma\delta G(SM(p)) & otherwise \end{cases} \tag{4}$$

where β and γ are weights balancing these two terms, set to $\beta = 1$, $\gamma = 2$ in our experiment. δM and δG denote the differences of magnitude and gradient and they are defined as the follows:

$$\begin{aligned} \delta M(SM(p)) &= ||I(n_{p'}) - I(q')|| + ||I(n_{q'}) - I(p')|| \\ \delta G(SM(p)) &= ||\nabla I(n_{p'}) - \nabla I(q')|| + ||\nabla I(n_{q'}) - \nabla I(p')|| \end{aligned} \tag{5}$$

where, I and ∇I are the magnitude and gradient at these locations. $p' = p + SM(p)$ and $q' = q + SM(q)$ are locations used to fill pixels p and q, respectively. $n_{p'}$ and $n_{q'}$ are two 4-connected neighbors of p' and q', respectively (Fig. 3d).

2.3 *Shift-Map Interpolation*

A full shift-map is first inferred from a completion at the lowest level of pyramid. Then it is interpolated to higher resolutions using a *nearest neighbor interpolation*, and the shift-map values are *doubled* to match the higher image resolution.

At the higher level, only small shifts relative to the initial guess are examined. It means that only some parent neighbors are considered instead of all possible labels. In our implementation, the shift relative for each coordinate varies in range [-a, a], so it takes $(2a+1)^2$ labels for both direction. It is important to note that the data and smoothness terms are always computed with respect to the actual shifts and not to the labels (Fig. 4).

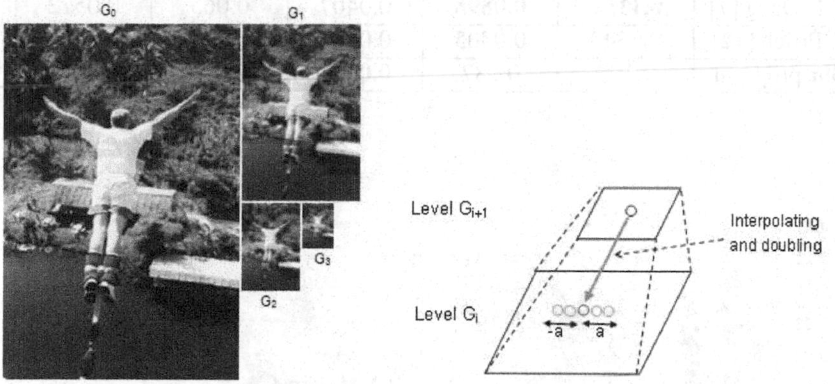

(a) Image gaussian pyramid (b) Interpolation between two adjacent levels

Fig. 4 Interpolation of Shift-Map

3 Experimental Results

This section is dedicated to the study of performance of the proposed algorithm using some typical real images that cover several major challenges for inpainting. In order to try and cover all inpainting methods would be infeasible. For the sake of comparison with literature, three inpainting methods corresponding to algorithms proposed by *A. Criminisi et al* [5] and *T. T. Dang et al* [7] for greedy strategy and *Y. Pritch et al* [12] for global optimization strategy have been implemented. Five images, given on Fig. 6 were chosen for this experiment (including *bungee* (206 × 308), *angle* (300 × 252), *silenus* (256 × 480), *boat* (300 × 225) and *seaman* (300 × 218)).

Figure 6 illustrates the results obtained with the proposed approach in comparison to the others. Fig. 6a gives images to be inpainted where damaged areas cover respectively *12.6%*, *5.83%*, *7.74%*, *10.73%* and *14.87%* of the whole image.

To evaluate the quality of inpainting output, some objective inpainted image quality metrics [17, 18] are considered and the metric in [18] is developed because all used images in our experiment are color. The metric values are shown in the table 1 and compared more visually in figure 5.

Table 1 The inpainted image quality metrics

Image	*bungee*	*angle*	*silenus*	*boat*	*seaman*
Size	(206 × 308)	(300 × 252)	(256 × 480)	(300 × 225)	(300 × 218)
Damaged Area	12.6%	5.83%	7.74%	10.73%	14.87%
A. Criminisi [5]	0.0685	0.0817	0.0358	0.061	0.0449
T. T. Dang [7]	**0.1157**	**0.0898**	0.0407	0.065	0.0572
Y. Pritch [12]	0.0343	0.0805	0.0289	0.0597	0.0407
Our proposal	0.107	0.087	**0.0407**	**0.069**	**0.0592**

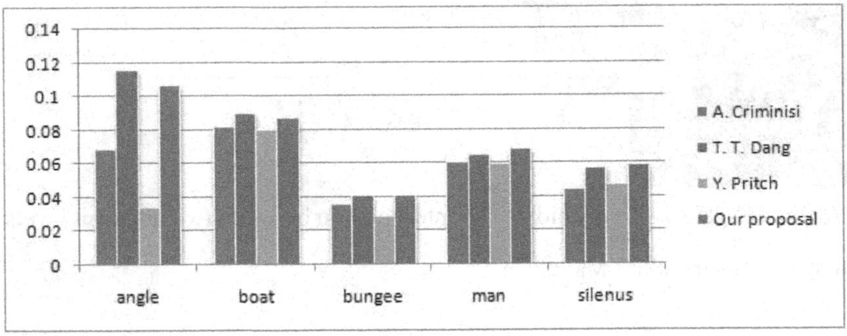

Fig. 5 A chart of quality performance

The performance of the proposed approach is quantitatively evaluated by implementation time in comparison with the other approaches. In order to avoid bias, all approaches are programmed by the same programming language, C/C++ programming language, and implemented on the same PC with the configuration of Intel Core i5, 2.8GHz CPU and 4GB RAM. The running time in seconds of each methods is given in table 2 and shown visually in figure 7. As it can be seen from these results, our method provides an acceptable visual quality, often outperforming the others, with a much faster implementation. Indeed, visual inspection of results shows that the completion performed by our approach looks more natural and more coherent than the other approaches.

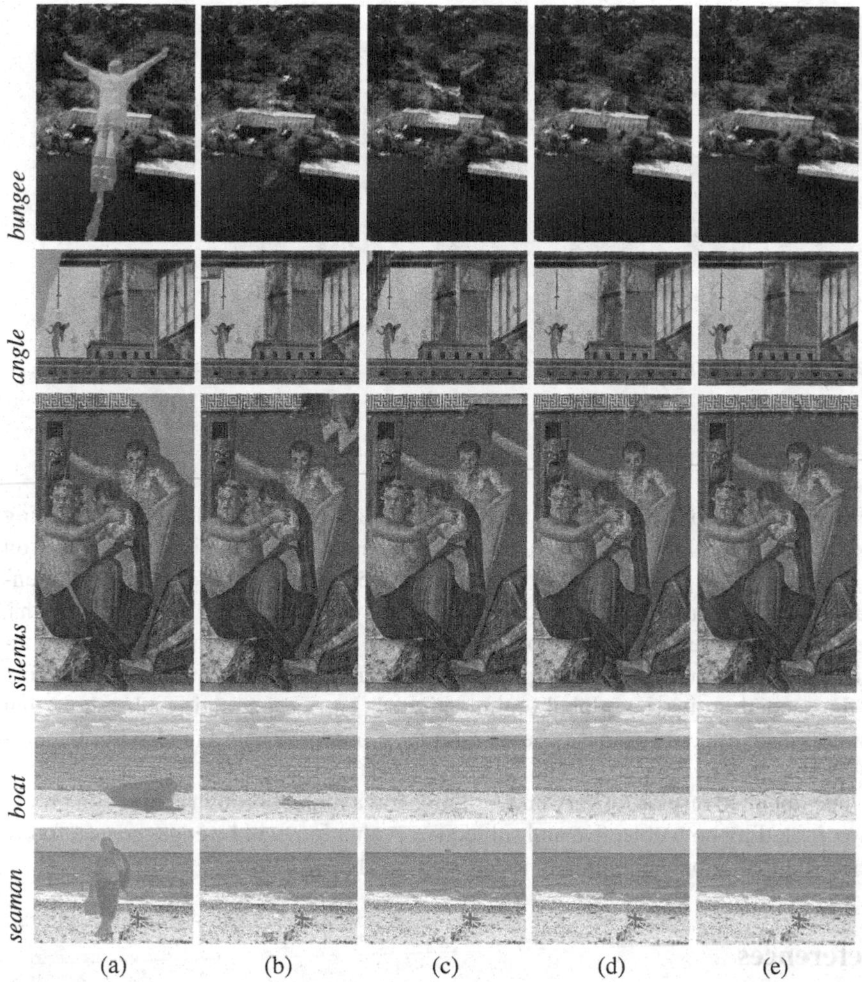

Fig. 6 The experimental results. (a) Image to be inpainted; The outputs when using the methods in (b) [5]; (c) [12]; (d) [7]; (e) our proposal.

Table 2 Computational time (in second) for implemented approaches and the set of used images

Image	*bungee*	*angle*	*silenus*	*boat*	*seaman*
Size	(206×308)	(300×252)	(256×480)	(300×225)	(300×218)
Damaged Area	12.6%	5.83%	7.74%	10.73%	14.87%
A. Criminisi [5]	16.30	8.20	38.29	24.54	27.31
T. T. Dang [7]	15.92	16.36	63.18	50.18	55.16
Y. Pritch [12]	35.39	13.24	57.68	21.18	15.50
Our proposal	**3,32**	**5.81**	**7.53**	**7.25**	**5.97**

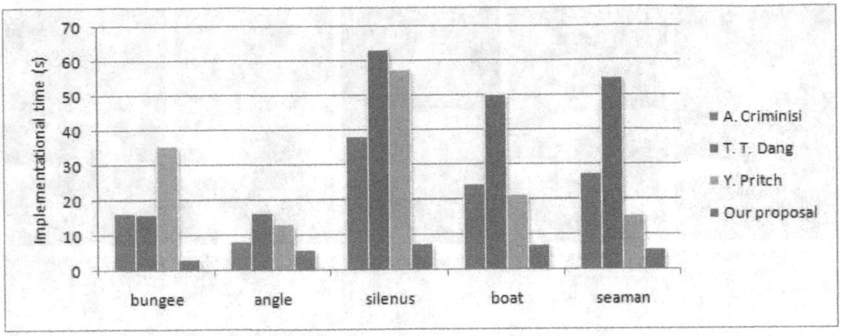

Fig. 7 A chart of implementation performance

4 Conclusions

In this paper, a novel framework of image completion is introduced by combining both greedy and global optimization strategies based on a pyramidal representation of the image. The greedy strategy is applied at the lowest resolution in order to generate a good initialization accounting for human perception. At higher resolutions, the shift map is refined by a global optimization algorithm and multi-label graph-cuts. A comparison with some representative approaches from literature belonging to the second group (i.e. global optimization) is carried out and results show that our approach not only produces better quality of output images but also implements noticeably faster.

The obtained results are very encouraging and a more thorough evaluation procedure, including both objective and subjective evaluation, will be engaged as a future work. Computational complexity issues will be also addressed.

References

1. Arias, P., Facciolo, G., Caselles, V., Sapiro, G.: A Variational Framework for Exemplar-Based Image Inpainting. International Journal of Computer Vision, 1–29 (2011)
2. Bertalmio, M., Sapiro, G., Caselles, V., Ballester, C.: Image inpainting. In: Proceedings of the 27th Annual Conference on Computer Graphics and Interactive Techniques, pp. 417–424 (2000)
3. Chan, T.F., Shen, J.: Non-texture inpainting by Curvature-Driven Diffusions (CCD). Journal of Visual Communication and Image Representation 4, 436–449 (2001)
4. Tschumperle, D.: Fast anisotropic smoothing of multi-valued images using curvature-preserving pdes. International Journal of Computer Vision 68, 65–82 (2006)
5. Criminisi, A., Perez, P., Toyama, K.: Region filling and object removal by exemplar-based image inpainting. IEEE Transaction of Image Process 13(9), 1200–1212 (2004)
6. Wu, J., Ruan, Q.: Object removal by cross isophotes exemplar based image inpainting. In: Proceeding of International Conference of Pattern Recognition, pp. 810–813 (2006)

7. Dang, T.T., Larabi, M.C., Beghdadi, A.: Multi-resolution patch and window-based priority for digital image inpainting problem. In: 3rd International Conference on Image Processing Theory, Tools and Applications, pp. 280–284 (2012)
8. Zhang, Q., Lin, J.: Exemplar-based image inpainting using color distribution analysis. Journal of Information Science and Engineering (2011)
9. Cheng, W., Hsieh, C., Lin, S., Wang, C., Wu, J.: Robust algorithm for exemplar-based image inpainting. In: Proceeding of International Conference on Computer Graphics, Imaging and Visualization (2005)
10. Wexler, Y., Shechtman, E., Irani, M.: Space-time video completion. IEEE Transactions Pattern Analysis and Machine Intelligence 29, 463–476 (2007)
11. Komodakis, G.T.N., Tziritas, G.: Image completion using global optimization. In: Proceeding of IEEE Computer Society Conference Computer Vision and Pattern Recognition, pp. 442–452 (2006)
12. Pritch, Y., Kav-Venaki, E., Peleg, S.: Shift-map image editing. In: IEEE Computer Society Conference on Computer Vision and Pattern Recognition, pp. 151–158 (2009)
13. Peter, J.B., Edward, H.A.: The Laplacian pyramid as a compact image code. IEEE Transactions on Communications 31, 532–540 (1983)
14. Boykov, Y., Veksler, O., Zabih, R.: Fast approximate energy minimization via graph cuts. IEEE Transactions on Pattern Analysis and Machine Intelligence 23(11), 1222–1239 (2001)
15. Agarwala, A., Dontcheva, M., Agrawala, M., Drucker, S., Colburn, A., Curless, B., Salesin, D., Cohen, M.: Interactive Digital Photomontage. In: Proceedings of SIGGRAPH, pp. 294–302 (2004)
16. Iordache, R., Beghdadi, A., de Lesegno, P.V.: Pyramidal perceptual filtering using Moon and Spencer contrast. In: International Conference on Image Processing, ICIP 2001, pp. 146–149 (2001)
17. Dang, T.T., Beghdadi, A., Larabi, M.C.: Perceptual evaluation of digital image completion quality. In: 21st European Signal Processing Conference, EUSIPCO 2013 (2013)
18. Dang, T.T., Beghdadi, A., Larabi, M.C.: Perceptual quality assessment for color image inpainting. In: IEEE International Conference on Image Processing, ICIP 2013 (2013)

The Un-normalized Graph p-Laplacian Based Semi-supervised Learning Method and Protein Function Prediction Problem

Loc Tran

Abstract. Protein function prediction is a fundamental problem in modern biology. In this paper, we present the un-normalized graph p-Laplacian semi-supervised learning methods. These methods will be applied to the protein network constructed from the gene expression data to predict the functions of all proteins in the network. These methods are based on the assumption that the labels of two adjacent proteins in the network are likely to be the same. The experiments show that that the un-normalized graph p-Laplacian semi-supervised learning methods are at least as good as the current state of the art method (the un-normalized graph Laplacian based semi-supervised learning method) but often lead to better classification accuracy performance measures.

1 Introduction

Protein function prediction is the important problem in modern biology. Identifying the function of proteins by biological experiments is very expensive and hard. Hence a lot of computational methods have been proposed to infer the functions of the proteins by using various types of information such as gene expression data and protein-protein interaction networks [1].

The classical way predicting protein function infers the similarity to function from sequence homologies among proteins in the databases using sequence similarity algorithms such as FASTA [2] and PSI-BLAST [3]. Next, to predict protein function, graph which is the natural model of relationship between proteins or genes can also be employed. This model can be protein-protein interaction network or gene co-expression network. In this model, the nodes represent proteins or genes and the edges represent for the possible interactions between nodes. Then, machine learning methods such as Support Vector Machine [5], Artificial Neural Networks

Loc Tran
University of Minnesota, USA
e-mail: tran0398@umn.edu

V.-N. Huynh et al. (eds.), *Knowledge and Systems Engineering, Volume 1*,
Advances in Intelligent Systems and Computing 244,
DOI: 10.1007/978-3-319-02741-8_5, © Springer International Publishing Switzerland 2014

[4], un-normalized graph Laplacian based semi-supervised learning method [6], the symmetric normalized and random walk graph Laplacian based semi-supervised learning methods [7], or neighbor counting method [8] can be applied to this graph to infer the functions of un-annotated protein. The neighbor counting method labels the protein with the function that occurs frequently in the protein's adjacent nodes in the protein-protein interaction network and hence does not utilized the full topology of the network. However, the Artificial Neural Networks, Support Vector Machine, un-normalized, symmetric normalized and random walk graph Laplacian based semi-supervised learning method utilizes the full topology of the network. The Artificial Neural Networks and Support Vector Machine are all supervised learning methods. The neighbor counting method, the Artificial Neural Networks, and the three graph Laplacian based semi-supervised learning methods are all based on the assumption that the labels of two adjacent proteins in graph are likely to be the same. However, SVM do not rely on this assumption. Unlike graphs used in neighbor counting method, Artificial Neural Networks, and the three graph Laplacian based semi-supervised learning methods are very sparse, the graph (i.e. kernel) used in SVM is fully-connected.

The Artificial Neural Networks method is applied to the single protein-protein interaction network. However, the SVM method and three graph Laplacian based semi-supervised learning methods try to use weighted combination of multiple networks (i.e. kernels) such as gene co-expression network and protein-protein interaction network to improve the accuracy performance measures. [5] (SVM method) determines the optimal weighted combination of networks by solving the semi-definite problem. [6] (un-normalized graph Laplacian based semi-supervised learning method) uses a dual problem and gradient descent to determine the weighted combination of networks. [7] uses the integrated network combined with equal weights, i.e. without optimization due to the integrated network combined with optimized weights has similar performance to the integrated network combined with equal weights and the high time complexity of optimization methods.

The un-normalized, symmetric normalized, and random walk graph Laplacian based semi-supervised learning methods are developed based on the assumption that the labels of two adjacent proteins or genes in the network are likely to be the same [6]. Hence this assumption can be interpreted as pairs of genes showing a similar pattern of expression and thus sharing edges in a gene co-expression network tend to have similar function. In [9], the single gene expression data is used for protein function prediction problem. However, assuming the pairwise relationship between proteins or genes is not complete, the information a group of genes that show very similar patterns of expression and tend to have similar functions [12] (i.e. the functional modules) is missed. The natural way overcoming the information loss of the above assumption is to represent the gene expression data as the hypergraph [10,11]. A hypergraph is a graph in which an edge (i.e. a hyper-edge) can connect more than two vertices. In [9], the un-normalized, random walk, and symmetric normalized hypergraph Laplacian based semi-supervised learning methods have been developed and successfully outperform the un-normalized, symmetric normalized, and random walk graph Laplacian based semi-supervised learning methods in protein function prediction problem.

In [13,14], the symmetric normalized graph p-Laplacian based semi-supervised learning method has been developed but has not been applied to any practical applications. To the best of my knowledge, the un-normalized graph p-Laplacian based semi-supervised learning method has not yet been developed and obviously has not been applied to protein function prediction problem. This method is worth investigated because of its difficult nature and its close connection to partial differential equation on graph field. Specifically, in this paper, the un-normalized graph p-Laplacian based semi-supervised learning method will be developed based on the un-normalized graph p-Laplacian operator definition such as the curvature operator of graph (i.e. the un-normalized graph 1-Laplacian operator). Please note that the un-normalized graph p-Laplacian based semi-supervised learning method is developed based on the assumption that the labels of two adjacent proteins or genes in the network are likely to be the same [6].

We will organize the paper as follows: Section 2 will introduce the preliminary notations and definitions used in this paper. Section 3 will introduce the definition of the gradient and divergence operators of graphs. Section 4 will introduce the definition of Laplace operator of graphs and its properties. Section 5 will introduce the definition of the curvature operator of graphs and its properties. Section 6 will introduce the definition of the p-Laplace operator of graphs and its properties. Section 7 will show how to derive the algorithm of the un-normalized graph p-Laplacian based semi-supervised learning method from regularization framework. In section 8, we will compare the accuracy performance measures of the un-normalized graph Laplacian based semi-supervised learning algorithm (i.e. the current state of art method applied to protein function prediction problem) and the un-normalized graph p-Laplacian based semi-supervised learning algorithms. Section 9 will conclude this paper and the future direction of researches of other practical applications in bioinformatics utilizing discrete operator of graph will be discussed.

2 Preliminary Notations and Definitions

Given a graph $G=(V,E,W)$ where V is a set of vertices with $|V| = n$, $E \subseteq V * V$ is a set of edges and W is a $n*n$ similarity matrix with elements $w_{ij} > 0$ $(1 \leq i,j \leq n)$.

Also, please note that $w_{ij} = w_{ji}$.

The degree function $d : V \to R^+$ is

$$d_i = \sum_{j \sim i} w_{ij}, \tag{1}$$

where $j \sim i$ is the set of vertices adjacent with i.

Define $D = diag(d_1, d_2, \ldots, d_n)$.

The inner product on the function space R^V is

$$< f,g >_V = \sum_{i \in V} f_i g_i \tag{2}$$

Also define an inner product on the space of functions R^E on the edges

$$<F,G>_E = \sum_{(i,j)\in E} F_{ij}G_{ij} \tag{3}$$

Here let $H(V) = (R^V, <.,.>_V)$ and $H(E) = (R^E, <.,.>_E)$ be the Hilbert space real-valued functions defined on the vertices of the graph G and the Hilbert space of real-valued functions defined in the edges of G respectively.

3 Gradient and Divergence Operators

We define the gradient operator $d : H(V) \to H(E)$ to be

$$(df)_{ij} = \sqrt{w_{ij}}(f_j - f_i), \tag{4}$$

where $f : V \to R$ be a function of $H(V)$.

We define the divergence operator $div : H(E) \to H(V)$ to be

$$<df, F>_{H(E)} = <f, -divF>_{H(V)}, \tag{5}$$

where $f \in H(V), F \in H(E)$

Next, we need to prove that

$$(divF)_j = \sum_{i\sim j} \sqrt{w_{ij}}(F_{ji} - F_{ij})$$

Proof:

$$
\begin{aligned}
<df,F> &= \sum_{(i,j)\in E} df_{ij}F_{ij} \\
&= \sum_{(i,j)\in E} \sqrt{w_{ij}}(f_j - f_i)F_{ij} \\
&= \sum_{(i,j)\in E} \sqrt{w_{ij}} f_j F_{ij} - \sum_{(i,j)\in E} \sqrt{w_{ij}} f_i F_{ij} \\
&= \sum_{k\in V}\sum_{i\sim k} \sqrt{w_{ik}} f_k F_{ik} - \sum_{k\in V}\sum_{j\sim k} \sqrt{w_{kj}} f_k F_{kj} \\
&= \sum_{k\in V} f_k \Big(\sum_{i\sim k} \sqrt{w_{ik}} F_{ik} - \sum_{i\sim k} \sqrt{w_{ki}} F_{ki}\Big) \\
&= \sum_{k\in V} f_k \sum_{i\sim k} \sqrt{w_{ik}}(F_{ik} - F_{ki})
\end{aligned}
$$

Thus, we have

$$(divF)_j = \sum_{i\sim j} \sqrt{w_{ij}}(F_{ji} - F_{ij}) \tag{6}$$

4 Laplace Operator

We define the Laplace operator $\triangle : H(V) \to H(V)$ to be

$$\triangle f = -\frac{1}{2} div(df) \tag{7}$$

Next, we compute

$$(\triangle f)_j = \frac{1}{2}\sum_{i\sim j}\sqrt{w_{ij}}((df)_{ij} - (df)_{ji})$$

$$= \frac{1}{2}\sum_{i\sim j}\sqrt{w_{ij}}(\sqrt{w_{ij}}(f_j - f_i) - \sqrt{w_{ij}}(f_i - f_j))$$

$$= \sum_{i\sim j} w_{ij}(f_j - f_i)$$

$$= \sum_{i\sim j} w_{ij} f_j - \sum_{i\sim j} w_{ij} f_i$$

$$= d_j f_j - \sum_{i\sim j} w_{ij} f_i$$

Thus, we have

$$(\triangle f)_j = d_j f_j - \sum_{i\sim j} w_{ij} f_i \tag{8}$$

The graph Laplacian is a linear operator. Furthermore, the graph Laplacian is self-adjoint and positive semi-definite.

Let $S_2(f) = <\triangle f, f>$, we have the following **theorem 1**

$$D_f S_2 = 2\triangle f \tag{9}$$

The proof of the above theorem can be found from [13,14].

5 Curvature Operator

We define the curvature operator $\kappa : H(V) \to H(V)$ to be

$$\kappa f = -\frac{1}{2} div\left(\frac{df}{||df||}\right) \tag{10}$$

Next, we compute

$$(\kappa f)_j = \frac{1}{2}\sum_{i\sim j}\sqrt{w_{ij}}\left(\left(\frac{df}{||df||}\right)_{ij} - \left(\frac{df}{||df||}\right)_{ji}\right)$$

$$= \frac{1}{2}\sum_{i\sim j}\sqrt{w_{ij}}(\frac{1}{\|d_if\|}\sqrt{w_{ij}}(f_j - f_i) - \frac{1}{\|d_jf\|}\sqrt{w_{ij}}(f_i - f_j))$$

$$= \frac{1}{2}\sum_{i\sim j} w_{ij}(\frac{1}{\|d_if\|} + \frac{1}{\|d_jf\|})(f_j - f_i)$$

Thus, we have

$$(\kappa f)_j = \frac{1}{2}\sum_{i\sim j} w_{ij}(\frac{1}{\|d_if\|} + \frac{1}{\|d_jf\|})(f_j - f_i) \tag{11}$$

From the above formula, we have

$$d_if = ((df)_{ij} : j \sim i)^T \tag{12}$$

The local variation of f at i is defined to be

$$\|d_if\| = \sqrt{\sum_{j\sim i}(df)^2_{ij}} = \sqrt{\sum_{j\sim i} w_{ij}(f_j - f_i)^2} \tag{13}$$

To avoid the zero denominators in (11), the local variation of f at i is defined to be

$$\|d_if\| = \sqrt{\sum_{j\sim i}(df)^2_{ij} + ?}, \tag{14}$$

where $? = 10^{-10}$.
The graph curvature is a non-linear operator.
Let $S_1(f) = \sum_i \|d_if\|$, we have the following **theorem 2**

$$D_fS_1 = \kappa f \tag{15}$$

The proof of the above theorem can be found from [13,14].

6 p-Laplace Operator

We define the p-Laplace operator $\triangle_p : H(V) \to H(V)$ to be

$$\triangle_p f = -\frac{1}{2}div(\|df\|^{p-2}df) \tag{16}$$

Clearly, $\triangle_1 = \kappa$ and $\triangle_2 = \triangle$. Next, we compute

$$(\triangle_p f)_j = \frac{1}{2}\sum_{i\sim j}\sqrt{w_{ij}}(\|df\|^{p-2}df_{ij} - \|df\|^{p-2}df_{ji})$$

$$= \frac{1}{2}\sum_{i\sim j}\sqrt{w_{ij}}(\|d_if\|^{p-2}\sqrt{w_{ij}}(f_j - f_i) - \|d_jf\|^{p-2}\sqrt{w_{ij}}(f_i - f_j))$$

$$= \frac{1}{2} \sum_{i \sim j} w_{ij} (\|d_i f\|^{p-2} + \|d_j f\|^{p-2}) (f_j - f_i)$$

Thus, we have

$$(\triangle_p f)_j = \frac{1}{2} \sum_{i \sim j} w_{ij} (\|d_i f\|^{p-2} + \|d_j f\|^{p-2}) (f_j - f_i) \tag{17}$$

Let $S_p(f) = \frac{1}{p} \sum_i \|d_i f\|^p$, we have the following **theorem 3**

$$D_f S_p = p \triangle_p f \tag{18}$$

7 Discrete Regularization on Graphs and Protein Function Classification Problems

Given a protein network *G=(V,E)*. *V* is the set of all proteins in the network and *E* is the set of all possible interactions between these proteins. Let *y* denote the initial function in *H(V)*. y_i can be defined as follows

$$y_i = \begin{cases} 1 \ if \ protein \ i \ belongs \ to \ the \ functional \ class \\ -1 \ if \ protein \ i \ does \ not \ belong \ to \ the \ functional \ class \\ 0 \ otherwise \end{cases}$$

Our goal is to look for an estimated function *f* in *H(V)* such that *f* is not only smooth on *G* but also close enough to an initial function *y*. Then each protein *i* is classified as $sign(f_i)$. This concept can be formulated as the following optimization problem

$$argmin_{f \in H(V)} \{S_p(f) + \frac{\mu}{2} \|f - y\|^2\} \tag{19}$$

The first term in (19) is the smoothness term. The second term is the fitting term. A positive parameter μ captures the trade-off between these two competing terms.

7.I) **2-smoothness**

When *p=2*, the optimization problem (19) is

$$argmin_{f \in H(V)} \{\frac{1}{2} \sum_i \|d_i f\|^2 + \frac{\mu}{2} \|f - y\|^2\} \tag{20}$$

By theorem 1, we have

Theorem 4: The solution of (20) satisfies

$$\triangle f + \mu (f - y) = 0 \tag{21}$$

Since $\triangle$ is a linear operator, the closed form solution of (21) is

$$f = \mu (\triangle + \mu I)^{-1} y, \tag{22}$$

Where I is the identity operator and $\triangle = D - W$. (22) is the algorithm proposed by [6].

7.II) **1-smoothness**

When *p=1*, the optimization problem (19) is

$$argmin_{f \in H(V)} \{\sum_i \|d_i f\| + \frac{\mu}{2} \|f - y\|^2\}, \tag{23}$$

By theorem 2, we have

<u>**Theorem 5:**</u> The solution of (23) satisfies

$$\kappa f + \mu (f - y) = 0, \tag{24}$$

The curvature κ is a non-linear operator; hence we do not have the closed form solution of equation (24). Thus, we have to construct iterative algorithm to obtain the solution. From (24), we have

$$\frac{1}{2} \sum_{i \sim j} w_{ij} \left(\frac{1}{\|d_i f\|} + \frac{1}{\|d_j f\|} \right) (f_j - f_i) + \mu (f_j - y_j) = 0 \tag{25}$$

Define the function $m : E \to R$ by

$$m_{ij} = \frac{1}{2} w_{ij} \left(\frac{1}{\|d_i f\|} + \frac{1}{\|d_j f\|} \right) \tag{26}$$

Then (25)

$$\sum_{i \sim j} m_{ij} (f_j - f_i) + \mu (f_j - y_j) = 0$$

can be transformed into

$$\left(\sum_{i \sim j} m_{ij} + \mu \right) f_j = \sum_{i \sim j} m_{ij} f_i + \mu y_j \tag{27}$$

Define the function $p : E \to R$ by

$$p_{ij} = \begin{cases} \frac{m_{ij}}{\sum_{i \sim j} m_{ij} + \mu} & if\ i \neq j \\ \frac{\mu}{\sum_{i \sim j} m_{ij} + \mu} & if\ i = j \end{cases} \tag{28}$$

Then

$$f_j = \sum_{i \sim j} p_{ij} f_i + p_{jj} y_j \tag{29}$$

Thus we can consider the iteration

$f_j^{(t+1)} = \sum_{i \sim j} p_{ij}^{(t)} f_i^{(t)} + p_{jj}^{(t)} y_j$ for all $j \in V$

to obtain the solution of (23).

7.III) **p-smoothness**

For any number p, the optimization problem (19) is

$$argmin_{f\in H(V)}\{\frac{1}{p}\sum_i \|d_i f\|^p + \frac{\mu}{2}\|f-y\|^2\}, \tag{30}$$

By theorem 3, we have

Theorem 6: The solution of (30) satisfies

$$\triangle_p f + \mu(f-y) = 0, \tag{31}$$

The *p-Laplace* operator is a non-linear operator; hence we do not have the closed form solution of equation (31). Thus, we have to construct iterative algorithm to obtain the solution. From (31), we have

$$\frac{1}{2}\sum_{i\sim j} w_{ij}\left(\|d_i f\|^{p-2} + \left\|d_j f\right\|^{p-2}\right)(f_j - f_i) + \mu(f_j - y_j) = 0 \tag{32}$$

Define the function $m : E \to R$ by

$$m_{ij} = \frac{1}{2} w_{ij}(\|d_i f\|^{p-2} + \left\|d_j f\right\|^{p-2}) \tag{33}$$

Then equation (32) which is

$$\sum_{i\sim j} m_{ij}(f_j - f_i) + \mu(f_j - y_j) = 0$$

can be transformed into

$$\left(\sum_{i\sim j} m_{ij} + \mu\right) f_j = \sum_{i\sim j} m_{ij} f_i + \mu y_j \tag{34}$$

Define the function $p : E \to R$ by

$$p_{ij} = \begin{cases} \frac{m_{ij}}{\sum_{i\sim j} m_{ij} + \mu} & if\ i \neq j \\ \frac{\mu}{\sum_{i\sim j} m_{ij} + \mu} & if\ i = j \end{cases} \tag{35}$$

Then

$$f_j = \sum_{i\sim j} p_{ij} f_i + p_{jj} y_j \tag{36}$$

Thus we can consider the iteration

$f_j^{(t+1)} = \sum_{i\sim j} p_{ij}^{(t)} f_i^{(t)} + p_{jj}^{(t)} y_j$ for all $j \in V$

to obtain the solution of (30).

8 Experiments and Results

8.1 Datasets

In this paper, we use the dataset available from [9,15] and the references therein. This dataset contains the gene expression data measuring the expression of 4062 S. cerevisiae genes under the set of 215 titration experiments. These proteins are annotated with 138 GO Biological Process functions. In the other words, we are given gene expression data ($R^{4062*215}$) matrix and the annotation (i.e. the label) matrix ($R^{4062*138}$). We filtered the datasets to include only those GO functions that had at least 150 proteins and at most 200 proteins. This resulted in a dataset containing 1152 proteins annotated with seven different GO Biological Process functions. Seven GO Biological Process functions are

1. Alcohol metabolic process
2. Proteolysis
3. Mitochondrion organization
4. Cell wall organization
5. rRNA metabolic process
6. Negative regulation of transcription, DNA-dependent, and
7. Cofactor metabolic process.

We refer to this dataset as **yeast**. There are three ways to construct the similarity graph from the gene expression data:

1. The ε-neighborhood graph: Connect all genes whose pairwise distances are smaller than ε.
2. k-nearest neighbor graph: Gene i is connected with gene j if gene i is among the k-nearest neighbor of gene j or gene j is among the k-nearest neighbor of gene i.
3. The fully connected graph: All genes are connected.

In this paper, the similarity function is the Gaussian similarity function

$$s(G(i,:),G(j,:) = e^{-\frac{d(G(i,:),G(j,:))}{t}}$$

In this paper, t is set to 1.25 and the 3-nearest neighbor graph is used to construct the similarity graph from **yeast**.

8.2 Experiments

In this section, we experiment with the above proposed un-normalized graph p-Laplacian methods with *p=1, 1.1, 1.2, 1.3, 1.4, 1.5, 1.6, 1.7, 1.8, 1.9* and the current state of the art method (i.e. the un-normalized graph Laplacian based semi-supervised learning method *p=2*) in terms of classification accuracy performance measure. The accuracy performance measure Q is given as follows

$$Q = \frac{True\ Positive + True\ Negative}{True\ Positive + True\ Negative + False\ Positive + False\ Negative}$$

All experiments were implemented in Matlab 6.5 on virtual machine. The three-fold cross validation is used to compute the average accuracy performance measures of all methods used in this paper. The parameter μ is set to 1.

The accuracy performance measures of the above proposed methods and the current state of the art method is given in the following table 1.

Table 1 The comparison of accuracies of proposed methods with different p-values

Functional classes		1	2	3	4	5	6	7
Accuracy Performance Measures (%)	p=1	**86.20**	**84.64**	**84.72**	**83.94**	**92.71**	**85.16**	**86.72**
	p=1.1	86.11	84.03	**84.72**	83.59	92.53	84.81	86.46
	p=1.2	85.94	84.03	**84.72**	83.68	92.62	84.72	86.46
	p=1.3	85.50	82.12	83.25	82.38	92.27	83.42	85.50
	p=1.4	85.59	83.25	84.11	82.90	92.88	84.64	86.28
	p=1.5	85.50	82.90	83.77	82.73	92.80	84.38	86.11
	p=1.6	85.42	82.64	83.68	82.64	92.88	83.94	85.94
	p=1.7	85.42	82.29	83.33	82.47	92.62	83.85	85.85
	p=1.8	85.42	82.12	83.33	82.55	92.53	83.51	85.59
	p=1.9	85.24	82.12	83.07	82.47	92.27	83.51	85.42
	p=2 (i.e. the current state of the art method)	85.50	82.12	83.25	82.38	92.27	83.42	85.50

From the above table, we easily recognized that the un-normalized graph 1-Laplacian semi-supervised learning method outperform other proposed methods and the current state of art method. The results from the above table shows that the un-normalized graph p-Laplacian semi-supervised learning methods are at least as good as the current state of the art method (*p=2*) but often lead to better classification accuracy performance measures.

9 Conclusions

We have developed the detailed regularization frameworks for the un-normalized graph p-Laplacian semi-supervised learning methods applying to protein function prediction problem. Experiments show that the un-normalized graph p-Laplacian semi-supervised learning methods are at least as good as the current state of the art method (i.e. $p=2$) but often lead to significant better classification accuracy performance measures.

Moreover, these un-normalized graph p-Laplacian semi-supervised learning methods can not only be used in classification problem but also in ranking problem. In specific, given a set of genes (i.e. the queries) making up a protein complex/-pathways or given a set of genes (i.e. the queries) involved in a specific disease (for e.g. leukemia), these methods can also be used to find more potential members of the complex/pathway or more genes involved in the same disease by ranking genes in gene co-expression network (derived from gene expression data) or the protein-protein interaction network or the integrated network of them. The genes with the highest rank then will be selected and then checked by biologist experts to see if the extended genes in fact belong to the same complex/pathway or are involved in the same disease. These problems are also called complex/pathway membership determination and biomarker discovery in cancer classification.

References

1. Shin, H.H., Lisewski, A.M., Lichtarge, O.: Graph sharpening plus graph integration: a synergy that improves protein functional classification. Bioinformatics 23, 3217–3224 (2007)
2. Pearson, W.R., Lipman, D.J.: Improved tools for biological sequence comparison. Proceedings of the National Academy of Sciences of the United States of America 85, 2444–2448 (1998)
3. Lockhart, D.J., Dong, H., Byrne, M.C., Follettie, M.T., Gallo, M.V., Chee, M.S., Mittmann, M., Wang, C., Kobayashi, M., Horton, H., Brown, E.L.: Expression monitoring by hybridization to high-density oligonucleotide arrays. Nature Biotechnology 14, 1675–1680 (1996)
4. Shi, L., Cho, Y., Zhang, A.: Prediction of Protein Function from Connectivity of Protein Interaction Networks. International Journal of Computational Bioscience 1(1) (2010)
5. Lanckriet, G.R.G., Deng, M., Cristianini, N., Jordan, M.I., Noble, W.S.: Kernel-based data fusion and its application to protein function prediction in yeast. In: Pacific Symposium on Biocomputing, PSB (2004)
6. Tsuda, K., Shin, H.H., Schoelkopf, B.: Fast protein classification with multiple networks. Bioinformatics (ECCB 2005) 21(suppl. 2), ii59–ii65 (2005)
7. Tran, L.: Application of three graph Laplacian based semi-supervised learning methods to protein function prediction problem. CoRR abs/1211.4289 (2012)
8. Schwikowski, B., Uetz, P., Fields, S.: A network of protein–protein interactions in yeast. Nature Biotechnology 18, 1257–1261 (2000)
9. Tran, L.: Hypergraph and protein function prediction with gene expression data. CoRR abs/1212.0388 (2012)

10. Zhou, D., Huang, J., Schoelkopf, B.: Beyond Pairwise Classification and Clustering Using Hypergraphs, Max Planck Institute Technical Report 143, Max Planck Institute for Biological Cybernetics, Tbingen, Germany (2005)
11. Zhou, D., Huang, J., Schoelkopf, B.: Learning with Hypergraphs: Clustering, Classification, and Embedding. In: Schoelkopf, B., Platt, J.C., Hofmann, T. (eds.) Advances in Neural Information Processing System (NIPS), pp. 1601–1608. MIT Press, Cambridge (2007)
12. Pandey, G., Atluri, G., Steinbach, M., Kumar, V.: Association Analysis Techniques for Discovering Functional Modules from Microarray Data. In: Proc. ISMB Special Interest Group Meeting on Automated Function Prediction (2008)
13. Zhou, D., Schölkopf, B.: Regularization on Discrete Spaces. In: Kropatsch, W.G., Sablatnig, R., Hanbury, A. (eds.) DAGM 2005. LNCS, vol. 3663, pp. 361–368. Springer, Heidelberg (2005)
14. Zhou, D., Schoelkopf, B.: Discrete Regularization. In: Chapelle, O., Schoelkopf, B., Zien, A. (eds.) Semi-Supervised Learning, pp. 221–232. MIT Press, Cambridge (2006)
15. Pandey, G., Myers, L.C., Kumar, V.: Incorporating Functional Inter-relationships into Protein Function Prediction Algorithms. BMC Bioinformatics 10, 142 (2009)

On Horn Knowledge Bases in Regular Description Logic with Inverse

Linh Anh Nguyen, Thi-Bich-Loc Nguyen, and Andrzej Szałas

Abstract. We study a Horn fragment called Horn-$\mathscr{R}eg^I$ of the regular description logic with inverse $\mathscr{R}eg^I$, which extends the description logic $\mathscr{ALC}$ with inverse roles and regular role inclusion axioms characterized by finite automata. In contrast to the well-known Horn fragments $\mathscr{EL}$, DL-Lite, DLP, Horn-$\mathscr{SHIQ}$ and Horn-$\mathscr{SROIQ}$ of description logics, Horn-$\mathscr{R}eg^I$ allows a form of the concept constructor "universal restriction" to appear at the left hand side of terminological inclusion axioms, while still has PTIME data complexity. Namely, a universal restriction can be used in such places in conjunction with the corresponding existential restriction. We provide an algorithm with PTIME data complexity for checking satisfiability of Horn-$\mathscr{R}eg^I$ knowledge bases.

1 Introduction

Description logics (DLs) are variants of modal logics suitable for expressing terminological knowledge. They represent the domain of interest in terms of individuals

Linh Anh Nguyen
Institute of Informatics, University of Warsaw, Banacha 2, 02-097 Warsaw, Poland, and
Faculty of Information Technology, VNU University of Engineering and Technology,
144 Xuan Thuy, Hanoi, Vietnam
e-mail: nguyen@mimuw.edu.pl

Thi-Bich-Loc Nguyen
Department of Information Technology, Hue University of Sciences,
77 Nguyen Hue, Hue City, Vietnam
e-mail: ntbichloc@hueuni.edu.vn

Andrzej Szałas
Institute of Informatics, University of Warsaw, Banacha 2, 02-097 Warsaw, Poland, and
Dept. of Computer and Information Science, Linköping University,
SE-581 83 Linköping, Sweden
e-mail: andsz@mimuw.edu.pl

V.-N. Huynh et al. (eds.), *Knowledge and Systems Engineering, Volume 1*,
Advances in Intelligent Systems and Computing 244,
DOI: 10.1007/978-3-319-02741-8_6, © Springer International Publishing Switzerland 2014

(objects), concepts and roles. A concept stands for a set of individuals, a role stands for a binary relation between individuals. The DL $\mathcal{SROIQ}$ [8] founds the logical base of the Web Ontology Language OWL 2, which was recommended by W3C as a layer for the architecture of the Semantic Web.

As reasoning in $\mathcal{SROIQ}$ has a very high complexity, W3C also recommended the profiles OWL 2 EL, OWL 2 QL and OWL 2 RL, which are based on the families of DLs $\mathcal{EL}$ [1, 2], DL-Lite [4] and DLP [6]. These families of DLs are monotonic rule languages enjoying PTIME data complexity. They are defined by selecting suitable Horn fragments of the corresponding full languages with appropriate restrictions adopted to eliminate nondeterminism. A number of other Horn fragments of DLs with PTIME data complexity have also been investigated (see [12] for references). The fragments Horn-$\mathcal{SHIQ}$ [9] and Horn-$\mathcal{SROIQ}$ [17] are notable, with considerable rich sets of allowed constructors and features.

To eliminate nondeterminism, all $\mathcal{EL}$ [1, 2], DL-Lite [4], DLP [6], Horn-$\mathcal{SHIQ}$ [9] and Horn-$\mathcal{SROIQ}$ [17] disallow (any form of) the universal restriction $\forall R.C$ at the left hand side of $\sqsubseteq$ in terminological axioms. The problem is that the general Horn fragment of the basic DL $\mathcal{ALC}$ allowing $\forall R.C$ at the left hand side of $\sqsubseteq$ has NP-complete data complexity [11]. Also, roles are not required to be serial (i.e., satisfying the condition $\forall x \exists y R(x,y)$), which complicates the construction of (logically) least models. For many application domains, the profiles OWL 2 EL, OWL 2 QL and OWL 2 RL languages and the underlying Horn fragments $\mathcal{EL}$, DL-Lite, DLP seem satisfactory. However, in general, forbidding $\forall R.C$ at the left hand side of $\sqsubseteq$ in terminological axioms is a serious restriction.

In [10] Nguyen introduced the deterministic Horn fragment of $\mathcal{ALC}$, where the constructor $\forall R.C$ is allowed at the left hand side of $\sqsubseteq$ in the combination with $\exists R.C$ (in the form $\forall R.C \sqcap \exists R.C$, denoted by $\forall\exists R.C$ [3]). He proved that such a fragment has PTIME data complexity by providing a bottom-up method for constructing a (logically) least model for a given deterministic positive knowledge base in the restricted language. In [11] Nguyen applied the method of [10] to regular DL $\mathcal{R}eg$, which extends $\mathcal{ALC}$ with regular role inclusion axioms characterized by finite automata. Let us denote the Horn fragment of $\mathcal{R}eg$ that allows the constructor $\forall\exists R.C$ at the left hand side of $\sqsubseteq$ by Horn-$\mathcal{R}eg$. As not every positive Horn-$\mathcal{R}eg$ knowledge base has a (logically) least model, Nguyen [11] proposed to approximate the instance checking problem in Horn-$\mathcal{R}eg$ by using its weakenings with PTIME data complexity.

The works [10, 11] found a starting point for the research concerning the universal restriction $\forall R.C$ at the left hand side of $\sqsubseteq$ in terminological axioms guaranteeing PTIME data complexity. However, a big challenge is faced: the bottom-up approach is used, but not every positive Horn-$\mathcal{R}eg$ knowledge base has a logically least model. As a consequence, the work [11] on Horn-$\mathcal{R}eg$ is already very complicated and the problem whether Horn-$\mathcal{R}eg$ has PTIME data complexity still remained open.

The goal of our research is to develop a Horn fragment of a DL (and therefore a rule language for the Semantic Web) that is substantially richer than all well-known Horn fragments $\mathcal{EL}$, DL-Lite, Horn-$\mathcal{R}eg$, Horn-$\mathcal{SHIQ}$, Horn-$\mathcal{SROIQ}$ as well as Horn-$\mathcal{R}eg$, while still has PTIME data complexity. Recently, we have succeeded to reach the goal by introducing such a Horn fragment, Horn-DL, and proving its PTIME data complexity [13]. In comparison with Horn-$\mathcal{SROIQ}$, Horn-DL additionally allows the universal role and assertions of the form *irreflexive*(s), $\neg s(a,b)$, $a \not\doteq b$. The most important feature of Horn-DL, however, is to allow the concept constructor $\forall\exists R.C$ to appear at the left hand side of $\sqsubseteq$ in terminological inclusion axioms. The (unpublished) manuscript [13] is too long for a conference paper. We are just publishing a short technical communication [14] without explanations and proofs.

In the current paper we present and explain the technique of [13]. Due to the lack of space, we simplify and shorten the presentation by discussing the technique only for the Horn fragment Horn-$\mathcal{R}eg^I$ of the DL $\mathcal{R}eg^I$, which extends $\mathcal{R}eg$ with inverse roles. Omitting the other additional features of Horn-DL allows us to concentrate on the concept constructor $\forall\exists R.C$. We provide an algorithm with PTIME data complexity for checking satisfiability of Horn-$\mathcal{R}eg^I$ knowledge bases. The key idea is to follow the top-down approach[1] and use a special technique to deal with non-seriality of roles.

The DL $\mathcal{R}eg^I$ is a variant of regular grammar logic with converse [5, 15]. The current work is based on the previous works [10, 11, 16]. Namely, [16] considers Horn fragments of serial regular grammar logics with converse. The current work exploits the technique of [16] in dealing with converse (like inverse roles), but the difference is that it concerns *non-serial* regular DL with inverse roles. The change from grammar logic (i.e., modal logic) to DL is syntactic, but may increase the readability for the DL community.

The main achievements of the current paper are that:

- it overcomes the difficulties encountered in [10, 11] by using the top-down rather than bottom-up approach, and thus enables to show that both Horn-$\mathcal{R}eg$ and Horn-$\mathcal{R}eg^I$ have PTIME data complexity, solving an open problem of [11];
- the technique introduced in the current paper for dealing with non-seriality leads to a solution for the important issue of allowing the concept constructor $\forall\exists R.C$ to appear at the left hand side of $\sqsubseteq$ in terminological inclusion axioms.

The rest of this paper is structured as follows. In Section 2 we present notation and semantics of $\mathcal{R}eg^I$ and recall automaton-modal operators. In Section 3 we define the Horn-$\mathcal{R}eg^I$ fragment. In Section 4 we present our algorithm of checking satisfiability of Horn-$\mathcal{R}eg^I$ knowledge bases and discuss our technique of dealing with $\forall\exists R.C$ at the left hand side of $\sqsubseteq$. We conclude this work in Section 5.

2 Preliminaries

2.1 Notation and Semantics of $\mathcal{R}eg^I$

Our language uses a finite set $\mathbf{C}$ of *concept names*, a finite set $\mathbf{R}_+$ of *role names* including a subset of *simple role names*, and a finite set $\mathbf{I}$ of *individual names*. We

[1] In the top-down approach, the considered query is negated and added into the knowledge base, and in general, a knowledge base may contain "negative" constraints.

use letters like a, b to denote individual names, letters like A, B to denote concept names, and letters like r, s to denote role names.

For $r \in \mathbf{R}_+$, we call the expression $\overline{r}$ the *inverse* of r. Let $\mathbf{R}_- = \{\overline{r} \mid r \in \mathbf{R}_+\}$ and $\mathbf{R} = \mathbf{R}_+ \cup \mathbf{R}_-$. For $R = \overline{r}$, let $\overline{R}$ stand for r. We call elements of $\mathbf{R}$ *roles* and use letters like R, S to denote them. We define a *simple role* to be either a simple role name or the inverse of a simple role name.

A *context-free semi-Thue system* $\mathscr{S}$ over $\mathbf{R}$ is a finite set of context-free production rules over alphabet $\mathbf{R}$. It is *symmetric* if, for every rule $R \to S_1 \dots S_k$ of $\mathscr{S}$, the rule $\overline{R} \to \overline{S}_k \dots \overline{S}_1$ is also in $\mathscr{S}$.[2] It is *regular* if, for every $R \in \mathbf{R}$, the set of words derivable from R using the system is a regular language over $\mathbf{R}$.

A context-free semi-Thue system is like a context-free grammar, but it has no designated start symbol and there is no distinction between terminal and non-terminal symbols. We assume that, for $R \in \mathbf{R}$, the word R is derivable from R using such a system.

A *role inclusion axiom* (RIA for short) is an expression of the form $S_1 \circ \dots \circ S_k \sqsubseteq R$, where $k \geq 0$. In the case $k = 0$, the left hand side of the inclusion axiom stands for the empty word ε.

A *regular RBox* $\mathscr{R}$ is a finite set of RIAs such that

$$\{R \to S_1 \dots S_k \mid (S_1 \circ \dots \circ S_k \sqsubseteq R) \in \mathscr{R}\}$$

is a symmetric regular semi-Thue system $\mathscr{S}$ over $\mathbf{R}$ such that if $R \in \mathbf{R}$ is a simple role then only words with length 1 or 0 are derivable from R using $\mathscr{S}$. We assume that $\mathscr{R}$ is given together with a mapping $\mathbf{A}$ that associates every $R \in \mathbf{R}$ with a finite automaton $\mathbf{A}_R$ recognizing the words derivable from R using $\mathscr{S}$. We call $\mathbf{A}$ the *RIA-automaton-specification* of $\mathscr{R}$.

Let $\mathscr{R}$ be a regular RBox and $\mathbf{A}$ be its RIA-automaton-specification. For $R, S \in \mathbf{R}$, we say that R is a *subrole* of S w.r.t. $\mathscr{R}$, denoted by $R \sqsubseteq_{\mathscr{R}} S$, if the word R is accepted by $\mathbf{A}_S$.

Concepts are defined by the following BNF grammar, where $A \in \mathbf{C}$, $R \in \mathbf{R}$:

$$C ::= \top \mid \bot \mid A \mid \neg C \mid C \sqcap C \mid C \sqcup C \mid \forall R.C \mid \exists R.C$$

We use letters like C, D to denote concepts (including complex concepts).

A *TBox* is a finite set of *TBox axioms* of the form $C \sqsubseteq D$. An *ABox* is a finite set of *assertions* of the form $C(a)$ or $r(a,b)$. A *knowledge base* is a tuple $\langle \mathscr{R}, \mathscr{T}, \mathscr{A} \rangle$, where $\mathscr{R}$ is a regular RBox, $\mathscr{T}$ is a TBox and $\mathscr{A}$ is an ABox.

An *interpretation* is a pair $\mathscr{I} = \langle \Delta^{\mathscr{I}}, \cdot^{\mathscr{I}} \rangle$, where $\Delta^{\mathscr{I}}$ is a non-empty set called the *domain* of $\mathscr{I}$ and $\cdot^{\mathscr{I}}$ is a mapping called the *interpretation function* of $\mathscr{I}$ that associates each individual name $a \in \mathbf{I}$ with an element $a^{\mathscr{I}} \in \Delta^{\mathscr{I}}$, each concept name $A \in \mathbf{C}$ with a set $A^{\mathscr{I}} \subseteq \Delta^{\mathscr{I}}$, and each role name $r \in \mathbf{R}_+$ with a binary relation $r^{\mathscr{I}} \subseteq \Delta^{\mathscr{I}} \times \Delta^{\mathscr{I}}$. Define $(\overline{r})^{\mathscr{I}} = (r^{\mathscr{I}})^{-1} = \{\langle y,x \rangle \mid \langle x,y \rangle \in r^{\mathscr{I}}\}$ (for $r \in \mathbf{R}_+$) and $\varepsilon^{\mathscr{I}} = \{\langle x,x \rangle \mid x \in \Delta^{\mathscr{I}}\}$. The interpretation function $\cdot^{\mathscr{I}}$ is extended to complex concepts as follows:

---

[2] In the case $k = 0$, the left hand side of the latter inclusion axiom also stands for ε.

$$
\begin{aligned}
&\top^{\mathscr{I}} = \Delta^{\mathscr{I}}, \quad \bot^{\mathscr{I}} = \emptyset, \quad (\neg C)^{\mathscr{I}} = \Delta^{\mathscr{I}} \setminus C^{\mathscr{I}}, \\
&(C \sqcap D)^{\mathscr{I}} = C^{\mathscr{I}} \cap D^{\mathscr{I}}, \quad (C \sqcup D)^{\mathscr{I}} = C^{\mathscr{I}} \cup D^{\mathscr{I}}, \\
&(\forall R.C)^{\mathscr{I}} = \{x \in \Delta^{\mathscr{I}} \mid \forall y(\langle x,y\rangle \in R^{\mathscr{I}} \Rightarrow y \in C^{\mathscr{I}})\}, \\
&(\exists R.C)^{\mathscr{I}} = \{x \in \Delta^{\mathscr{I}} \mid \exists y(\langle x,y\rangle \in R^{\mathscr{I}} \land y \in C^{\mathscr{I}})\}.
\end{aligned}
$$

Given an interpretation $\mathscr{I}$ and an axiom/assertion φ, the satisfaction relation $\mathscr{I} \models \varphi$ is defined as follows, where $\circ$ at the right hand side of "if" stands for composition of relations:

$$
\begin{array}{ll}
\mathscr{I} \models S_1 \circ \cdots \circ S_k \sqsubseteq R & \text{if } S_1^{\mathscr{I}} \circ \cdots \circ S_k^{\mathscr{I}} \subseteq R^{\mathscr{I}} \\
\mathscr{I} \models \varepsilon \sqsubseteq R & \text{if } \varepsilon^{\mathscr{I}} \sqsubseteq R^{\mathscr{I}} \\
\mathscr{I} \models C \sqsubseteq D & \text{if } C^{\mathscr{I}} \subseteq D^{\mathscr{I}} \\
\mathscr{I} \models C(a) & \text{if } a^{\mathscr{I}} \in C^{\mathscr{I}} \\
\mathscr{I} \models r(a,b) & \text{if } \langle a^{\mathscr{I}}, b^{\mathscr{I}}\rangle \in r^{\mathscr{I}}.
\end{array}
$$

If $\mathscr{I} \models \varphi$ then we say that $\mathscr{I}$ *validates* φ.

An interpretation $\mathscr{I}$ is a *model* of an RBox $\mathscr{R}$, a TBox $\mathscr{T}$ or an ABox $\mathscr{A}$ if it validates all the axioms/assertions of that "box". It is a *model* of a knowledge base $\langle \mathscr{R}, \mathscr{T}, \mathscr{A} \rangle$ if it is a model of all $\mathscr{R}$, $\mathscr{T}$ and $\mathscr{A}$.

A knowledge base is *satisfiable* if it has a model. For a knowledge base KB, we write $KB \models \varphi$ to mean that every model of KB validates φ. If $KB \models C(a)$ then we say that a is an *instance* of C w.r.t. KB.

2.2 Automaton-Modal Operators

Recall that a *finite automaton* A over alphabet $\mathbf{R}$ is a tuple $\langle \mathbf{R}, Q, q_0, \delta, F \rangle$, where Q is a finite set of states, $q_0 \in Q$ is the initial state, $\delta \subseteq Q \times \mathbf{R} \times Q$ is the transition relation, and $F \subseteq Q$ is the set of accepting states. A *run* of A on a word $R_1 \ldots R_k$ over alphabet $\mathbf{R}$ is a finite sequence of states $q_0, q_1, \ldots, q_k$ such that $\delta(q_{i-1}, R_i, q_i)$ holds for every $1 \leq i \leq k$. It is an *accepting run* if $q_k \in F$. We say that A *accepts* a word w if there exists an accepting run of A on w.

Given an interpretation $\mathscr{I}$ and a finite automaton A over alphabet $\mathbf{R}$, define $\mathsf{A}^{\mathscr{I}} = \{\langle x,y\rangle \in \Delta^{\mathscr{I}} \times \Delta^{\mathscr{I}} \mid$ there exist a word $R_1 \ldots R_k$ accepted by A and elements $x_0 = x$, $x_1, \ldots, x_k = y$ of $\Delta^{\mathscr{I}}$ such that $\langle x_{i-1}, x_i \rangle \in R_i^{\mathscr{I}}$ for all $1 \leq i \leq k\}$.

We will use auxiliary modal operators $[\mathsf{A}]$ and $\langle \mathsf{A} \rangle$, where A is a finite automaton over alphabet $\mathbf{R}$. We call $[\mathsf{A}]$ (resp. $\langle \mathsf{A} \rangle$) a *universal* (resp. *existential*) *automaton-modal operator*. Automaton-modal operators were used earlier, among others, in [7, 11].

In the *extended language*, if C is a concept then $[\mathsf{A}]C$ and $\langle \mathsf{A} \rangle C$ are also concepts. The semantics of $[\mathsf{A}]C$ and $\langle \mathsf{A} \rangle C$ are defined as follows:

$$
\begin{aligned}
([\mathsf{A}]C)^{\mathscr{I}} &= \{x \in \Delta^{\mathscr{I}} \mid \forall y(\langle x,y\rangle \in \mathsf{A}^{\mathscr{I}} \text{ implies } y \in C^{\mathscr{I}})\} \\
(\langle \mathsf{A} \rangle C)^{\mathscr{I}} &= \{x \in \Delta^{\mathscr{I}} \mid \exists y(\langle x,y\rangle \in \mathsf{A}^{\mathscr{I}} \text{ and } y \in C^{\mathscr{I}})\}.
\end{aligned}
$$

For a finite automaton A over $\mathbf{R}$, assume that $\mathsf{A} = \langle \mathbf{R}, Q_\mathsf{A}, q_\mathsf{A}, \delta_\mathsf{A}, F_\mathsf{A} \rangle$.

If q is a state of a finite automaton A then by A_q we denote the finite automaton obtained from A by replacing the initial state by q.

The following lemma can be proved in a straightforward way.

Lemma 0.1. *Let $\mathcal{I}$ be a model of a regular RBox $\mathcal{R}$, $\mathbf{A}$ be the RIA-automaton-specification of $\mathcal{R}$, C be a concept, and $R \in \mathbf{R}$. Then $(\forall R.C)^\mathcal{I} = ([\mathbf{A}_R]C)^\mathcal{I}$, $(\exists R.C)^\mathcal{I} = (\langle \mathbf{A}_R \rangle C)^\mathcal{I}$, $C^\mathcal{I} \subseteq ([\mathbf{A}_{\overline{R}}]\langle \mathbf{A}_R \rangle C)^\mathcal{I}$ and $C^\mathcal{I} \subseteq ([\mathbf{A}_{\overline{R}}]\exists R.C)^\mathcal{I}$.*

3 The Horn-$\mathcal{R}eg^I$ Fragment

Let $\forall\exists R.C$ stand for $\forall R.C \sqcap \exists R.C$. *Left-hand-side Horn-$\mathcal{R}eg^I$ concepts*, called *LHS Horn-$\mathcal{R}eg^I$ concepts* for short, are defined by the following grammar, where $A \in \mathbf{C}$ and $R \in \mathbf{R}$:

$$C ::= \top \mid A \mid C \sqcap C \mid C \sqcup C \mid \forall\exists R.C \mid \exists R.C$$

Right-hand-side Horn-$\mathcal{R}eg^I$ concepts, called *RHS Horn-$\mathcal{R}eg^I$ concepts* for short, are defined by the following BNF grammar, where $A \in \mathbf{C}$, D is an LHS Horn-$\mathcal{R}eg^I$ concept, and $R \in \mathbf{R}$:

$$C ::= \top \mid \bot \mid A \mid \neg D \mid C \sqcap C \mid \neg D \sqcup C \mid \forall R.C \mid \exists R.C$$

A *Horn-$\mathcal{R}eg^I$ TBox axiom*, is an expression of the form $C \sqsubseteq D$, where C is an LHS Horn-$\mathcal{R}eg^I$ concept and D is an RHS Horn-$\mathcal{R}eg^I$ concept.

A *Horn-$\mathcal{R}eg^I$ TBox* is a finite set of *Horn-$\mathcal{R}eg^I$ TBox axioms*.

A *Horn-$\mathcal{R}eg^I$ clause* is a Horn-$\mathcal{R}eg^I$ TBox axiom of the form $C_1 \sqcap \ldots \sqcap C_k \sqsubseteq D$ or $\top \sqsubseteq D$, where:

- each C_i is of the form A, $\forall\exists R.A$ or $\exists R.A$,
- D is of the form $\bot$, A, $\forall R.A$ or $\exists R.A$,
- $k \geq 1$, $A \in \mathbf{C}$ and $R \in \mathbf{R}$.

A *clausal Horn-$\mathcal{R}eg^I$ TBox* is a TBox consisting of Horn-$\mathcal{R}eg^I$ clauses.

A *Horn-$\mathcal{R}eg^I$ ABox* is a finite set of assertions of the form $C(a)$ or $r(a,b)$, where C is an RHS Horn-$\mathcal{R}eg^I$ concept. A *reduced ABox* is a finite set of assertions of the form $A(a)$ or $r(a,b)$.

A knowledge base $\langle \mathcal{R}, \mathcal{T}, \mathcal{A} \rangle$ is called a *Horn-$\mathcal{R}eg^I$ knowledge base* if $\mathcal{T}$ is a Horn-$\mathcal{R}eg^I$ TBox and $\mathcal{A}$ is a Horn-$\mathcal{R}eg^I$ ABox. When $\mathcal{T}$ is a clausal Horn-$\mathcal{R}eg^I$ TBox and $\mathcal{A}$ is a reduced ABox, we call such a knowledge base a *clausal Horn-$\mathcal{R}eg^I$ knowledge base*.

Example 0.1. This example is about Web pages. Let $\mathbf{R}_+ = \{link, path\}$ and let $\mathcal{R}$ be the regular RBox consisting of the following role axioms:

$$link \sqsubseteq path, \qquad \overline{link} \sqsubseteq \overline{path},$$
$$link \circ path \sqsubseteq path, \qquad \overline{path} \circ \overline{link} \sqsubseteq \overline{path}.$$

This RBox "defines" *path* to be the transitive closure of *link*. As the RIA-automaton-specification of $\mathscr{R}$ we can take the mapping $\mathbf{A}$ such that:

$$\begin{aligned}
\mathbf{A}_{link} &= \langle \mathbf{R}, \{1,2\}, 1, \{(1, link, 2)\}, \{2\}\rangle \\
\mathbf{A}_{\overline{link}} &= \langle \mathbf{R}, \{1,2\}, 2, \{(2, \overline{link}, 1)\}, \{1\}\rangle \\
\mathbf{A}_{path} &= \langle \mathbf{R}, \{1,2\}, 1, \{(1, link, 1), (1, link, 2), (1, path, 2)\}, \{2\}\rangle \\
\mathbf{A}_{\overline{path}} &= \langle \mathbf{R}, \{1,2\}, 2, \{(1, \overline{link}, 1), (2, \overline{link}, 1), (2, \overline{path}, 1)\}, \{1\}\rangle.
\end{aligned}$$

Let $\mathscr{T}$ be the TBox consisting of the following program clauses:

$$\begin{aligned}
&perfect \sqsubseteq interesting \sqcap \forall path.interesting \\
&interesting \sqcap \forall\exists path.interesting \sqsubseteq perfect \\
&interesting \sqcup \forall\exists link.interesting \sqsubseteq worth_surfing.
\end{aligned}$$

Let $\mathscr{A}$ be the ABox specified by the concept assertions $perfect(b)$ and the following role assertions of *link*:

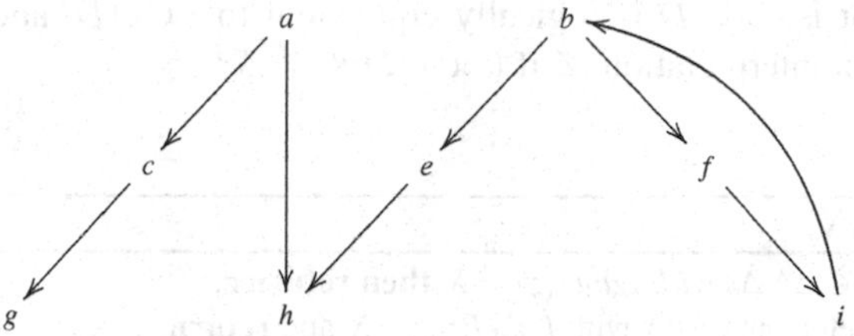

Then $KB = \langle \mathscr{R}, \mathscr{T}, \mathscr{A}\rangle$ is a Horn-$\mathscr{R}eg^I$ knowledge base. (Ignoring $\overline{link}$ and $\overline{path}$, which are not essential in this example, *KB* can be treated as a Horn-$\mathscr{R}eg$ knowledge base.) It can be seen that b, e, f, i are instances of the concepts *perfect*, *interesting*, *worth_surfing* w.r.t. *KB*. Furthermore, h is also an instance of the concept *interesting* w.r.t. *KB*. ◁

The *length* of a concept, an assertion or an axiom φ is the number of symbols occurring in φ. The *size* of an ABox is the sum of the lengths of its assertions. The *size* of a TBox is the sum of the lengths of its axioms.

The *data complexity* class of Horn-$\mathscr{R}eg^I$ is defined to be the complexity class of the problem of checking satisfiability of a Horn-$\mathscr{R}eg^I$ knowledge base $\langle \mathscr{R}, \mathscr{T}, \mathscr{A}\rangle$, measured in the size of $\mathscr{A}$ when assuming that $\mathscr{R}$ and $\mathscr{T}$ are fixed and $\mathscr{A}$ is a reduced ABox.

Proposition 0.1. *Let $KB = \langle \mathscr{R}, \mathscr{T}, \mathscr{A}\rangle$ be a Horn-$\mathscr{R}eg^I$ knowledge base.*

1. *If C is an LHS Horn-$\mathscr{R}eg^I$ concept then $KB \models C(a)$ iff the Horn-$\mathscr{R}eg^I$ knowledge base $\langle \mathscr{R}, \mathscr{T} \cup \{C \sqsubseteq A\}, \mathscr{A} \cup \{\neg A(a)\}\rangle$ is unsatisfiable, where A is a fresh concept name.*
2. *KB can be converted in polynomial time in the sizes of $\mathscr{T}$ and $\mathscr{A}$ to a Horn-$\mathscr{R}eg^I$ knowledge base $KB' = \langle \mathscr{R}, \mathscr{T}', \mathscr{A}'\rangle$ with $\mathscr{A}'$ being a reduced ABox such that KB is satisfiable iff KB' is satisfiable.*

3. *KB can be converted in polynomial time in the size of $\mathscr{T}$ to a Horn-$\mathscr{R}eg^I$ knowledge base $KB' = \langle \mathscr{R}, \mathscr{T}', \mathscr{A} \rangle$ with $\mathscr{T}'$ being a clausal Horn-$\mathscr{R}eg^I$ TBox such that KB is satisfiable iff KB' is satisfiable.*

See the long version [12] for the proof of this proposition.

Corollary 0.1. *Every Horn-$\mathscr{R}eg^I$ knowledge base $KB = \langle \mathscr{R}, \mathscr{T}, \mathscr{A} \rangle$ can be converted in polynomial time in the sizes of $\mathscr{T}$ and $\mathscr{A}$ to a clausal Horn-$\mathscr{R}eg^I$ knowledge base $KB' = \langle \mathscr{R}, \mathscr{T}', \mathscr{A}' \rangle$ such that KB is satisfiable iff KB' is satisfiable.*

4 Checking Satisfiability of Horn-$\mathscr{R}eg^I$ Knowledge Bases

In this section we present an algorithm that, given a clausal Horn-$\mathscr{R}eg^I$ knowledge base $\langle \mathscr{R}, \mathscr{T}, \mathscr{A} \rangle$ together with the RIA-automaton-specification $\mathbf{A}$ of $\mathscr{R}$, checks whether the knowledge base is satisfiable. The algorithm has PTIME data complexity.

We will treat each TBox axiom $C \sqsubseteq D$ from $\mathscr{T}$ as a concept standing for a global assumption. That is, $C \sqsubseteq D$ is logically equivalent to $\neg C \sqcup D$, and it is a global assumption for an interpretation $\mathscr{I}$ if $(\neg C \sqcup D)^{\mathscr{I}} = \Delta^{\mathscr{I}}$.

Function Find(X)

if *there exists* $z \in \Delta \setminus \Delta_0$ *with* $Label(z) = X$ **then return** z;
else add a new element z to Δ with $Label(z) := X$ and **return** z;
;

Procedure ExtendLabel(z, X)

if $X \subseteq Label(z)$ **then return**;
;
if $z \in \Delta_0$ **then** $Label(z) := Label(z) \cup \mathsf{Satr}(X)$;
else
  $z_* := \mathtt{Find}(Label(z) \cup \mathsf{Satr}(X))$;
  foreach y, R, C *such that* $Next(y, \exists R.C) = z$ **do** $Next(y, \exists R.C) := z_*$;

Function CheckPremise(x, C)

if $C = \top$ **then return** *true*;
else let $C = C_1 \sqcap \ldots \sqcap C_k$;
;
foreach $1 \leq i \leq k$ **do**
  if $C_i = A$ *and* $A \notin Label(x)$ **then return** *false*;
  else if $C_i = \forall \exists R.A$ *and* ($\exists R.\top \notin Label(x)$ *or* $Next(x, \exists R.\top)$ *is not defined or* $A \notin Label(Next(x, \exists R.\top))$) **then return** *false*;
  else if $C_i = \exists R.A$ *and* $\langle \mathbf{A}_R \rangle A \notin Label(x)$ **then return** *false*;
return *true*;

Algorithm 1: checking satisfiability of a clausal Horn-$\mathscr{R}eg^I$ knowledge base

Input: a clausal Horn-$\mathscr{R}eg^I$ knowledge base $\langle \mathscr{R}, \mathscr{T}, \mathscr{A} \rangle$ and the RIA-automaton-specification $\mathbf{A}$ of $\mathscr{R}$.
Output: *true* if $\langle \mathscr{R}, \mathscr{T}, \mathscr{A} \rangle$ is satisfiable, or *false* otherwise.

let Δ_0 be the set of all individuals occurring in $\mathscr{A}$;
if $\Delta_0 = \emptyset$ **then** $\Delta_0 := \{\tau\}$;
;
$\Delta := \Delta_0$, $\mathscr{T}' := \mathsf{Satr}(\mathscr{T})$, and set *Next* to the empty mapping;
foreach $a \in \Delta_0$ **do** $Label(a) := \mathsf{Satr}(\{A \mid A(a) \in \mathscr{A}\}) \cup \mathscr{T}'$;
;
repeat
    foreach $r(a,b) \in \mathscr{A}$ **do** `ExtendLabel`$(b, \mathsf{Trans}(Label(a), r))$;
    ;
    foreach x, $\exists R.C$, y *s.t.* x *is reachable from* Δ_0 *and* $Next(x, \exists R.C) = y$ **do**
        $Next(x, \exists R.C) :=$ `Find`$(Label(y) \cup \mathsf{Satr}(\mathsf{Trans}(Label(x), R)))$;
    foreach $\langle x, R, y \rangle \in Edges$ *such that* x *is reachable from* Δ_0 **do**
        `ExtendLabel`$(x, \mathsf{Trans}(Label(y), \overline{R}))$
    foreach $x \in \Delta$ *reachable from* Δ_0 *and* $\exists R.C \in Label(x)$ *with* $R \in \mathbf{R}$ **do**
        if $Next(x, \exists R.C)$ *is not defined* **then**
            $Next(x, \exists R.C) :=$ `Find`$(\mathsf{Satr}(\{C\} \cup \mathsf{Trans}(Label(x), R)) \cup \mathscr{T}')$
    foreach $x \in \Delta$ *reachable from* Δ_0 *and* $(C \sqsubseteq D) \in Label(x)$ **do**
        if `CheckPremise`(x, C) **then** `ExtendLabel`$(x, \{D\})$;
    if *there exists* $x \in \Delta$ *such that* $\bot \in Label(x)$ **then return** *false*;
    ;
until *no changes occurred in the last iteration*;
return *true*;

Let X be a set of concepts. The *saturation* of X (w.r.t. $\mathbf{A}$ and $\mathscr{T}$), denoted by $\mathsf{Satr}(X)$, is defined to be the least extension of X such that:

- if $\forall R.C \in \mathsf{Satr}(X)$ then $[\mathbf{A}_R]C \in \mathsf{Satr}(X)$,
- if $[\mathsf{A}]C \in \mathsf{Satr}(X)$ and $q_{\mathsf{A}} \in F_{\mathsf{A}}$ then $C \in \mathsf{Satr}(X)$,
- if $\forall\exists R.A$ occurs in $\mathscr{T}$ for some A then $[\mathbf{A}_{\overline{R}}]\exists R.\top \in \mathsf{Satr}(X)$,
- if $A \in \mathsf{Satr}(X)$ and $\exists R.A$ occurs at the left hand side of $\sqsubseteq$ in some clause of $\mathscr{T}$ then $[\mathbf{A}_{\overline{R}}]\langle \mathbf{A}_R \rangle A \in \mathsf{Satr}(X)$.

Remark 0.1. Note the third item in the above list. It is used for dealing with non-seriality and the concept constructor $\forall\exists R.A$. Another treatment for the problem of non-seriality and $\forall\exists R.A$ is the step 6 of Function `CheckPremise` (used in our algorithm) and will be explained later.

For $R \in \mathbf{R}$, the *transfer* of X through R is

$$\mathsf{Trans}(X, R) = \{[\mathsf{A}_q]C \mid [\mathsf{A}]C \in X \text{ and } \langle q_{\mathsf{A}}, R, q \rangle \in \delta_{\mathsf{A}}\}.$$

Algorithm 2 (on page 305) checks satisfiability of $\langle \mathscr{R}, \mathscr{T}, \mathscr{A} \rangle$. It uses the following data structures:

- Δ_0 : the set consisting of all individual names occurring in $\mathscr{A}$,
- Δ : a set of objects including Δ_0,
- *Label* : a function mapping each $x \in \Delta$ to a set of concepts,
- *Next* : $\Delta \times \{\exists R.\top, \exists R.A \mid R \in \mathbf{R}, A \in \mathbf{C}\} \to \Delta$ is a partial mapping.

For $x \in \Delta$, $Label(x)$ is called the *label* of x. A fact $Next(x, \exists R.C) = y$ means that $\exists R.C \in Label(x)$, $C \in Label(y)$, and $\exists R.C$ is "realized" at x by going to y. When defined, $Next(x, \exists R.\top)$ denotes the "logically smallest" R-successor of x.

Define $Edges = \{\langle x, R, y\rangle \mid R(x,y) \in \mathscr{A}$ or $Next(x, \exists R.C) = y$ for some $C\}$.

We say that $x \in \Delta$ is *reachable* from Δ_0 if there exist $x_0, \ldots, x_k \in \Delta$ and elements $R_1, \ldots, R_k$ of $\mathbf{R}$ such that $k \geq 0$, $x_0 \in \Delta_0$, $x_k = x$ and $\langle x_{i-1}, R_i, x_i\rangle \in Edges$ for all $1 \leq i \leq k$.

Algorithm 2 attempts to construct a model of $\langle \mathscr{R}, \mathscr{T}, \mathscr{A}\rangle$. The intended model extends $\mathscr{A}$ with disjoint trees rooted at the named individuals occurring in $\mathscr{A}$. The trees may be infinite. However, we represent such a semi-forest as a graph with global caching: if two nodes that are not named individuals occur in a tree or in different trees and have the same label, then they should be merged. In other words, for every finite set X of concepts, the graph contains at most one node $z \in \Delta \setminus \Delta_0$ such that $Label(z) = X$. The function $\texttt{Find}(X)$ (on page 44) returns such a node z if it exists, or creates such a node z otherwise. A tuple $\langle x, R, y\rangle \in Edges$ represents an edge $\langle x, y\rangle$ with label R of the graph. The notions of *predecessor* and *successor* are defined as usual.

For each $x \in \Delta$, $Label(x)$ is a set of requirements to be "realized" at x. To realize such requirements at nodes, sometimes we have to extend their labels. Suppose we want to extend the label of $z \in \Delta$ with a set X of concepts. Consider the following cases:

- Case $z \in \Delta_0$ (i.e., z is a named individual occurring in $\mathscr{A}$): as z is "fixed" by the ABox $\mathscr{A}$, we have no choice but to extend $Label(z)$ directly with $\mathsf{Satr}(X)$.
- Case $z \notin \Delta_0$ and the requirements X are directly caused by z itself or its successors: if we directly extend the label of z (with $\mathsf{Satr}(X)$) then z will possibly have the same label as another node not belonging to Δ_0 and global caching is not fulfilled. Hence, we "simulate" changing the label of z by using $z_* := \texttt{Find}(Label(z) \cup \mathsf{Satr}(X))$ for playing the role of z. In particular, for each y, R and C such that $Next(y, \exists R.C) = z$, we set $Next(y, \exists R.C) := z_*$.

Extending the label of z for the above two cases is done by Procedure $\texttt{ExtendLabel}(z, X)$ (on page 44). The third case is considered below.

Suppose that $Next(x, \exists R.C) = y$. Then, to realize the requirements at x, the label of y should be extended with $X = \mathsf{Satr}(\mathsf{Trans}(Label(x), R))$. How can we realize such an extension? Recall that we intend to construct a forest-like model for $\langle \mathscr{R}, \mathscr{T}, \mathscr{A}\rangle$, but use global caching to guarantee termination. There may exist another $Next(x', \exists R'.C') = y$ with $x' \neq x$. That is, we may use y as a successor for two different nodes x and x', but the intention is to put x and x' into disjoint trees. If we directly modify the label of y to realize the requirements of x, such a modification may affect x'. The solution is to delete the edge $\langle x, R, y\rangle$ and reconnect x to

$y_* := \texttt{Find}(Label(y) \cup X)$ by setting $Next(x, \exists R.C) := y_*$. The extension is formally realized by steps 10-11 of Algorithm 2.

Consider the other main steps of Algorithm 2:

- Step 8: If $r(a,b) \in \mathscr{A}$ then we extend $Label(b)$ with $\mathsf{Satr}(\mathsf{Trans}(Label(a), R))$.
- Steps 12-13: If $\langle x, R, y \rangle \in Edges$ then we extend the label of x with $\mathsf{Trans}(Label(y), \overline{R})$ by using the procedure `ExtendLabel` discussed earlier.
- Steps 14-16: If $\exists R.C \in Label(x)$ and $Next(x, \exists R.C)$ is not defined yet then to realize the requirement $\exists R.C$ at x we connect x via R to a node with label $X = \mathsf{Satr}(\{C\} \cup \mathsf{Trans}(Label(x), R) \cup \mathscr{T})$ by setting $Next(x, \exists R.C) := \texttt{Find}(X)$.
- Steps 17-18: If $(C \sqsubseteq D) \in Label(x)$ and C "holds" at x then we extend the label of x with $\{D\}$ by using the procedure `ExtendLabel` discussed earlier. Suppose $C = C_1 \sqcap \ldots \sqcap C_k$. How to check whether C "holds" at x? It "holds" at x if C_i "holds" at x for each $1 \leq i \leq k$. There are the following cases:
  - Case $C_i = A$: C_i "holds" at x if $A \in Label(x)$.
  - Case $C_i = \forall\exists R.A$: C_i "holds" at x if both $\forall R.A$ and $\exists R.\top$ "hold" at x. If $\exists R.\top$ "holds" at x by the evidence of a path connecting x to a node z with (forward or backward) "edges" labeled by $S_1, \ldots, S_k$ such that the word $S_1 \ldots S_k$ is accepted by the automaton $\mathsf{A} = \mathbf{A}_R$, that is:
    - there exist nodes $x_0, \ldots, x_k$ such that $x_0 = x$, $x_k = z$ and, for each $1 \leq j \leq k$, either $\langle x_{j-1}, S_j, x_j \rangle \in Edges$ or $\langle x_j, \overline{S}_j, x_{j-1} \rangle \in Edges$,
    - there exist states $q_0, \ldots, q_k$ of A such that $q_0 = q_{\mathsf{A}}$, $q_k \in Q_{\mathsf{A}}$ and, for each $1 \leq j \leq k$, $\langle q_{j-1}, S_j, q_j \rangle \in \delta_{\mathsf{A}}$,

    then, with $\overline{\mathsf{A}} = \mathbf{A}_{\overline{R}}$, we have that:
    - since $Label(z)$ is saturated, $[\mathbf{A}_{\overline{R}}]\exists R.\top \in Label(z)$, which means $[\overline{\mathsf{A}}_{q_k}]\exists R.\top \in Label(x_k)$,
    - by the steps 8–13 of Algorithm 2, for each j from $k-1$ to 0, we can expect that $[\overline{\mathsf{A}}_{q_j}]\exists R.\top \in Label(x_j)$,
    - consequently, since $q_0 = q_{\mathsf{A}} \in Q_{\overline{\mathsf{A}}}$, due to the saturation we can expect that $\exists R.\top \in Label(x_0)$.

    That is, we can expect that $\exists R.\top \in Label(x)$ and $Next(x, \exists R.\top)$ is defined. To check whether C_i "holds" at x we just check whether $\exists R.\top \in Label(x)$, $Next(x, \exists R.\top)$ is defined and $A \in Label(Next(x, \exists R.\top))$. The intuition is that, $y = Next(x, \exists R.\top)$ is the "least R-successor" of x, and if $A \in Label(y)$ then A will occur in all R-successors of x.
  - Case $C_i = \exists R.A$: If $\exists R.A$ "holds" at x by the evidence of a path connecting x to a node z with (forward or backward) "edges" labeled by $S_1, \ldots, S_k$ such that the word $S_1 \ldots S_k$ is accepted by $\mathbf{A}_R$ and $A \in Label(z)$ then, since $[\mathbf{A}_{\overline{R}}]\langle \mathbf{A}_R \rangle A$ is included in $Label(z)$ by saturation, we can expect that $\langle \mathbf{A}_R \rangle A \in Label(x)$. To check whether $C_i = \exists R.A$ "holds" at x, we just check whether $\langle \mathbf{A}_R \rangle A \in Label(x)$. (Semantically, $\langle \mathbf{A}_R \rangle A$ is equivalent to $\exists R.A$.) The reason for using this technique is due to the use of global caching (in order to guarantee termination).

We do global caching to represent a possibly infinite semi-forest by a finite graph possibly with cycles. As a side effect, direct checking "realization" of existential automaton-modal operators is not safe. Furthermore, we cannot allow universal modal operators to "run" along such cycles. "Running" universal modal operators backward along an edge is safe, but "running" universal modal operators forward along an edge is done using a special technique, which may replace the edge by another one as in the steps 10-11 of Algorithm 2. Formally, checking whether the premise C of a Horn-$\mathcal{R}eg^I$ clause $C \sqsubseteq D$ "holds" at x is done by Function `CheckPremise`(x,C) (on page 44).

Expansions by modifying the label of a node and/or setting the mapping *Next* are done only for nodes that are reachable from Δ_0. Note that, when a node z is simulated by z_* as in Procedure `ExtendLabel`, the node z becomes unreachable from Δ_0. We do not delete such nodes z because they may be reused later.

When some $x \in \Delta$ has $Label(x)$ containing $\perp$, Algorithm 2 returns *false*, which means that the knowledge base $\langle \mathcal{R}, \mathcal{T}, \mathcal{A} \rangle$ is unsatisfiable. When the graph cannot be expanded any more, the algorithm terminates in the normal mode with result *true*, which means $\langle \mathcal{R}, \mathcal{T}, \mathcal{A} \rangle$ is satisfiable.

Theorem 0.1. *Algorithm 2 correctly checks satisfiability of clausal Horn-$\mathcal{R}eg^I$ knowledge bases and has* PTIME *data complexity.*

See the long version [12] of the current paper for the proof of this theorem. The following corollary follows from this theorem and Proposition 0.1.

Corollary 0.2. *The problem of checking satisfiability of Horn-$\mathcal{R}eg^I$ knowledge bases has* PTIME *data complexity.*

5 Conclusions

We have explained our technique of dealing with non-seriality that leads to a solution for the important issue of allowing the concept constructor $\forall\exists R.C$ to appear at the left hand side of $\sqsubseteq$ in terminological inclusion axioms. We have developed an algorithm with PTIME data complexity for checking satisfiability of Horn-$\mathcal{R}eg^I$ knowledge bases. This shows that both Horn-$\mathcal{R}eg$ and Horn-$\mathcal{R}eg^I$ have PTIME data complexity, solving an open problem of [11].

Acknowledgements. This work was supported by the Polish National Science Centre (NCN) under Grants No. 2011/01/B/ST6/02769 and 2011/01/B/ST6/02759.

References

1. Baader, F., Brandt, S., Lutz, C.: Pushing the EL envelope. In: Kaelbling, L.P., Saffiotti, A. (eds.) Proceedings of IJCAI 2005, pp. 364–369. Morgan-Kaufmann Publishers (2005)
2. Baader, F., Brandt, S., Lutz, C.: Pushing the EL envelope further. In: Proceedings of the OWLED 2008 DC Workshop on OWL: Experiences and Directions (2008)
3. Brandt, S.: Polynomial time reasoning in a description logic with existential restrictions, GCI axioms, and - what else? In: de Mántaras, R.L., Saitta, L. (eds.) Proceedings of ECAI 2004, pp. 298–302. IOS Press (2004)

4. Calvanese, D., De Giacomo, G., Lembo, D., Lenzerini, M., Rosati, R.: Tractable reasoning and efficient query answering in description logics: The DL-Lite family. J. Autom. Reasoning 39(3), 385–429 (2007)
5. Demri, S., de Nivelle, H.: Deciding regular grammar logics with converse through first-order logic. Journal of Logic, Language and Information 14(3), 289–329 (2005)
6. Grosof, B.N., Horrocks, I., Volz, R., Decker, S.: Description logic programs: combining logic programs with description logic. In: Proceedings of WWW 2003, pp. 48–57 (2003)
7. Harel, D., Kozen, D., Tiuryn, J.: Dynamic Logic. MIT Press (2000)
8. Horrocks, I., Kutz, O., Sattler, U.: The even more irresistible SROIQ. In: Doherty, P., Mylopoulos, J., Welty, C.A. (eds.) Proceedings of KR 2006, pp. 57–67. AAAI Press (2006)
9. Hustadt, U., Motik, B., Sattler, U.: Reasoning in description logics by a reduction to disjunctive Datalog. J. Autom. Reasoning 39(3), 351–384 (2007)
10. Nguyen, L.A.: A bottom-up method for the deterministic horn fragment of the description logic $\mathcal{ALC}$. In: Fisher, M., van der Hoek, W., Konev, B., Lisitsa, A. (eds.) JELIA 2006. LNCS (LNAI), vol. 4160, pp. 346–358. Springer, Heidelberg (2006)
11. Nguyen, L.A.: Horn knowledge bases in regular description logics with PTime data complexity. Fundamenta Informaticae 104(4), 349–384 (2010)
12. Nguyen, L.A., Nguyen, T.-B.-L., Szałas, A.: A long version of the current paper, `http://www.mimuw.edu.pl/~nguyen/HornRegI-long.pdf`
13. Nguyen, L.A., Nguyen, T.-B.-L., Szałas, A.: A long version of the paper [14], `http://www.mimuw.edu.pl/~nguyen/horn_dl_long.pdf`
14. Nguyen, L.A., Nguyen, T.-B.-L., Szałas, A.: HornDL: An expressive horn description logic with pTime data complexity. In: Faber, W., Lembo, D. (eds.) RR 2013. LNCS, vol. 7994, pp. 259–264. Springer, Heidelberg (2013)
15. Nguyen, L.A., Szałas, A.: ExpTime tableau decision procedures for regular grammar logics with converse. Studia Logica 98(3), 387–428 (2011)
16. Nguyen, L.A., Szałas, A.: On the Horn fragments of serial regular grammar logics with converse. In: Proceedings of KES-AMSTA 2013. Frontiers of Artificial Intelligence and Applications, vol. 252, pp. 225–234. IOS Press (2013)
17. Ortiz, M., Rudolph, S., Simkus, M.: Query answering in the Horn fragments of the description logics SHOIQ and SROIQ. In: Walsh, T. (ed.) Proceedings of IJCAI 2011, pp. 1039–1044 (2011)

On the Semantics of Defeasible Reasoning for Description Logic Ontologies

Viet-Hoai To, Bac Le, and Mitsuru Ikeda

Abstract. Research in nonmonotonic reasoning for description logics to handle incomplete knowledge on the Semantic Web has attracted much attention in recent years. Among proposed approaches, preferential description logic has a well-formed semantics while defeasible reasoning shows its efficiency in the propositional case. In this paper, we propose a method to define formal definition of semantics of defeasible reasoning for description logic based on the framework of preferential DL. The semantics of defeasible DL theory is defined via its simulated theory constructed by two proposed transformations. This proposal fills the gap between these two approaches and may achieve great benefit by utilizing the advantages of both approaches.

1 Introduction

The achievement of research on ontology has widely spread its application in many areas. However, the monotonic characteristic of underlying description logics of ontology has shown to be not suitable in many practical application. Let consider an example inspired by interesting discovery in [8] to illustrate the situation. An ontology of mad cow disease has a finding that the cause of disease is eating sheep brains. It refers to animal ontology that contains knowledge cow is vegetarian thus does not eat animal. The reference is invalid because of the inconsistency of mad cow concept itself, so no further knowledge about mad cow can be inferred.

Nonmonotonic reasoning has been introduced to description logic to provide a more tolerant solution to this problem by many related works. Among those

Viet-Hoai To · Mitsuru Ikeda
Japan Advanced Institute of Science and Technology, Nomi, Ishikawa, Japan
e-mail: {viet.to,ikeda}@jaist.ac.jp

Bac Le
University of Science, Vietnamese National University, Ho Chi Minh City, Vietnam
e-mail: lhbac@fit.hcmus.edu.vn

V.-N. Huynh et al. (eds.), *Knowledge and Systems Engineering, Volume 1*, Advances in Intelligent Systems and Computing 244,
DOI: 10.1007/978-3-319-02741-8_7, © Springer International Publishing Switzerland 2014

approaches, preferential description logic [4] lifts the semantics of preferential logic from propositional case [9] to description logic case and thus inherits the elegant and well-formed semantics of the original one. Application of this approach is also implemented as a plugins of common ontology editor Protégé [12]. In this research, we focus on another approach for nonmonotonic reasoning named defeasible logic [13] due to its efficiency and flexibility [3]. Defeasible inference in propositional form of logic can be performed in linear time as proved in [10]. However, defeasible logic relies on a proof-based semantics, which is not fully compatible with model-based semantics of description logic. This limitation prevents the application of defeasible reasoning on the ontology reasoning. Current work on defeasible description logic [7, 14] converts DL axioms into propositional rules and perform reasoning with logic programming techniques. This approach shows some restrictions on the expressiveness and efficiency of those works as will be discussed in the final section.

We study the semantics of defeasible reasoning for description logic in this paper. The contribution of this paper is to propose a method to define a model-based semantics of defeasible reasoning for description logics based on preferential and classical semantics. We introduce two transformation to construct an *simulated* theory of defeasible logic theory in preferential description logic. A simulated theory of a logic theory is the one that return the same answer to those of original theory for every possible query in original language[11]. Therefore, we can define the satisfibility of every sentence in defeasible DL theory via its simulated theory. The model-based semantics allows us to apply defeasible reasoning algorithm for DL ontologies, which may result a greater benefit than the current approaches. The followings explain our work in detail. Section 2 summarizes preferential description logic with its formal semantics. Section 3 introduces basic knowledge about defeasible logic and its proof theory. A connecting example is introduced through those above sections to show the differences of these approaches. Section 4 introduces our proposal to formalize the semantics of defeasible reasoning in the framework of preferential DL. The last section presents our discussion about related works and conclude this paper with future work.

2 Preferential Description Logic

2.1 Preliminary

This section summarizes some important notions of preferential DL which relate to its semantics. We assume that the reader is familiar with the basic definitions of description logic ontology and keep this part from the summarization.

Preferential DL enriches the classical $\mathscr{ALC}$ DL with the *defeasible subsumption operator* ($\sqsubsetsim$). This operator is 'supraclassical' of the classical subsumption operator ($\sqsubseteq$). The sentence $\alpha \sqsubsetsim \beta$ is a defeasible subsumption sentence which intuitively means that the most *typical* α's are also β's (as opposed to *all* α's being β's in the classical case). In preferential DL, a defeasible KB contains some defeasible axioms. In this paper, we consider defeasible reasoning for DL TBox only, according to the limitation of underlying work in [4].

Preferential DL defines model-based semantics by specifying a preferential order among axioms and utilizing this order to remove (some) axiom(s) when KB is inconsistent. Specificity, or exceptionality (these two terms are used interchangeably), of an axiom is used to determine the preferential order. For an example of specificity, consider $\mathscr{K} = \{\mathsf{A} \mathrel{\underset{\sim}{\sqsubset}} \mathsf{B}, \mathsf{B} \mathrel{\underset{\sim}{\sqsubset}} \mathsf{C}, \mathsf{A} \mathrel{\underset{\sim}{\sqsubset}} \neg\mathsf{C}\}$. A is more semantically specific than B since $\mathsf{A} \mathrel{\underset{\sim}{\sqsubset}} \mathsf{B}$ and among axioms that cause conflict, $\mathsf{B} \mathrel{\underset{\sim}{\sqsubset}} \mathsf{C}$ and $\mathsf{A} \mathrel{\underset{\sim}{\sqsubset}} \neg\mathsf{C}$, the latter is more specific than the former.

Procedure to compute preferential order is summarized as follows. Firstly, all defeasible axioms $\alpha \mathrel{\underset{\sim}{\sqsubset}} \beta$ are transformed into their classical counterparts $\alpha \sqsubseteq \beta$. Mean while, classical (or strict) axioms $\alpha \sqsubseteq \beta$ are transformed into $\alpha \sqcap \neg\beta \sqsubseteq \bot$. Example 0.1 taken from [12] shows a defeasible KB and its classical counterpart. Then, a *ranking* is computed to arrange axioms according to their *exceptionality*. From the classical counterpart $\mathscr{K}^{\sqsubseteq}$, exceptionality of every concept and axiom in $\mathscr{K}$ is determined by the following conditions:

- C is *exceptional* w.r.t $\mathscr{K}$ iff $\mathscr{K}^{\sqsubseteq} \models C \sqsubseteq \bot$
- $C \sqsubseteq D$ is *exceptional* w.r.t $\mathscr{K}$ iff $C \sqcap \neg D$ is *exceptional* w.r.t $\mathscr{K}$
- $C \mathrel{\underset{\sim}{\sqsubset}} D$ is *exceptional* w.r.t $\mathscr{K}$ iff C is *exceptional* w.r.t $\mathscr{K}$
- $\mathscr{K}' \subseteq \mathscr{K}$, is *exceptional* w.r.t $\mathscr{K}$ iff every element of $\mathscr{K}'$ and *only* the elements of $\mathscr{K}'$ are exceptional w.r.t $\mathscr{K}$ (we say that $\mathscr{K}'$ is *more exceptional* than $\mathscr{K}$).

Example 0.1. Let $\mathscr{K}$ is the defeasible description logic KB contains some defeasible subsumption axioms

$$\mathscr{K} = \left\{\begin{array}{l} \mathsf{Meningitis} \mathrel{\underset{\sim}{\sqsubset}} \neg\mathsf{FatalInfection} \\ \mathsf{BacterialMeningitis} \sqsubseteq \mathsf{Meningitis} \\ \mathsf{ViralMeningitis} \sqsubseteq \mathsf{Meningitis} \\ \mathsf{BacterialMeningitis} \mathrel{\underset{\sim}{\sqsubset}} \mathsf{FatalInfection} \\ \mathsf{Meningitis} \mathrel{\underset{\sim}{\sqsubset}} \mathsf{ViralDisease} \end{array}\right\}$$

The transformed KB of $\mathscr{K}$ is:

$$\mathscr{K}^{\sqsubseteq} = \left\{\begin{array}{l} \mathsf{Meningitis} \sqsubseteq \neg\mathsf{FatalInfection} \\ \mathsf{BacterialMeningitis} \sqcap \neg\mathsf{Meningitis} \sqsubseteq \bot \\ \mathsf{ViralMeningitis} \sqcap \neg\mathsf{Meningitis} \sqsubseteq \bot \\ \mathsf{BacterialMeningitis} \sqsubseteq \mathsf{FatalInfection} \\ \mathsf{Meningitis} \sqsubseteq \mathsf{ViralDisease} \end{array}\right\}$$

The *ranking* of the KB is computed as a collection of sets of sentences $\mathscr{D}$. Each sentence in a particular set from $\mathscr{D}$ shares the same level of exceptionality. Given the collection of sets $\mathscr{D} = \{\mathscr{D}_1, \ldots, \mathscr{D}_n\}$, $\mathscr{D}_1$ represents the *lowest* rank containing the *least* exceptional (specific) sentences in the KB. Then they are the ones which can be "disregarded" with the most confidence when contradicting information is found. Meanwhile, the highest ranking subset $\mathscr{D}_\infty$ contains all non-defeasible or classical axioms that will not be disregarded and should always remain in the KB.

These ranks is computed iteratively, starting from the lowest rank [12]: axioms that are *not* exceptional w.r.t the current KB are removed from the KB and added to the current rank. For example, two axioms $\mathsf{Meningitis} \sqsubseteq \mathsf{ViralDisease}$ and $\mathsf{Meningitis} \sqsubseteq \neg\mathsf{FatalInfection}$ are not exceptional w.r.t original KB due to the third condition so they are moved to $\mathscr{D}_1$. The remaining KB is used to compute next higher ranks. The iteration continues until current KB is empty or all axioms are *exceptional* w.r.t to current KB. Example 0.2 shows the complete ranking of the Meningitis example.

Example 0.2. The ranking of KB from Example 0.1

$$\mathscr{D}_\infty = \left\{ \begin{array}{ll} \mathsf{BacterialMeningitis} \sqcap \neg\mathsf{Meningitis} \sqsubseteq \bot, & (1) \\ \mathsf{ViralMeningitis} \sqcap \neg\mathsf{Meningitis} \sqsubseteq \bot & (2) \end{array} \right\}$$

$$\mathscr{D}_2 = \left\{ \mathsf{BacterialMeningitis} \sqsubseteq \mathsf{FatalInfection} \quad (3) \right\}$$

$$\mathscr{D}_1 = \left\{ \begin{array}{ll} \mathsf{Meningitis} \sqsubseteq \neg\mathsf{FatalInfection}, & (4) \\ \mathsf{Meningitis} \sqsubseteq \mathsf{ViralDisease} & (5) \end{array} \right\}$$

2.2 Semantics of Preferential DL

Preferential DL reasoning uses the ranking of a defeasible DL to verify the satisfiability of a defeasible subsumption sentence, that is to answer whether a defeasible KB entails a defeasible subsumption sentence. For a given query, preferential DL reasoning procedure finds the maximal subset of the ranking that satisfies the antecedent and returning the reasoning result according to this subset. There are two strategies two find the maximal subset corresponding to two type of reasoning in preferential DL: prototypical and presumptive reasoning. Formal definition of semantics of preferential DLs is given as below.

Definition 1 (Preferential Semantics). *Given a defeasible DL Tbox $\mathscr{K}$, $\mathscr{K}$ satisfies a defeasible subsumption statement $C \precsim D$ w.r.t prototypical reasoning or presumptive reasoning, denoted by $\mathscr{K} \models_{Prot} C \precsim D$ or $\mathscr{K} \models_{Pres} C \precsim D$, respectively, if $\bigcup_{i \geq r} \mathscr{D}_i \models C \sqsubseteq D$ where*

- *$\mathscr{D} = \{\mathscr{D}_1, \ldots, \mathscr{D}_\infty\}$ is the ranking of $\mathscr{K}$ correspondent to prototypical reasoning or presumptive reasoning, respectively, and*
- *$\bigcup_{i \geq r} \mathscr{D}_i$ satisfies C and $\forall r' < r$, $\bigcup_{i \geq r'} \mathscr{D}_i \models C \sqsubseteq \bot$.*

Prototypical reasoning uses the original ranking. That means it will remove the *entire* rank when antecedent is not satisfiable. In Example 0.2, for the query $\mathsf{BacterialMeningitis} \precsim \neg\mathsf{FatalInfection}$, the maximal subset of $\mathscr{D}$ is $\mathscr{D}_\infty \cup D_2$, $\mathscr{D}_1$ is removed since $(1) \sqcap (3) \sqcap (4) \models \mathsf{BacterialMeningitis} \sqsubseteq \bot$. Therefore, the query is not satisfiable. The same result is returned for the query $\mathsf{BacterialMeningitis} \precsim \mathsf{ViralDiease}$.

Presumptive reasoning provides more lenient answer than prototypical reasoning by trying to remove *one by one* axiom before removing the whole rank. It performs this computation by adding some ranks above every rank $\mathscr{D}_i (i \neq \infty)$ in the prototypical ranking. Every additional rank corresponds to the union of subsets of $\mathscr{D}_i$ after removing one axiom. Example 0.3 shows the converted ranking for presumptive reasoning. The additional rank $\mathscr{D}'_{12}$ is added and it is the union of the subset of $\mathscr{D}_1$ after remove one axiom. With this converted ranking, reasoning result for query $\mathsf{BacterialMeningitis} \precsim \mathsf{ViralDiease}$ is satisfiable. Since presumptive reasoning gives more informative answer than prototypical reasoning, preferential DL with presumptive reasoning is used to construct the simulated theory of defeasible DL in the following sections.

Example 0.3. The converted ranking $\mathscr{D}'$ for presumptive reasoning is as follows:

$$\mathscr{D}'_\infty = \left\{ \begin{array}{l} \mathsf{BacterialMeningitis} \sqcap \neg\mathsf{Meningitis} \sqsubseteq \bot, \\ \mathsf{ViralMeningitis} \sqcap \neg\mathsf{Meningitis} \sqsubseteq \bot \end{array} \right\}$$

$$\mathscr{D}'_2 = \{ \mathsf{BacterialMeningitis} \sqsubseteq \mathsf{FatalInfection} \}$$

$$\mathscr{D}'_{12} = \{ \mathsf{Meningitis} \sqsubseteq \neg\mathsf{FatalInfection} \sqcup \mathsf{ViralDisease} \}$$

$$\mathscr{D}'_{11} = \left\{ \begin{array}{l} \mathsf{Meningitis} \sqsubseteq \neg\mathsf{FatalInfection}, \\ \mathsf{Meningitis} \sqsubseteq \mathsf{ViralDisease} \end{array} \right\}$$

3 Defeasible Logic

3.1 Basic Definition

Defeasible logic [13] is a rule-based logic which can handle the subjective preference among contradicting conclusions. A *defeasible theory D* is a tripple $(F, R, >)$ where F is a finite set of literals (called *facts*), R a finite set of rules, and ¿ a superiority relation on R. A *rule* has three parts: a finite set of literals on the left (the *antecedent*), an arrow in the middle, and a literal on the right (the *consequence*). There are three kinds of rules: *strict rules* contain *strict arrow* $\rightarrow$ and are interpreted in the classical sense, *defeasible rules* contain *plausible arrow* $\Rightarrow$ and can be defeated by contrary evidence, and defeaters contain $\leadsto$ and are used to prevent some conclusions.

In this paper, we will consider defeasible proof theory in the basis form of defeasible logic. A *basic defeasible theory* contains neither superiority relationship nor defeater. Any defeasible theory can be converted into an equivalent basic defeasible theory using three transformations proposed by [2]. Basic defeasible theory therefore provides us with a simple framework to analyze and formalize semantics of defeasible reasoning for DL.

3.2 Defeasible Proof Theory

A *conclusion* of D is a tagged literal that may be proved by D and can have one of the following four forms: $+\Delta q$, $-\Delta q$, $+\partial q$, and $-\partial q$. $+\Delta q$ shows the provability of q using forward chaining on the strict rules only and $-\Delta q$ is the negation $+\Delta q$. They represent the satisfiability of literal q according to the classical sense. For basic defeasible theory, a literal q is defeasibly provable by a theory D, denoted by $+\partial q$, if the one of the following conditions is satisfied: (1) q is definitely provable $(+\Delta q)$, or (2) there is a rule which has q as consequence and its antecedent is defeasibly provable, and there is no rule which has $\sim q$ as consequence and its antecedent is defeasibly provable.

We consider the application of defeasible proof theory on DL ontologies. However, this proof theory is merely syntactical. In order to define a model-based semantics for defeasible DL, we construct an simulated theory in preferential DL, which has a model-based definition semantics as shown in the previous section. Firstly, we investigate reasoning result of defeasible proof theory on the converted KB from defeasible DL KB presented in Example 0.1. Notice that for every classical axiom $C \sqsubseteq D$ two strict rules $C \rightarrow D$ and $\neg D \rightarrow \neg C$ are generated, following the conversion approach proposed by [7]. Meanwhile, every defeasible axiom corresponds to a defeasible rule in the KB. The converted theory is shown in Example 0.4.

Example 0.4. The propositional conversion of defeasible DL KB in Example 0.1 is represented as follows:

$$D = (F, R, \emptyset), \text{ where } F = \emptyset \text{ and}$$

$$R = \left\{ \begin{array}{ll} BacterialMeningitis & \rightarrow Meningitis, \\ \neg Meningitis & \rightarrow \neg BacterialMeningitis, \\ ViralMeningitis & \rightarrow Meningitis \\ \neg Meningitis & \rightarrow \neg ViralMeningitis \\ BacterialMeningitis & \Rightarrow FatalInfection, \\ Meningitis & \Rightarrow \neg FatalInfection, \\ Meningitis & \Rightarrow ViralDisease \end{array} \right\}$$

From the converted KB, we compute classical and defeasible superclasses of BacterialMeningitis by adding this concept to F, i.e. $F = \{BacterialMeningitis\}$, and performing reasoning procedure to obtain the following result sets ($+\Delta = \{q \mid D \vdash +\Delta q\}$ and $+\partial = \{q \mid D \vdash +\partial q\}$):
$+\Delta = \{BacterialMeningitis, Meningitis\}$
$+\partial = \{BacterialMeningitis, Meningitis, ViralDisease\}$

A *simple proof graph* as in Figure 1 is used to illustrate the defeasible DL ontology and explain reasoning result of defeasible proof theory. In this graph, *node* stands for one concept. We use abbreviation of the concept in label of the node. *Edge* stands for a axiom: double line edge for classical axiom, and single line edge for defeasible axiom. A *slash* on the end of the edge to indicate negative concept in the axiom. A *proof path* on this graph corresponds with a chain of axioms that allow

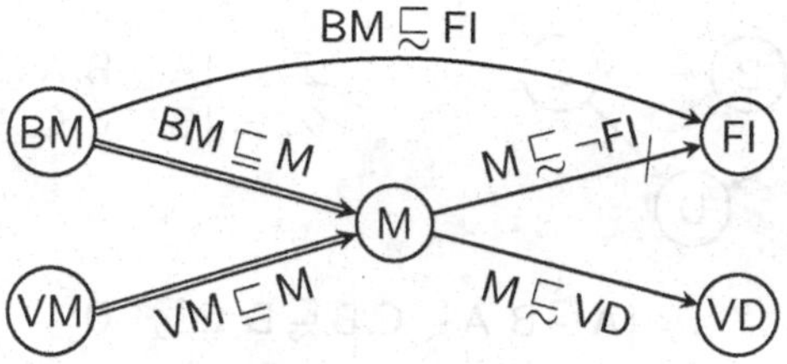

Fig. 1 Proof graph for Meningitis ontology

to infer one concept from other concept. A *conflict situation* happens when there are two proof paths from one node to another node with contrary conclusions.

We compare the answers of defeasible reasoning and presumptive reasoning on the defeasible queries to identify the difference between these approaches. In reasoning results w.r.t defeasible proof theory, hence we call defeasible reasoning, concepts in $+\Delta$ and $+\partial$ correspond to the classical and defeasible superclasses of the interested concept in defeasible DL, respectively. For the defeasible subsumption relations, those preliminary comparison can be observed:

- Defeasible reasoning does not entail $\mathsf{BacterialMeningitis} \precsim \mathsf{FatalInfection}$ while presumptive reasoning does. In Figure 1, there are two proof path from BM leading to contradicting conclusions FI and ¬FI, defeasible reasoning rejects both conclusions. Presumptive reasoning entails this sentence because it belongs to $\mathscr{D}_2$, which satisfies the antecedent.
- Both reasoning approaches entail $\mathsf{BacterialMeningitis} \precsim \mathsf{ViralDiease}$.

We will give a detail comparison and analysis on the reasoning results of defeasible reasoning and presumptive reasoning in next section. From that analysis, we can propose a method to construct the simulated theory of defeasible DL in the framework of preferential DL and define semantics of defeasible reasoning for DL via this simulated theory.

4 Semantics of Defeasible Reasoning for DL

4.1 Skepticism vs. Principle of Specificity

Firstly, we consider the transformation that resolves skeptical problem of defeasible reasoning in the case ontology contains defeasible axioms only.

Example 0.5. Consider the following two defeasible KBs and their correspondent proof graphs and prototypical rankings respectively:

$$\mathscr{K}^1 = \{\mathsf{A} \precsim \mathsf{B}, \mathsf{A} \precsim \neg\mathsf{D}, \mathsf{B} \precsim \mathsf{C}, \mathsf{C} \precsim \mathsf{D}\}$$

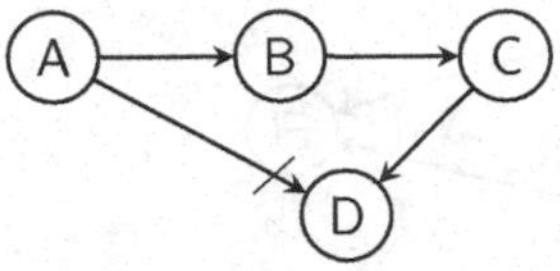

$$\mathscr{D}_2^1 = \{\mathsf{A} \sqsubseteq \mathsf{B}, \mathsf{A} \sqsubseteq \neg\mathsf{D}\}$$
$$\mathscr{D}_1^1 = \{\mathsf{B} \sqsubseteq \mathsf{C}, \mathsf{C} \sqsubseteq \mathsf{D}\}$$

$$\mathscr{K}^2 = \{\mathsf{A} \underset{\sim}{\sqsubset} \mathsf{B}, \mathsf{A} \underset{\sim}{\sqsubset} \mathsf{C}, \mathsf{B} \underset{\sim}{\sqsubset} \mathsf{D}, \mathsf{C} \underset{\sim}{\sqsubset} \neg\mathsf{D}\}$$

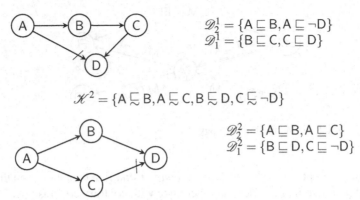

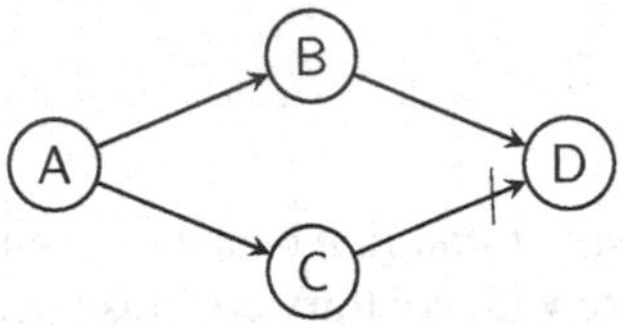

$$\mathscr{D}_2^2 = \{\mathsf{A} \sqsubseteq \mathsf{B}, \mathsf{A} \sqsubseteq \mathsf{C}\}$$
$$\mathscr{D}_1^2 = \{\mathsf{B} \sqsubseteq \mathsf{D}, \mathsf{C} \sqsubseteq \neg\mathsf{D}\}$$

The reasoning results according to defeasible reasoing are identical in both examples. Neither $\mathsf{A} \underset{\sim}{\sqsubset} \mathsf{D}$ nor $\mathsf{A} \underset{\sim}{\sqsubset} \neg\mathsf{D}$ is satisfiable by $\mathscr{K}^1$ or $\mathscr{K}^2$ since from A there are two proof paths lead to contradicting concepts D and $\neg\mathsf{D}$ in both graphs. This is the *skeptical principle* in defeasible proof theory: we can not conclude a consequence with the existence of contradicting evidence. Notice that in defeasible proof theory, we can easily choose the preferential conclusion by explicitly specifying the superiority relations among those axioms. Other defeasible subsupmtion relations such as $\mathsf{A} \underset{\sim}{\sqsubset} \mathsf{C}$ and $\mathsf{B} \underset{\sim}{\sqsubset} \mathsf{D}$ are satisfiable because there is no conflict evidence.

Meanwhile, presumptive reasoning results vary from $\mathscr{K}^1$ to $\mathscr{K}^2$ due to the exceptionality of the concepts and axioms. Axioms with antecedent A belongs to the higher rank than others in both $\mathscr{K}^1$ and $\mathscr{K}^2$, i.e. $\mathscr{D}_2^1$ and $\mathscr{D}_2^2$ respectively. In $\mathscr{K}^1$, $\mathsf{A} \sqsubseteq \neg\mathsf{D}$ and $\mathsf{C} \sqsubseteq \mathsf{D}$ are in different ranks: $\mathsf{A} \underset{\sim}{\sqsubset} \neg\mathsf{D}$ is preferred to $\mathsf{C} \underset{\sim}{\sqsubset} \mathsf{D}$ when conflict occurs and thus $\mathscr{K}^1 \models_{Pres} \mathsf{A} \underset{\sim}{\sqsubset} \neg\mathsf{D}$. The reasoning result in $\mathscr{K}^1$ agrees with the principle of specificity (or exceptionality) of Lehmann and Magidor's approach [9]. In $\mathscr{K}^2$, $\mathsf{B} \sqsubseteq \mathsf{D}$ and $\mathsf{C} \sqsubseteq \neg\mathsf{D}$ are in the same rank: no defeasible sentence is preferred to the other when conflict occurs. Conclusions in $\mathscr{K}^2$ are same as those of defeasible reasoning.

Another conclusion of presumptive reasoning we should notice is that $\mathscr{K}^1 \not\models_{Pres} \mathsf{A} \underset{\sim}{\sqsubset} \mathsf{C}$. The sentence $\mathsf{B} \sqsubseteq \mathsf{C}$ which is necessary to entail $\mathsf{A} \sqsubseteq \mathsf{C}$ belongs to removed rank $\mathscr{D}_1^1$. But either $\mathsf{B} \sqsubseteq \mathsf{C}$ or $\mathsf{C} \sqsubseteq \mathsf{D}$ can be removed to satisfy A, no axiom is preferred to the other.

From above analysis, the differences in reasoning results of defeasible reasoning and presumptive reasoning are caused by the exceptionality of the axioms. In order to simulate reasoning results of defeasible reasoning in preferential DL, we introduce the transformation to process the exceptionality of the axioms. The goal of this transformation is to decrease the exceptionality of axioms that may cause conflict, i.e. axiom whose consequence contradicts those of other axiom in KB, and increase the exceptionality of the others. From here, we call them conflict axioms and non-conflict axioms, respectively. The formal definition of this transformation is as below.

Definition 2 (tranException). *Given a defeasible DL Tbox* $\mathscr{K}$ *with defeasible axioms only. Define* $tranException(\mathscr{K}) = \bigcup_{\varphi \in \mathscr{K}} tranException(\varphi)$ *where*

$$tranException(\varphi) = \begin{cases} \{\alpha \precsim \beta_\alpha, \beta_\alpha \precsim \beta & \mid (\varphi = \alpha \precsim \beta) \wedge (\gamma \precsim {\sim}\beta \in \mathscr{K})\} \\ \{\alpha \precsim \beta, \alpha \precsim \beta_\alpha, \beta_\alpha \precsim {\sim}\beta & \mid (\varphi = \alpha \precsim \beta) \wedge (\gamma \precsim {\sim}\beta \notin \mathscr{K})\} \end{cases}$$

and $\sim\beta$ is the complement of β.

Transformation *tranException* changes the exceptionality by adding virtual concepts and axioms to KB. For the conflict axiom, *tranException* creates an intermediate concept between the antecedent and consequence and replace the original axiom by two axioms go through this intermediate concept. Since the newly created concept does not entail any contradiction, it has the lowest exceptionality. All conflict axioms from original KB are moved to the lowest rank and presumptive reasoning do not entail the consequence with contradicting evidence since there is no preference between those axioms. For non-conflict axiom, *tranException* creates an virtual concept and two virtual axioms from the antecedent to the contradicting consequence via this virtual concept. With two new virtual axioms, the original axiom's exceptionality is increased, if it was not exceptional formerly, since its antecedent becomes exceptional w.r.t the transformed KB.

Example 0.6. The following is the transformed KB $\mathscr{K}^{1e}$ of $\mathscr{K}^{1}$ using *tranException*:

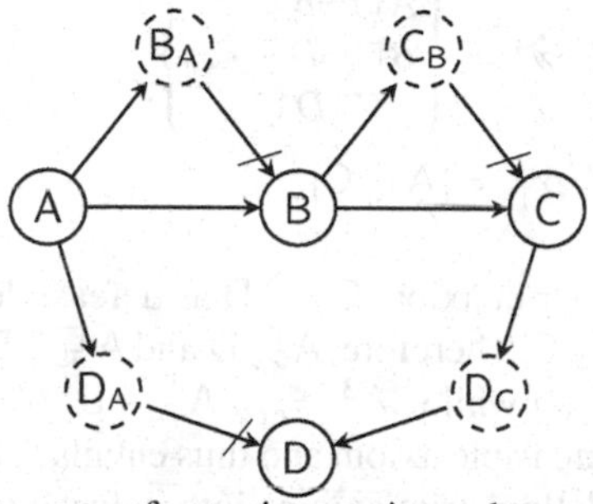

$$\mathscr{D}_2^{1e} = \left\{ \begin{array}{c} A \sqsubseteq B, A \sqsubseteq B_A, \\ B \sqsubseteq C, B \sqsubseteq C_B, \\ A \sqsubseteq D_A \end{array} \right\}$$

$$\mathscr{D}_1^{1e} = \left\{ \begin{array}{c} B_A \sqsubseteq \neg B, C_B \sqsubseteq \neg C, \\ D_A \sqsubseteq \neg D, D_C \sqsubseteq D, \\ C \sqsubseteq D_C \end{array} \right\}$$

After transformation, axioms have contradicting conclusions, D and $\neg$D, are in the lowest rank and other axioms in the higher rank. The result of presumptive reasoning on this transformed KB is same as those of defeasbile reasoning: $\mathscr{K}_{1e} \not\models_{Pres} \mathsf{A} \precsim \mathsf{D}$, $\mathscr{K}_{1e} \not\models_{Pres} \mathsf{A} \precsim \neg\mathsf{D}$, and $\mathscr{K}_{1e} \models_{Pres} (\mathsf{A} \precsim \mathsf{C}) \sqcap (\mathsf{B} \precsim \mathsf{D})$.

4.2 *Interaction between Defeasible and Strict Axioms*

Now we consider a defeasible KB contains both strict and defeasible subsumption sentences. When $\mathscr{K}$ consists of both strict and defeasible sentences, there is a special rank $\mathscr{D}_\infty$ that contains strict axioms. Axioms $\mathscr{D}_\infty$ have ultimate preference to axioms in other ranks. Adding virtual defeasible axioms in this case may not generate the desired result because of the existence of this special rank. Therefore, we need to translate all strict axioms into defeasible axioms. According to defeasible proof theory, strict subsumption relation can be computed using strict axioms only so we can separate the computation of strict subsumption and defeasible subsumption relations. The following investigate the transformation that convert strict axioms into defeasible axioms to compute defeasible subsumption relations.

In defeasible proof theory, strict sentence can be used in two different ways: as a strict axiom or as a defeasible axiom. The former case occurs when antecedent is given or entailed by a chain of strict axioms only. The latter occurs when antecedent is entailed defeasibly. Example 0.7 shows two cases of interaction between strict and defeasible axioms.

Example 0.7. Consider the following two KBs which contain both stricts and defeasible axioms

$$\mathscr{K}^3 = \{\mathsf{A} \sqsubseteq \mathsf{B}, \mathsf{B} \underset{\sim}{\sqsubset} \mathsf{D}, \mathsf{A} \underset{\sim}{\sqsubset} \mathsf{C}, \mathsf{C} \sqsubseteq \neg\mathsf{D}\}$$

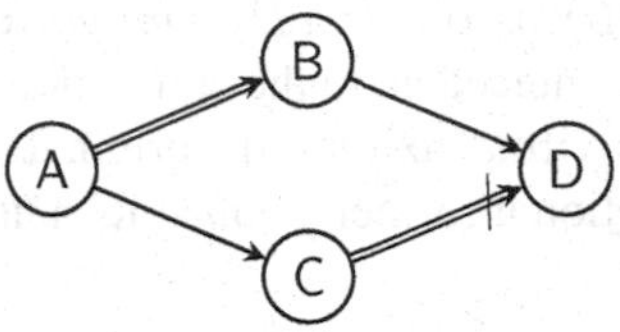

$$\mathscr{D}^3_\infty = \{\mathsf{A} \sqcap \neg\mathsf{B} \sqsubseteq \bot, \mathsf{C} \sqcap \mathsf{D} \sqsubseteq \bot\}$$
$$\mathscr{D}^3_2 = \{\mathsf{A} \sqsubseteq \mathsf{C}\}$$
$$\mathscr{D}^3_1 = \{\mathsf{B} \sqsubseteq \mathsf{D}\}$$

$$\mathscr{K}^4 = \{\mathsf{A} \sqsubseteq \mathsf{B}, \mathsf{B} \sqsubseteq \mathsf{D}, \mathsf{A} \underset{\sim}{\sqsubset} \mathsf{C}, \mathsf{C} \sqsubseteq \neg\mathsf{D}\}$$

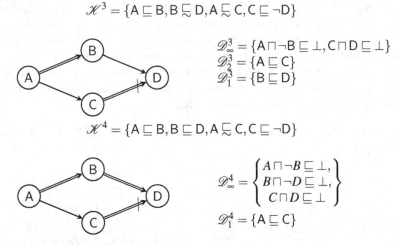

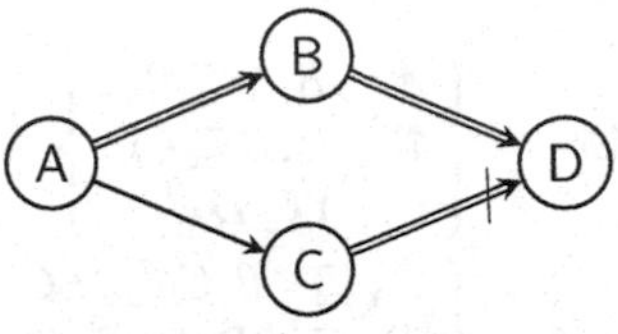

$$\mathscr{D}^4_\infty = \left\{\begin{matrix} A \sqcap \neg B \sqsubseteq \bot, \\ B \sqcap \neg D \sqsubseteq \bot, \\ C \sqcap D \sqsubseteq \bot \end{matrix}\right\}$$
$$\mathscr{D}^4_1 = \{\mathsf{A} \sqsubseteq \mathsf{C}\}$$

In $\mathscr{K}^3$, defeasible reasoning will consider strict axiom $\mathsf{C} \sqsubseteq \neg\mathsf{D}$ as defeasible axiom because it follows a defeasible axiom $\mathsf{A} \underset{\sim}{\sqsubset} \mathsf{C}$. Therefore, $\mathsf{A} \underset{\sim}{\sqsubset} \mathsf{D}$ and $\mathsf{A} \underset{\sim}{\sqsubset} \neg\mathsf{D}$ are not satisfiable by $\mathscr{K}^3$ w.r.t defeasible reasoning while $\mathscr{K}^3 \models_{Pres} \mathsf{A} \underset{\sim}{\sqsubset} \neg\mathsf{D}$. In $\mathscr{K}^4$, defeasible reasoning also treats $\mathsf{C} \sqsubseteq \neg\mathsf{D}$ as defeasible axiom and thus entails $\mathsf{A} \sqsubseteq \mathsf{D}$, same as presumptive reasoning. When translating strict axioms into defeasible axioms, we need to keep the preference of the chain of strict axioms as described in these examples.

We propose the transformation to translate all strict axioms into defeasible axioms in Definition 3. The idea of this transformation is to separate reasoning process into two parallel processes, one for previously strict axioms and one for all axioms. Then we can set the superiority of chaining of strict axioms over defeasible one.

Definition 3 (elimStrict). *Given a defeasible DL Tbox $\mathscr{K}$ containing both strict and defeasible axioms. Define $elimStrict(\mathscr{K}) = \bigcup_{\varphi \in \mathscr{K}} elimStrict(\varphi)$ where*

$$elimStrict(\varphi) = \begin{cases} tranStrict(\alpha \sqsubseteq \beta) \cup tranStrict(\sim\beta \sqsubseteq \sim\alpha) & , \text{if } \varphi = \alpha \sqsubseteq \beta \\ tranDefeasible(\alpha \underset{\sim}{\sqsubset} \beta) & , \text{if } \varphi = \alpha \underset{\sim}{\sqsubset} \beta \end{cases}$$

$$tranStrict(\alpha \sqsubseteq \beta) = \{\alpha \underset{\sim}{\sqsubset} \beta, \alpha \underset{\sim}{\sqsubset} \alpha^d, \alpha^d \underset{\sim}{\sqsubset} \beta^d\}$$

$$tranDefeasible(\alpha \underset{\sim}{\sqsubset} \beta) = \{\alpha \underset{\sim}{\sqsubset} \beta^d, \alpha \underset{\sim}{\sqsubset} \alpha^d, \alpha^d \underset{\sim}{\sqsubset} \beta^d\}.$$

Transformation *elimStrict* creates a auxiliary concept for every concept. The defeasible axioms are applicable for those auxiliary concepts only while strict axioms are applicable for both concepts. The transformation of strict axioms can affect the original and auxiliary concepts, but the transformation of defeasible axioms only affect the auxiliary concepts. For strict axiom, its reverse is also added by the transformation. Figure 2 shows a part of proof graph, which does not contain reverse axioms for the purpose of simplification, of transformation *elimStrict* on $\mathcal{K}^4$. From this proof graph, we can see that transformed KB entails $\mathsf{A} \precsim \mathsf{D}$ even it does not entail $\mathsf{A} \precsim \mathsf{D}'$. By setting D preferential to D' in the result set, we can obtain the desired result.

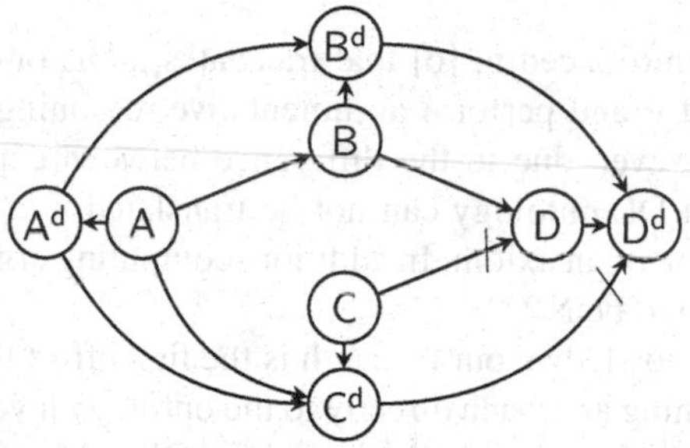

Fig. 2 A part of transformation *elimStrict* on $\mathcal{K}_4$

4.3 Semantics of Defeasible Reasoning for DLs

With two transformations introduced above, we propose a formal definition of the semantics of defeasible reasoning for DLs as following.

Definition 4 (Defeasible Semantics). *Given a defeasible DL Tbox $\mathcal{K}$, Σ is the language of $\mathcal{K}$. $\mathcal{K}$ satisfies a defeasible subsumption statements $C \precsim D$ s.t. $C, D \in \Sigma$, denoted by $\mathcal{K} \models_{Def} C \precsim D$, iff $\mathcal{K}^D \models_{Pres} C \precsim D$ or $\mathcal{K}^D \models_{Pres} C \precsim D^d$ where $\mathcal{K}^D = tranException(elimStrict(K))$.*

The satisfiability of a defeasible sentence w.r.t defeasible reasoning is defined via its simulated theory constructed by two transformations in preferential DL framework. As shown in above analysis, the transformed theory generates the same answer for every possible query with defeasible proof theory in those cases. Due to the space limitation, we plan to introduce the proof of equivalence for the general case in the extension of this paper. Another note is that transformations significantly increase the size of KB. However it is not our aim to use those transformations for defeasible reasoning. Recall that we have an efficient reasoning algorithm for defeasible propositional logic. Lifting this algorithm to defeasible DL case is a promising direction in our future research, as concluded in the next, final section.

5 Related Work and Final Discussion

Owning to its simplicity, flexibility and efficiency, defeasible proof theory has been introduced to DLs by other research such as in [7, 14, 1]. Those studies apply logical conversion approach to translate DL ontology axioms into propositional rules and add defeasible rules to this converted KB. Defeasbile reasoning is then performed on the rule layer of the ontology. This approach relies on the semantics of propositional logic, which is much "weaker" than those of current DLs such as $\mathscr{SROIQ}$ DL [5], causing much difficulty in extending the semantics of proposed defeasible DL. It also suffers from the complexity of the propositionalization of DL constructors. For example, [14] shows that instantiating universal quantifier of DL has the complexity $\mathscr{O}(n^4)$.

A similar work is introduced in [6] that translates a DL ontology into defeasible logic program DeLP and perform argumentative reasoning by an justification abstract machine. However, due to the difference between expressiveness of two logics, some axiom in DL ontology can not be translated into DeLP, for example existential in conclusion of an axiom. In addition, computing argumentation in logic program is also known to be NP.

To the best of our knowledge, our research is the first effort to introduce defeasible proof theory reasoning approach directly to the ontology layer. The model-based definition of semantics of defeasible reasoning for DLs proposed in this paper allows us to lift the reasoning approach from propositional case to DL case. The advantage of this lifting can be two-fold. Firstly, defeasible reasoning for DL can be more flexible and efficient than preferential DLs. Secondly, semantics of defeasible reasoning for DL can be more extensible than previously proposed defeasile DL. It is our plan to realize defeasible reasoning algorithm for DL ontology and verify those hypotheses in future research.

References

1. Antoniou, G.: Nonmonotonic rule systems on top of ontology layers. In: Horrocks, I., Hendler, J. (eds.) ISWC 2002. LNCS, vol. 2342, pp. 394–398. Springer, Heidelberg (2002)
2. Antoniou, G., Billington, D., Governatori, G., Maher, M.J.: Representation results for defeasible logic. ACM Trans. Comput. Logic. 2(2), 255–287 (2001)
3. Billington, D., Antoniou, G., Governatori, G., Maher, M.: An inclusion theorem for defeasible logics. ACM Trans. Comput. Logic 12(1), 1–27 (2010)
4. Britz, K., Meyer, T., Varzinczak, I.: Semantic Foundation for Preferential Description Logics. In: Wang, D., Reynolds, M. (eds.) AI 2011. LNCS, vol. 7106, pp. 491–500. Springer, Heidelberg (2011)
5. Grau, B.C., Horrocks, I., Motik, B., Parsia, B., Patel-Schneider, P., Sattler, U.: OWL 2: The next step for OWL. Journal of Web Semantics 6(4), 309–322 (2008)
6. Gómez, S.A., Chesñevar, C.I., Simari, G.R.: Reasoning with inconsistent ontologies through argumentation. Applied Artificial Intelligence 24(1&2), 102–148 (2010)
7. Governatori, G.: Defeasible Description Logics. In: Antoniou, G., Boley, H. (eds.) RuleML 2004. LNCS, vol. 3323, pp. 98–112. Springer, Heidelberg (2004)

8. Horrocks, I.: Ontologies and the semantic web. Communications of the ACM 51(12), 58–67 (2008)
9. Lehmann, D., Magidor, M.: What does a conditional knowledge base entail? Artificial Intelligence 55(1), 1–60 (1992)
10. Maher, M.J.: Propositional defeasible logic has linear complexity. Theory Pract. Log. Program. 1(6), 691–711 (2001)
11. Maher, M.J.: Relative expressiveness of defeasible logics. Theory and Practice of Logic Programming 12, 793–810 (2012)
12. Moodley, K., Meyer, T., Varzinczak, I.J.: A defeasible reasoning approach for description logic ontologies. In: Proceedings of the SAICSIT 2012, pp. 69–78. ACM, New York (2012)
13. Nute, D.: Defeasible logic. In: Handbook of Logic in Artificial Inteligence and Logic Programming, vol. 3, pp. 353–395. Oxford University Press (1987)
14. Pothipruk, P., Governatori, G.: $\mathcal{ALE}$ defeasible description logic. In: Sattar, A., Kang, B.-H. (eds.) AI 2006. LNCS (LNAI), vol. AI 2006, pp. 110–119. Springer, Heidelberg (2006)

SudocAD: A Knowledge-Based System for the Author Linkage Problem*

Michel Chein, Michel Leclère, and Yann Nicolas

Abstract. SudocAD is a system concerning the author linkage problem in a bibliographic database context. Having a bibliographic database $\mathscr{E}$ and a (new) bibliographic notice d, r being an identifier of an author in $\mathscr{E}$ and r' being an identifier of an author in d: is that r and r' refer to the same author ? The system, which is a prototype, has been evaluated in a real situation. Compared to results given by expert librarians, the results of SudocAD are interesting enough to plan a transformation of the prototype into a production system. SudocAD is based on a method combining numerical and knowledge based techniques. This method is abstractly defined and even though SudocAD is devoted to the author linkage problem the method could be adapted for other kinds of linkage problems especially in the semantic web context.

1 Introduction

The issue addressed in this paper is a linkage problem that arises in bibliographic or document databases. The fundamental problem to be solved can be abstractly stated as follows. Let B and B' be two computer systems (e.g. bibliographic databases), let r be a referent in B to an exterior world entity and r' be a referent in B' to an exterior world entity (e.g. identifiers of authors of documents), then are r and r' referring to the same entity (e.g. the same author)? For a long time (cf. e.g. the seminal paper [20]), various problems having linkage problems at their core have been extensively studied under different names (cf. [29] and [30]), such as:

Michel Chein · Michel Leclère
LIRMM-GraphIK (CNRS, INRIA, UM2), 161 Rue Ada F-34392, Montpellier Cedex 5, France

Yann Nicolas
ABES, 25 rue Guillaume Dupuytren, BP 4367, 34196, Montpellier Cedex 5, France

* This work benefited from the support of ANR, the French Research National Agency (ANR-12-CORD-0012).

V.-N. Huynh et al. (eds.), *Knowledge and Systems Engineering, Volume 1*, Advances in Intelligent Systems and Computing 244,
DOI: 10.1007/978-3-319-02741-8_8, © Springer International Publishing Switzerland 2014

- *record linkage* (e.g. [15],[16], [9])
- *entity resolution, reference reconciliation, co-reference resolution* (e.g., [3], [16], [26])
- *de-duplication* (e.g. [8])
- *merge/purge* (e.g. [17])
- *entity alignment*(e.g. [25], [19])
- *object identification* (e.g. [24])

These problems have been considered for many kinds of databases, especially data warehouses where it is important to know if two identifiers refer to the same object or to different entities in the exterior world. These problems have increased in importance due to the web, especially with respect to Linked Open Data, where a key issue is to gather all available information concerning the same entity.

Let us mention some reasons that highlight the importance of linkage problems in bibliographic or document databases. Firstly, all of the previously mentioned problems are important in libraries: some of them are induced by the evolution of libraries (e.g. adding records to a base, merging bibliographic bases, maintenance of different bases), others concern the quality of the record bases (e.g. consistency inside and between bases, relevance of the subject). Secondly, a bibliographic database is a very rich source of information on the documents themselves and also on authorities. International work is under way to standardize the metadata (FRBR [11], CIDOC CRM [5], RDA [23]), to build shared ontologies. This allows the transformation of a document base into a knowledge base and then knowledge-based techniques can be used to make the most of the information present in the base. Thirdly, whenever semantic web languages are used for solving object identification problems and due to the large size of document bases, the techniques developed are not only interesting *per se* but also as a testbed for linkage problems in the web of data context.

Most solutions to linkage problems are based on classification techniques (e.g. [13], [21], [16]). In such an approach, an entity is described by a list of attributes; attribute values are simple data (e.g. names, dates, numbers, etc.) and approximate similarity measures are assigned for each kind of attribute; a similarity measure is built for lists of attributes (often it is a weighted combination of the attribute similarities); finally, a decision procedure allows to state when two lists of attributes represent the same entity. Recently, logical approaches using knowledge representation techniques (e.g. [26]) and combinations of these two kinds of methods have been developed (e.g. [14], [2]).

Contribution. As far as we know, the methods used for solving author linkage problems are essentially based on name comparisons. SudocAD has a lot of original facets since it uses:

- **Combination of numerical and logical computations.** Numerical computations are used for comparing low level attributes, and logical rules are applied for higher level attributes. The semantic web languages RDFS and OWL, as well as knowledge representation techniques (cf. [6]), have been used for developing the system.

- **Super-authorities.** A notion of super-authority is introduced, which is an authority enriched by information present in the whole bibliographic database. Furthermore, if the document database considered does not contain authority records, the super-authority building technique can be used to build such authority records.
- **Qualitative partition.** The result obtained is an **ordered qualitative partition** of the set of candidates. This ordered partition can be used in different ways, thus allowing different strategies for the linkage problem (in automatic mode as well as in decision-aided mode).
- **Genericity.** The method underlying SudocAD is generic and even though it deals with authors it can be used for other authorities as well, e.g. collective entities.

Another key contribution of the present paper is the evaluation method used (and the results of this evaluation).

The paper is organized as follows. In Section 2, the bibliographic database context of the system SudocAD is presented. The hybrid method, combining numerical and logical aspects, underlying SudocAD is presented in Section 3. The methodology and the results of the evaluation of SudocAD are described in Section 4. Cogui, the tool used for implementing SudocAD, is presented in Section 5. The Conclusion and further work end the paper.

2 The Bibliographic Context

In this section, the bibliographic context of SudocAD is briefly described.

2.1 *Bibliographic and Authority Notices, Sudoc and IdRef*

Sudoc (cf.[27]) is the national bibliographic infrastructure for French higher education developed and maintained by ABES (National Agency in charge of Sudoc cf.[1]). As this infrastructure, Sudoc is both:

- a shared database where bibliographic records are created once, but used by all;
- a network of professional catalogers, with common tools, and guidelines for using the tools.

The core function of this collective endeavour is to create and maintain database records that describe the documents held or licensed by French Universities and other higher education institutions. Sudoc contains more than ten million of such records (2012 figure). Documents described by Sudoc are mainly electronic or printed books and journals, but also manuscripts, pictures, maps, etc.

A Sudoc record consists of three kinds of information:

- Meta-information
- Descriptive information
- Access points

Meta-information is information about the record itself. It is beyond the scope of this paper.Descriptive information is mere transcription of information that is found in the document to be catalogued. For instance, the descriptive field *Title* contains the text string that on the title page. The same is true for the descriptive field *Author*, the record has to keep the author's name as it is found in the document, even if the librarian knows that the title page misspelled this name. This strictly descriptive approach aims to identify the publication without any ambiguity. The transcribed information has to be sufficient to distinguish two editions of the same work. But this descriptive approach may make the document harder to find. If the title page, hence the bibliographic record, assumes that the author's name is 'Jean Dupond' whereas the actual name is 'Jean Dupont', library users who are only aware of the actual name will fail to find and access the record in the catalog and then to find and access the document in the library.To overcome this kind of problem, the cataloguing rules prescribe to add the actual name in the record, not instead of but rather besides the one found on the title page. This kind of additional non-descriptive information is called "access point".

Access points constitute the third kind of information that is to be found in a bibliographic record. An access point is a piece of information that focuses on some aspect of the described document that may be relevant for finding the record in a database and that is not directly observable in the document itself. As we have seen above, it can be the actual author's name. It can also be the author's name written according to a specific convention (e.g. 'Dupont, Jean' or 'Dupont, Jean (1956-)').But an access point is not necessarily an alternative textual form of a piece of information that was observed on the described object. It can be the result of the analysis of the document content. For instance, an access point can be a keyword expressing the subject of the document (e.g. 'cars').

Any bibliographic record is the result of the description of the document and the selection of relevant access points. But when is a bibliographic record deemed to be achieved? Theoretically, the description could last forever but cataloguing guidelines set a finite list in which document properties must or may be described. But regarding access points, when to stop? If access points are available to help the user find the record, there can be plenty of them. Should the record describing the book by Jean Dupont and about cars have access points mentioning all variants of Jean Dupont's name and as many access points as there are synonyms for 'cars'? And should this be the case for each book by Jean Dupont or about cars?

In order not to repeat all the variants of a name or a concept in all bibliographic records using this name or this concept as an access point, these variants are grouped in specific records, namely authority records (cf. [18]). For instance, the authority record for Jean Dupont will contain his name variants. One of these variants will be distinguished as the preferred form. Some guidelines will help librarians choose the preferred form. Some guidelines expect the preferred form to be unique. The preferred form is the only form to be used as an access point in the bibliographic records. If it is unique, it works like an identifier for the authority record that contains the rest of the variants: the preferred form used as an access point links the bibliographic record to the authority record.

In many recent catalogs, the link from the bibliographic record to the authority record is not the preferred form but the authority number. This is the case in Sudoc. Sudoc bibliographic record access points are just authority record identifiers. When the record is displayed or exported, the Sudoc system follows the link and substitutes a name or a term for the identifier. Sudoc has more than ten million bibliographic records and two million authority records. An authority record is not an entry in a biographical dictionary. It is supposed to contain nothing but information sufficient to distinguish one person from another one described by another authority record. Authority records for people mainly contain information about their names, dates and country (cf. [18]). Authority records for concepts contain information about their labels and relationships to other concepts (broader, narrower, etc).

2.2 Semantization of Bibliographic Metadata

Bibliographic and authority records have been expressed in RDF. The RDF vocabulary has to meet some expectations:

- to be able to express precise bibliographic assertions,
- to be minimally stable and maintained by a community.

The FRBRoo vocabulary, which is an object-oriented version of FRBR (cf. [12]), meets these expectations:

- It has been developed for fine-grained bibliographic requirements,
- It is well documented,
- It is maintained and still developed by an active community,
- It is connected to other major modeling initiatives.

Firstly, FRBRoo takes its main concepts from FRBR, a prominent model that was developed by the International Federation of Library Associations during the 1990s. FRBRoo keeps core FRBR concepts but claims to overcome some of its alleged limitations. Secondly, FRBRoo is built as an extension of another model for cultural objects, namely CIDOC CRM. CIDOC CRM's main scope is material cultural heritage, as curated by art or natural history museums. It is focused on expressing the various events that constitute an object life, before or after its accession to the museum. FRBRoo imports many of its classes and properties from CIDOC CRM, but needs to forge a lot of new ones, in order to cope with more abstract entities such as texts and works.

The native Sudoc catalog is in UNIMARC. For SudocAD's needs, it was not necessary to convert all UNIMARC fields in FRBRoo. But even for conversion of the fields needed for the reasoning, we had to extend FRBRoo and forge some new classes and properties. The whole ontology used in SudocAD, which is composed of a domain ontology and a linkage ontology, contains a hierarchy of (313) concepts and a hierarchy of (1179) relations. It is presented in the Examples part on Cogui's website (cf. [7]). A preliminary report about the system SudocAD has been published, in French, on the ABES website (cf. [28]).

2.3 SudocAD

In a very general setting, the input and results of SudocAD can be stated as follows:

- **The data.** The input of the system consists of the sudoc catalog [27] and the person authority base IdRef [18], as well as a part of the Persee bibliographic database [22]. The part of Persee considered for SudocAD evaluation contains bibliographic records of papers in social science journals.
- **The problem.** Given a Persee bibliographic record d, link reference E to an author of a the paper described by d to an authority A in IdRef, if E and A refer to the same author.
- **The result.** The result is an ordered list of seven pairwise subsets of IdRef: $S(trong)$, $M(edium)$, $W(eak)$, $P(oor)$, $N(eutral)$, $U(nrelated)$, $I(mpossible)$. The order is a qualitative ordering of authorities which could be linked to E. $S(trong)$ contains authorities for which there is strong evidence that they refer to the same author as E. For $M(edium)$, the evidence is less strong etc., and $I(mpossible)$ contains authorities for which there is strong evidence that they do not refer to the same author as E. This result can be used in an automatic mode, the system makes a decision to link or not link r, or in an aided-decision mode, the system presents the ordered list to a person who has to make a decision.

3 The Method

3.1 Principles

The important tasks of the implemented method can be briefly stated as follows (they are detailed in the forthcoming subsections).

- **Linkage knowledge:** comparison criteria and rules.
  A preliminary fundamental task consists of building linkage knowledge whose main components are comparison criteria and logical rules. An *elementary comparison criterion* is a function c_δ that assigns to (E,A), where E is a referent to an author in a bibliographic record d and A is a referent to a person authority, a qualitative value representing the similarity of$E$ and A with respect to δ. The elementary comparison criteria built for SudocAD are: c_{denom} (dealing with denominations of authors), c_{dom} (dealing with scientific domains of documents), c_{date} (dealing with date information), and c_{lang} (dealing with the languages in which the documents are written). For instance, for computing $c_{date}(E,A)$, knowledge such as "if the publication date of d is t then E cannot be identical to a person authority A whose birth date is posterior to t."
  The set of values of a comparison criterion c_δ is a totally ordered set of qualitative values representing the similarity between E and A with respect to δ (typically: *similar, intermediate, weak, dissimilar*). See Section 3.4 for details.
  The (global) comparison between E and A is also expressed as a qualitative value, $S(trong)$ or $M(edium)$ or $W(eak)$ or $P(oor)$ or $N(eutral)$ or $U(nrelated)$ or $I(mpossible)$), which is the conclusion of a logical rule. These logical rules

are as follows: If $c_{denom}(E,A) = H_1$ and $c_{dom}(E,A) = H_2$ and $c_{date}(E,A) = H_3$ and $c_{lang}(E,A) = H_4$ then $linkage(E,A) = C$. (cf. Section 3.5).

The notions needed for expressing these rules are gathered in the linkage ontology $\mathcal{O}_L$.

Once this linkage knowledge has been built, the input data is processed by three successive steps as follows.

- **Working knowledge base.** The whole bibliographic database (sudoc catalog + IdRef) is very large. In the first processing step, it is restricted to a knowledge base $\mathcal{W}$ expressed with the formal ontology (composed of the domain ontology $\mathcal{O}_B$, based on FRBRoo, and on the linkage ontology $\mathcal{O}_L$). $\mathcal{W}$ should have two main properties: it should contain all authorities which may be linked with E in d, and these authorities are called authority candidates. The second property it that $\mathcal{W}$ should be small enough to efficiently perform the computations needed by the linkage problems. See Section 3.2 for the construction of $\mathcal{W}$.
- **Authority enrichment.** In the second step, each authority candidate in $\mathcal{W}$ is enriched, and the result of such enrichment is called a *super-authority*. The links between document records and authority records are used to build these super-authorities. For instance, one can add an attribute 'area of competence' of an authority A and if A is the author of many documents dealing with medicine it is probably relevant to add that medicine is within the area of competence of A. Instead of adding new attributes, it is also possible to specialize information already existing in an authority record. These super-authorities are compared to the information about E, i.e. the entity to be linked, obtained from d. See Sectioncf. 3.3) for more details.
- **Linkage computations.** The linkage task itself is decomposed into two subtasks as follows. For each couple (E,A) and for each comparison criterion c_δ, the value $c_\delta(E,A)$ is computed. This computation uses numerical computations but the resulting value $c_\delta(E,A)$ is a qualitative value (e.g. *similar* or *intermediate* or *weak* or *dissimilar*). These qualitative values are used as hypotheses in logical rules (see Section 3.5).

Finally, logical rules, having values of comparison criteria (between E and A) as conditions and a value of the global qualitative comparison criterion, *link*, as conclusion are fired. The result is a partition of the authority candidates ordered by decreasing relevance with respect to the possibility of linking E and A. For instance, if $link(E,A) = Strong$, there is strong evidence that E and A can be linked, i.e. that they represent the same entity in the exterior world.

3.2 Working Base and Authority Candidates

The construction of the working base $\mathcal{W}$ is detailed in this section. The main steps are as follows.

1. For each author name in d, the first task is to represent an author name in d in the same way, say *name*, as denominations are represented in authority records

(this may need an alignment between 'author name' in d and 'denomination' in authority records).

2. A set $\mathscr{A} = sim(name)$ of authorities having a denomination close to *name* is computed. Note that the variants of the name present in the authority record are used for this computation.
3. For each authority A in $\mathscr{A}$, the set $Bib(A)$ of bibliographic records having A as an author or as a contributor with a significant role is computed.
4. The working base $\mathscr{W}$ is obtained by making the union of the authority records in $\mathscr{A}$ and the document records $Bib(A)$ for all A in $\mathscr{A}$.

A fundamental condition is that the function *sim* is sufficiently robust to author name variations so that if there is an authority in the authority base corresponding to the author whose name is *name* in d, then this authority is (almost surely) in $\mathscr{A}$. Another reason for considering, in this step, a generous similarity function is that $\mathscr{W}$ should contain sufficient contextual knowledge concerning the authorities in order to remove the ambiguities, i.e. to solve the linkage problem.

3.3 Super-Authority

An authority record does not contain a lot of information. Indeed, an authority record for people is not a biographical record, its only goal is to distinguish one person from another one (also described by an authority record). Authority records for people essentially contain information about their names, dates and country (cf. [18]). Enriching an authority record with information concerning this authority that can be obtained by searching the bibliographic records is a key idea of SudocAD.

In d, for a bibliographic record of a paper in a scientific journal, one has only, besides the names of the authors, the publication date, paper title, language in which the paper is written and a list of scientific domains. For each of these notions, one can possibly enrich an authority record by using information in the document records, $Bib(A)$, in which A is a contributor. For instance, aggregating all domains of records in $Bib(A)$ is generally more precise than a piece of information concerning the competence in A. In the same way, one can compute, and then assign to A, an interval of publication dates. Note that, due to the nature of d (scientific paper), the kinds of contributor considered are restricted to those having a scientific role, e.g. author, PhD supervisor, scientific editor, preface writer. Note that, as in a bibliographic record d within the bibliographic base Persee, there is only, besides the names of the authors, the publication date, paper title and, language. Thus, super-authorities deal only with information that can be compared with information in d, i.e., domain, date, language. The next section stipulates how this new information is computed and defines the comparison criteria.

Remark. Note that if a document base does not contain authority records but only bibliographic records then the method for building super-authorities can be used, starting with an authority record containing only a name attribute, to build authority records.

3.4 Comparison Criteria

Due to the nature of Persee bibliographic records, the only information taken into account in SudocAD is: *denomination, domain, date, language*. The corresponding elementary comparison criteria and how their values are computed for a given couple (E,A) are described in this section.

3.4.1 Denomination

An author authority record (in the Sudoc base) contains all known variants of an author name, thus the attribute denomination is not used for computing a super-authority.

The name n_d of an author E in d, and a denomination n_A of an authority A, are split into two strings, respectively (n,p) and (n',p'). The first string is the most discriminant part of the name, in our case it corresponds to the family name, and the second string (less important) is composed of first names or first name initials. The strings n,n',p,p' are normalized (transformation of uppercases into lowercases, deletion of accents and redundant spaces, etc.). Two functions, denoted c and c', respectively compare n,n' and p, p'. For both, their result is one of the following qualitative values $I(dentical)$, $S(trong)$, $C(ompatible)$, $D(istant)$, $Dif(erent)$. c and c' use the same distance algorithm (Levenshtein's algorithm) with the same threshold, but due to their different nature (e.g. possibly initials in p's) the two functions are rather different. For instance, they differently use the prefix notion (these functions are completely described on Cogui's website).

The results of these two functions are aggregated as follows, where the qualitative values of $c_{denom}(E,A)$ are coded as follows: +++ stands for *same denomination*, ++ for *close denomination*, + for *distant denomination*, - for *dissimilar denomination*.

```
if (c(n, n')  = I or S) then
  if c'(p,p')= I or S return +++;
  if c'(p,p')= C or D return ++;
  otherwise return +);
if (c(n, n')  = C) then
  if c'(p,p')= I or S or C return ++;
  if c'(p,p')=  D return +;
  otherwise return -);
if (c(n, n')  = D) then
  if c'(p,p')= I or S or C or D return +;
  otherwise return -);
otherwise return -
```

Finally, since an authority A may have a set of denominations, $Denom(A)$, the value of the denomination criterion $c_{denom}(E,A)$ is equal to the maximum value over the set of denominations of A, i.e. if n_d is the name of E,

$c_{denom}(E,A) = max\{c_{denom}(n_d,n_A)|n_A \in Denom(A)\}$.

3.4.2 Domain

For each authority candidate A, a domain profile, which is a set of weighted domains $\{(d_1, p_1), \ldots, (d_k, p_k)\}$, is computed as follows.

A bibliographic record has a domain attribute which is a multi-set of domains (the same domain can occur several times). To a bibliographic record B which is in $Bib(A)$, i.e., for which A is a scientific contributor, is assigned a weighted set $\{(d_1, q_1), \ldots, (d_m, q_m)\}$, where $\{d_1, \ldots, d_m\}$ is the set of domains in B and $q_i = 1/\#d_i$ where $\#d_i$ is the number of occurrences of d_i in the domain attribute of B. This set is considered as a vector on the set of domains and the domain profile of A is the sum of these vectors for all documents in $Bib(A)$. Note that this domain profile is a new piece of information in the super-authority of A.

In the same way, a domain profile can also be assigned to document d, with the weight of a domain of d being equal to $1/\#dd$, where $\#dd$ is the number of domains associated with d. The domains associated with d is the domains of the Journal in which d has been published. We compare the set of domains of the Journal in which d has been published with the weighted list of domains in the authority A.

The similarity measure between two domain profiles $P = \{(d_1, p_1), \ldots, (d_m, p_m)\}$ and $P' = \{(d'_1, p'_1), \ldots, (d'_n, p'_n)\}$ that have been used in SudocAD is

$$\sigma(P,P') = \sum_{i=1}^{m} min(p_i, \sum_{j=1}^{n} p'_j \sigma(d_i, d'_j))$$

In this formula,

$$\sigma(d, d') \in [0, 1]$$

is a similarity between domains. $\sigma(d, d') = 1$ means that $d = d'$ or that d and d' are synonyms and $\sigma(d, d') = 0$ means that it is quite impossible that a given person has a scientific role in a document about d and a document about d'.

The qualitative values of the *domain* criterion are coded as follows : +++ stands for *strong correspondence*, ++ for *intermediate correspondence*, + for *weak correspondence*, - for *without correspondence*, and ? for *unknown* (Bibliographic records do not always have domain attributes, thus it is possible that domain profiles cannot be computed). The qualitative values for the domain criterion are defined as follows, $c_{dom}(d, A)$ is equal to:

- +++ whenever $0.8 < \sigma(P,P') \leq 1$,
- ++ whenever $0.5 < \sigma(P,P') \leq 0.8$,
- + whenever $0.2 < \sigma(P,P') \leq 0.5$,
- - whenever $0 \geq \sigma(P,P') \leq 0.2$.

Note that $c_{dom}(E, A) = c_{dom}(d, A)$ for any author in d since we only use d to compute the domain profile of an author of d.

3.4.3 Date

Two notions are used to define the date criterion. The first one expresses compatibility between the publication date p_d of d and the interval of publication dates (*beginPeriod*, *endPeriod*) of A and the other expresses compatibility between the publication date of d and the life interval (*birthDate*, *deathDate*) of A when these dates are known. If only one of these dates exists, the second one is approximated w.r.t. a (reasonable) maximal lifespan.

Three parameters are used: T_1 is the age at which a person can begin to publish, T_2 is the maximal lifespan whenever either the birth date or the death date of an author is unknown and T_3 is used to express that the publication date of d is close to the interval of publication dates. In SudocAD, T_1 is set at 20, T_2t at 100 and T_3 at 10.

The following definitions are used in the table specifying the date criterion.

$inPer = (beginPeriod \leq p_d)$ and $(p_d \leq endPeriod)$;

$closePer =$ not $inPer$ and $(beginPeriod - T_3 \leq p_d)$ and $(p_d \leq endPeriod + T_3)$;

$outPer =$ not $inPer$ and not $closePer$;

$beforeLife = (p_d \leq birthDate + T_1)$;

$inLife = (birthDate + T_1 \leq p_d)$ and $(p_d \leq deathDate))$;

$afterLife = (p_d \geq deathDate)$;

The qualitative values of the *date* criterion are coded as follows: +++ stands for *strong correspondence*, ++ for *intermediate correspondence*, + for *weak correspondence*, - for *without correspondence*, and ? for *unknown*.

The value of the date criterion is given by the following table.

date criterion	*inPer*	*closePer*	*outPer*
inLife	+++	++	++
afterLife	++	+	+
no LifeDates	++	+	+
beforeLife	-	-	-

3.4.4 Language

For our experiment, the language is not discriminant and we have chosen a very simple language criterion: if in $Bib(A)$ there is a document written in the same language as d then the value of $c_{lang}(d,A)$ is + and otherwise it is -. Note that, as for the domain criterion, the language criterion actually deals with d and is transferred to each author E in d.

3.5 Linkage

3.5.1 Principles

Let us recall some notations. E is an author in d and $\mathscr{A}$ is the set of authority candidates obtained in the working base computation step (see Section3.2). For each (E,A), A in $\mathscr{A}$, and each criterion c_δ, the value $c_\delta(E,A)$ is computed as explained in Section 3.4. These values are used for computing the possibility of linkage between E and A. This possibility is expressed by a (global) comparison criterion called *linkage*. The value of $linkage(E,A)$ is a value in the ordered set: $S(trong)$, $M(edium)$, $W(eak)$,$P(oor)$, $N(eutral)$, $U(nrelated)$, $I(mpossible)$s.

$linkage(E,A)$ is obtained as the conclusion of logical rules whose premises are the values of $c_{denom}(E,A)$, $c_{date}(E,A)$, $c_{dom}(E,A)$ and $c_{lang}(E,A)$. Here is an example of a rule (noted *LS*2 in the table in Section 3.5.2).

```
If $c_{denom}(E,A) = +++$ and $c_{date}(E,A)=++$ and
   $c_{dom}(E,A)=+++$ and $c_{lang}(E,A)=+$
then $linkage(E,A)=S$.
```

This rule is used as follows. If for a given pair (E,A) all premises are true, then the system concludes that there is strong evidence that E and A represent the same entity. Said otherwise, in this situation, the system cannot distinguish E from A. At the other end, if $linkage(E,A) = I$, this means that the system considers that there is strong evidence that E and A do not refer to the same author.

The values of $linkage(E,A)$ partition $\mathscr{A}$, a class being composed of all A in$\mathscr{A}$ having the same value $linkage(E,A)$. This partition is ordered by decreasing relevance with respect to the possibility of linking E and A.

This partition can be used in different ways, and the choices made in SudocAD for an automatic linkage mode and for a decision-aided mode are presented in Section 4.

3.5.2 In SudocAD

The 22 rules used in SudocAD are listed in the following table in which the joker * stands for any value of a criterion as well as the absence of value.

These rules are fired in a specific order and $linkage(E,A)$ is the first value obtained, i.e. as soon as a rule is fired for a pair (E,A) the others are not fired. The chosen order is as follows: LI1, LI2, LU3, LP4, LS1, LS2, LM1, LM2, LM3, LM4, LM5, LW1, LW2, LW3, LW4, LW5, LP1, LP2, LP3, LU1, LU2, Another case.

The order on the set of values of a comparison criterion is also used. If the value of a criterion is positive, say p, then it is assumed that it also has all the positive values $p' \leq p$. The way the rules are fired ensures that the result obtained is equivalent to the result given by 300 rules fired in any order, i.e. used in a declarative way.

Rule	Denom	Date	Dom	Lang	Linkage
LS1	+++	+++	++	+	S
LS2	+++	++	+++	+	S
LM1	+++	*	+++	*	M
LM2	+++	+	++	+	M
LM3	++	+++	+++	*	M
LM4	++	++	++	+	M
LM5	+++	++	+	+	M
LW1	++	++	+	*	W
LW2	++	+	++	+	W
LW3	+	+++	+++	*	W
LW4	++	+	+	*	W
LW5	+++	*	++	+	W
LP1	+++	*	-	*	P
LP2	++	*	*	*	P
LP3	+	++	*	*	P
LP4	*	*	*	-	P
LU1	+	*	*	-	U
LU2	*	+	-	*	U
LU3	*	-	*	*	U
LI1	*	-	-	*	I
LI2	-	*	*	*	I
Other case	*	*	*	*	N

4 Evaluation

4.1 Methodology

It seems difficult to assess the quality of the partition obtained by SudocAD and even to define the quality of such a partition. Thus, we consider two ways of using this partition which can be evaluated by human experts: an automatic mode and a decision-aided mode. In the automatic mode, the system either proposes an authority A to be linked to E or proposes no link. In the decision-aided mode, the system proposes a list of authorities in a decreasing relevance order until the operator chooses one authority or stops using the system.

From the database Persee (cf. [22]), 150 bibliographic records, referencing 212 authors, were chosen at random. For these records, professional librarians had to do their usual work, that is to try to link authors in Persee records to authorities in IdRef [18] in their usual work environment. That is to say, librarians had the Persee records and usual on-line access to the sudoc catalog and to the IdRef authorities. Librarians had also a limited time, no more than 5 min for linking an author, and they also had to respect the usual constraint to avoid creating erroneous links. For an author E in a Persee record, a librarian could make one of the following decisions:

- link with certainty E to an authority A
- link with uncertainty E to an authority A and suggest other possible authorities
- refrain with certainty for linking
- refrain with uncertainty for linking and suggest possible authorities

Note that the last two situations should normally prompt the librarian to create a new authority record.

An expert librarian analyzed the results of this first step. He was not involved in this step and could use any source of information (e.g. the web) and he had no time limited. The results obtained after this step are hoped to be the best possible linkages in such a way that the comparison of these results with those obtained by SudocAD is meaningful.

Expert Results	Number
Link with certainty	146
Link with uncertainty and other choices	3
Link with uncertainty and no other choices	19
No link with certainty	37
No link without certainty and other choices	7
No link without certainty and no other choices	0

4.2 Automatic Linkage

We considered four different ways of automatic linkage, listed as follows from the most restrictive to the least restrictive linkage.

- AL_1: If the best class, i.e. $S(trong)$, contains only one authority, then E is linked to this authority;
- AL_2: If the union of the two best classes, i.e. $S(trong)$ and $M(edium)$, contains only one authority, then E is linked to this authority;
- AL_3: If the union of the three best classes, i.e. $S(trong)e$ and $M(edium)$ and $W(eak)$, contains only one authority, then E is linked to this authority;
- AL_4: If the union of the four best classes, i.e. $S(trong)$ and $M(edium)$ and $W(eak)$ and $P(oor)$, contains only one authority, then E is linked to this authority.

For the comparison between the expert choice and one of these methods we considered that the answer given by the method corresponded to one of the following decisions:

- *Good decision:* either when the expert links with certainty E to A and the method links E to A or when the expert does not link with certainty E and the method does not propose a link;

- *Acceptable decision:* either when the expert links with uncertainty E to A and the method links E to A or when the expert does not link with certainty E and the method proposes a link to a possible candidate proposed by the expert;
- *Bad decision:* either when the expert links with certainty E to A and the method links E to $A' \neq A$ or when the expert does not link with certainty E and the method proposes a link;
- *Prudent decision:* when the expert links with or without certainty and the method does not propose a link.

The means of these parameters for the 212 authors occurring in the 150 Persee bibliographic records are as follows.

Method	Good	Acceptable	Bad	Prudent
AL_1	54.7%	0%	1.89%	43.4%
AL_2	77.36%	0.47%	1.89%	20.28%
AL_3	80.19%	0.47%	3.77%	15.57%
AL_4	86.79%	0.94%	6.6%	5.66%

The results were very positive. The choice of a method AL_i could be guided by the importance given to the quality of the decision made by the system. If bad decisions have to be strictly avoided while having a significant number of good decisions, then AL_2 is a good choice. If the number of good or acceptable decisions is the main criterion, then AL_4 can be chosen. Note furthermore, that most bad decisions of any method AL_i arise from errors in the bibliographic database and are not related to intrinsic flaws in the system itself. Indeed, when there are false links from bibliographic records to authority records, these erroneous links can lead to building incorrect super-authorities.

4.3 Decision-Aided

In a decision-aided mode, the system presents an ordered list of candidate authorities to a human operator. This list is presented in decreasing order of relevance $S(trong)$, $M(edium)$, $W(eak)$, etc. until the operator chooses one authority to be linked to E or stops and concludes that there is no authority which can be linked to E.

Three classical Information Retrieval parameters were considered to evaluate the use of our system in a decision-aided mode. They use the following sets of authorities:

- *Cand* is the set of authority candidates in the working base,
- *Impos* is the set of authorities related by $I(mpossible)$ to E,
- *Selec* is the union of the authority linked to E by the operator and, when there is uncertainty, the set of possible authorities proposed by the operator.

The parameters are then defined as follows.

- *Recall*
  $recall = |Selec \cap (Cand \setminus Impos)| / |Selected|$

- *Precision*
  $precision = |Selec \cap (Cand \setminus Impos)| / |(Cand \setminus Impos)|$
- *Relevance*
  $relevance = |Selec| / MaxPos$,
  where $MaxPos$ is the last position in the ordered result of an authority in $Selected$.

The means of these parameters for the 212 authors occurring in the 150 Persee records and without considering the candidates in $I(mpossible)$ or $U(nrelated)$ are as follows.

Expert choice	Recall	Precision	Relevance
No link with uncertainty	92.86%	43.94%	56.12%
Link with certainty	100%	78.76%	94.32%
Link with uncertainty	98.48%	68.25%	91.71%

Note that when the librarian has chosen to link, the relevance is higher than 90%. This means that when the librarian searches the candidates in the order given by SudocAD, then the first one is, with few exceptions, the good one.

More importantly, note also that these parameters allow only a partial evaluation of the decision-aided mode. Indeed, there are two different situations: the operator can decide either to link or not to link E to an authority. If he decides to link, the recall is not relevant, because in this case the recall is equal to 1 unless $Selected \cap Impossible \neq \emptyset$, which is unlikely. If he decides not to link, *Selected* is empty. It is possible to propose other parameters but for a significant evaluation one should carry out an experiment comparing two systems. This is further work that is planned for a decision-aided use of our system.

5 COGUI

The system SudocAD has been developed on COGUI (see [7]) and uses web services for accessing an RDF version of the Sudoc catalog and Sudoc authority records. COGUI is a platform for designing, building and reasoning –with logical rules (cf. [4])– on conceptual graph knowledge bases. COGUI was used at each stage of the project (cf. [6]). Information concerning SudocAD can be found on the Cogui website, on the examples page (cf. [10]).

- *Ontology.* The ontology used in SudocAD contains a hierarchy of (313) concepts and a hierarchy of (1179) relations. The root of the linkage relations is *liage : binding vocabulary relation (Resource,Resource).* A relation has a signature which indicates its arity and the maximal types an attribute can have (e.g. the two attributes of the relation *liage : liageAuthority (Person,Person)* have to be of type (less than or equal to) Person.
- *Input Data.* Input data consists of a part of the Sudoc database, which consists of a catalog and an authority base, and of a Persee record. As explained in section 2 the Sudoc database was translated into RDF. A part of it, containing

authority records whose denominations are close to the name of an author in the Persee record and the bibliographic records in which these authorities have a scientific role, is accessed through web services and then imported into Cogui through the import RDF/S natural mode tool.
- *Scripts.* The method described in section 3 is implemented through scripts which use queries for searching the data graph.

6 Conclusion and Further Work

As far as we know, SudocAD is an original system whose main facets are: the combination of numerical and logical computations; the notion and construction of super-authorities; the type of the result, a qualitatively ordered partition of the set of candidates, which allows to build different strategies for the linkage problem (in an automatic mode as well as in a decision-aided mode); and the genericity of the method, SudocAD deals with authors but can be used for other authorities as well, e.g. collective entities. Note also that the super-authority step can be used to build authority records for a document database only containing bibliographic records. The results obtained by SudocAD have been evaluated by a rigorous and demanding method and are very positive. ABES plan to develop SudocAD into a system usable in production conditions.

Improvements can be made when using the same methodology. Here are some of them we plan to investigate in the near future:

- **Improve criteria.** For instance, the domain criterion is the only attribute concerning the content of a document. This can be improved by using the title of a document, which gives more precise information concerning the content of a document than the domain attribute. The date criterion can also be improved by considering the time in which data are relevant intervals (e.g. period during which a person is affiliated to an institution or has a particular function).
- **Enrich the super-authority and add new criteria.** A bibliographic or document database is a rich resource that can be more intensively scanned. For instance, neither co-authors nor the affiliation relationships between institutions and authors have been used.
- **Narrow the search results.** The genericity of the method can allow to iteratively use it in order to reinforce the results. First, use a methodology similar to SudocAd for linking institutions, use these links for linking authors (using the affiliation of authors to institutions) then check or modify the links between institutions.

The algorithms used in SudocAD assume that the bases are correct. Each record, either bibliographic record or authority record, can indeed be assumed to be correct (especially when they have been built by librarians and not by an automatic system). However, links between records can be erroneous, especially since some of them have been automatically computed when the considered base has been obtained by automatically merging several bases. Further work are needed to define the quality of document bases, for improving the quality of the links existing in a document base, and for taking into account the quality of the considered bases in linkage problems.

References

1. ABES, http://www.abes.fr/
2. Arasu, A., Christopher, R., Suciu, D.: Large-scale deduplication with constraints using dedupalog. In: Proceedings of the 25th International Conference on Data Engineering (ICDE), pp. 952–963 (2009)
3. Benjelloun, O., Garcia-Molina, H., Menestrina, D., Su, Q., Euijong Whang, S., Widom, J.: Swoosh: a generic approach to entity resolution. The VLDB Journal 18(9-10), 255–276 (2009)
4. Baget, J.-F., Leclère, M., Mugnier, M.-L., Salvat, E.: On rules with existential variables: Walking the decidability line. Artif. Intell. 175(9-10), 1620–1654 (2011)
5. The CIDOC CRM, http://www.cidoc-crm.org/
6. Chein, M., Mugnier, M.-L.: Graph-based Knowledge Representation. Springer, London (2009)
7. COGUI, http://www.lirmm.fr/cogui/
8. de Carvalho, M.G., Laender, A.H.F., Goncalves, M.A., da Silva, A.S.: Genetic programming approach to record deduplication. IEEE Transactions on Knowledge and Data Engineering 24(3), 399–412 (2012)
9. Elmagarmid, A.K., Ipeirotis, P.G., Verykios, V.S.: Duplicate record detection: A survey. Transactions on Knowledge and Data Engineering, p. 2007 (2007)
10. COGUI examples, http://www.lirmm.fr/cogui/examples.php#sudocad
11. Functional Requirements for Bibliographic Records, http://www.ifla.org/publications/functional-requirements-for-bibliographic-records
12. Object Formulation of FRBR, http://www.cidoc-crm.org/frbr_inro.html
13. Fellegi, I.P., Sunter, A.B.: A theory for record linkage. Journal of the American Statistical Association (1969)
14. Fatiha Sais, F., Pernelle, N., Rousset, M.-C.: Combining a logical and a numerical method for reference reconciliation. Journal of Data Semantics, 66–94 (2009)
15. Gu, L., Baxter, R., Vickers, D., Rainsford, C.: Record linkage: current practice and future directions. Technical Report 03/83, CSIRO Mathematical and Information Sciences (2003)
16. Gomatam, S.: An empirical comparison of record linkage procedures. Statist. Med. 21(1), 1485–1496 (2002)
17. Hernández, M.A., Stolfo, S.J.: Real-world data is dirty: data cleansing and the merge/purge problem. Data Min. Knowl. Discov. 20(2(1)), 9–37 (1998)
18. IdRef:authority files of the Sudoc database, http://en.abes.fr/Other-services/IdRef
19. Suchanek, F.M., Abiteboul, S., Senellart, P.: Paris: Probabilistic alignment of relations, instances, and schema. Proceedings of the VLDB Endowment 5(3), 157–168 (2012)
20. Newcombe, H.B., Kennedy, J.M., Axford, S.J., James, A.P.: Automatic linkage of vital records. Science (1959)
21. Neil, N.R., Smalheiser, R., Torvik, V.I.: Author name disambiguation. Annual Review of Information Science and Technology (ARIST) 43 (2009)

22. PerseeD, `http://www.persee.fr/web/guest/home`
23. RDA: Resource Description and Access, `http://www.rda-jsc.org/rda.html`
24. Singla, P., Domingos, P.: Object identification with attribute-mediated dependences. In: Jorge, A.M., Torgo, L., Brazdil, P.B., Camacho, R., Gama, J. (eds.) PKDD 2005. LNCS (LNAI), vol. 3721, pp. 297–308. Springer, Heidelberg (2005)
25. Shvaiko, P., Euzenat, J.: Ontology matching: State of the art and future challenges. IEEE Trans. Knowl. Data Eng. 25(1), 158–176 (2013)
26. Sais, F., Pernelle, N., Rousset, M.-C.: L2r: a logical method for reference reconciliation. In: Proc. of AAAI 2007, pp. 329–334 (2007)
27. SUDOC, `http://www.abes.fr/Sudoc/Sudoc-public`
28. sudocAD, `http://www.abes.fr/Sudoc/Projets-en-cours/SudocAD`
29. Winkler, W.E.: Overview of record linkage and current research directions. Technical report, U.S. Census Bureau (2006)
30. Winkler, W.E.: Record linkage references. Technical report, U.S. Census Bureau (2008)

Word Confidence Estimation and Its Integration in Sentence Quality Estimation for Machine Translation

Ngoc-Quang Luong, Laurent Besacier, and Benjamin Lecouteux

Abstract. This paper proposes some ideas to build an effective estimator, which predicts the quality of words in a Machine Translation (MT) output. We integrate a number of features of various types (system-based, lexical, syntactic and semantic) into the conventional feature set, for our baseline classifier training. After the experiments with all features, we deploy a "Feature Selection" strategy to filter the best performing ones. Then, a method that combines multiple "weak" classifiers to build a strong "composite" classifier by taking advantage of their complementarity allows us to achieve a better performance in term of F score. Finally, we exploit word confidence scores for improving the estimation system at sentence level.

1 Introduction

Statistical Machine Translation (SMT) systems in recent years have marked impressive breakthroughs with numerous fruitful achievements, as they produced more and more user-acceptable outputs. Nevertheless the users still face with some open questions: are these translations ready to be published as they are? Are they worth to be corrected or do they require retranslation? It is undoubtedly that building a method which is capable of pointing out the correct parts as well as detecting the translation errors in each MT hypothesis is crucial to tackle these above issues. If we limit the concept "parts" to "words", the problem is called Word-level Confidence Estimation (WCE). The WCE's objective is to judge each word in the MT hypothesis as correct or incorrect by tagging it with an appropriate label. A classifier which has been trained beforehand calculates the confidence score for the MT output

Ngoc-Quang Luong · Laurent Besacier · Benjamin Lecouteux
Laboratoire d'Informatique de Grenoble, Campus de Grenoble, 41, Rue des Mathématiques, BP53, F-38041 Grenoble Cedex 9, France
e-mail: {Ngoc-Quang.Luong,Laurent.Besacier, Benjamin.Lecouteux}@imag.fr
http://www.liglab.fr

V.-N. Huynh et al. (eds.), *Knowledge and Systems Engineering, Volume 1*, Advances in Intelligent Systems and Computing 244,
DOI: 10.1007/978-3-319-02741-8_9, © Springer International Publishing Switzerland 2014

word, and then compares it with a pre-defined threshold. All words with scores that exceed this threshold are categorized in the *Good* label set; the rest belongs to the *Bad* label set.

The contributions of WCE for the other aspects of MT are incontestable. First, it assists the post-editors to quickly identify the translation errors, determine whether to correct the sentence or retranslate it from scratch, hence improve their productivity. Second, the confidence score of words is a potential clue to re-rank the SMT N-best lists. Last but not least, WCE can also be used by the translators in an interactive scenario [2].

This article conveys ideas towards a better word quality prediction, including: novel features integration, feature selection and Boosting technique. It also investigates the usefulness of using WCE in a sentence-level confidence estimation (SCE) system. After reviewing some related researches in Section 2, we depict all the features used for the classifier construction in Section 3. The settings and results of our preliminary experiments are reported in Section 4. Section 5 explains our feature selection procedure. Section 6 describes the Boosting method to improve the system performance. The role of WCE in SCE is discussed in Section 7. The last section concludes the paper and points out some perspectives.

2 Previous Work Review

To cope with WCE, various approaches have been proposed, aiming at two major issues: features and model to build the classifier. In [1], the authors combine a considerable number of features, then train them by the Neural Network and naive Bayes learning algorithms. Among these features, Word Posterior Probability (henceforth WPP) proposed by [3] is shown to be the most effective system-based features. Moreover, its combination with IBM-Model 1 features is also shown to overwhelm all the other ones, including heuristic and semantic features [4].

A novel approach introduced in [5] explicitly explores the phrase-based translation model for detecting word errors. A phrase is considered as a contiguous sequence of words and is extracted from the word-aligned bilingual training corpus. The confidence value of each target word is then computed by summing over all phrase pairs in which the target part contains this word. Experimental results indicate that the method yields an impressive reduction of the classification error rate compared to the state-of-the-art on the same language pairs.

Xiong et al. [6] integrate the POS of the target word with another lexical feature named "Null Dependency Link" and train them by Maximum Entropy model. In their results, linguistic features sharply outperform WPP feature in terms of F-score and classification error rate. Similarly, 70 linguistic features guided by three main aspects of translation: accuracy, fluency and coherence are applied in [9]. Results reveal that these features are helpful, but need to be carefully integrated to reach better performance.

Unlike most of previous work, the authors in [7] apply solely external features with the hope that their classifier can deal with various MT approaches, from statistical-based to rule-based. Given a MT output, the BLEU score is predicted by their regression model. Results show that their system maintains consistent performance across various language pairs.

Nguyen et al. [8] study a method to calculate the confidence score for both words and sentences relied on a feature-rich classifier. The novel features employed include source side information, alignment context, and dependency structure. Their integration helps to augment marginally in F-score as well as the Pearson correlation with human judgment.

3 Features

This section depicts 25 features exploited to train the classifier. Some of them are already used in our previous paper [18]. Among them, those marked with a Ⓟ symbol are proposed by us, and the remaining comes from the other researches.

3.1 System-Based Features (Directly Extracted from SMT System)

3.1.1 Target Side Features

We take into account the information of every word (at position i in the MT output), including:

- The word itself.
- The sequences formed between it and a word before $(i-1/i)$ or after it $(i/i+1)$.
- The trigram sequences formed by it and two previous and two following words (including: $i-2/i-1/i$; $i-1/i/i+1$; and $i/i+1/i+2$).
- The number of occurrences in the sentence.

3.1.2 Source Side Features

Using the alignment information, we can track the source words which the target word is aligned to. To facilitate the alignment representation, we apply the BIO[1] format: if multiple target words are aligned with one source word, the first word's alignment information will be prefixed with symbol "B-" (means "Begin"); meanwhile "I-" (means "Inside") will be added at the beginning of the alignment information for each of the remaining ones. The target words which are not aligned with any source word will be represented as "O" (means "Outside"). Table 1 shows an example for this representation, in case of the hypothesis is *"The public will soon have the opportunity to look again at its attention."*, given its source: *"Le public aura bientôt l'occasion de tourner à nouveau son attention."*. Since two target words *"will"* and *"have"* are aligned to *"aura"* in the source sentence, the alignment information for them will be "B-aura" and "I-aura" respectively. In case a target word has multiple

[1] `http://www-tsujii.is.s.u-tokyo.ac.jp/GENIA/tagger/`

Table 1 Example of using BIO format to represent the alignment information

Target words	Source aligned words	Target words	Source aligned words
The	B-le	to	B-de
public	B-public	look	B-tourner
will	B-aura	again	B-à|nouveau
soon	B-bientôt	at	B-son
have	I-aura	its	I-son
the	B-l'	attention	B-attention
opportunity	B-occasion	.	B-.

aligned source words (such as *"again"*), we separate these words by the symbol "|" after putting the prefix "B-" at the beginning.

3.1.3 Alignment Context Features

These features are proposed by [8] in regard with the intuition that collocation is a believable indicator for judging if a target word is generated by a particular source word. We also apply them in our experiments, containing:

- Source alignment context features: the combinations of the target word and one word before (left source context) or after (right source context) the source word aligned to it.
- Target alignment context features: the combinations of the source word and each word in the window ±2 (two before, two after) of the target word.

For instance, in case of *"opportunity"* in Table 1, the source alignment context features are: *"opportunity/l' "* and *"opportunity/de"*; while the target alignment context features are: *"occasion/have"*, *"occasion/the"*, *"occasion/opportunity"*, *"occasion/to"* and *"occasion/look"*.

3.1.4 Word Posterior Probability

WPP [3] is the likelihood of the word occurring in the target sentence, given the source sentence. To calculate it, the key point is to determine sentences in N-best lists that contain the word e under consideration in a fixed position i. In this work, we exploit the graph that represents MT hypotheses [10]. From this, the WPP of word e in position i (denoted by WPP *exact*) can be calculated by summing up the probabilities of all paths containing an edge annotated with e in position i of the target sentence. Another form is "WPP *any*" in case we sum up the probabilities of all paths containing an edge annotated with e in any position of the target sentence. In this paper, both forms are employed.

3.1.5 Graph Topology Features

They are based on the N-best list graph merged into a confusion network. On this network, each word in the hypothesis is labelled with its WPP, and belongs to one *confusion set*. Every completed path passing through all nodes in the network represents one sentence in the N-best, and must contain exactly one link from each confusion set. Looking into a confusion set (which the hypothesis word belongs to), we find some information that can be the useful indicators, including: the *number of alternative paths* it contains (called *Nodes*Ⓟ), and the distribution of posterior probabilities tracked over all its words (most interesting are *maximum and minimum probabilities*, called *Max*Ⓟ and *Min*Ⓟ). We assign three above numbers as features for the hypothesis word.

3.1.6 Language Model Based Features

Applying SRILM toolkit [11] on the bilingual corpus, we build 4-gram language models for both target and source side, which permit to compute two features: the *"longest target n-gram length"*Ⓟ and *"longest source n-gram length"*Ⓟ (length of the longest sequence created by the current word and its previous ones in the language model). For example, with the target word w_i: if the sequence $w_{i-2}w_{i-1}w_i$ appears in the target language model but the sequence $w_{i-3}w_{i-2}w_{i-1}w_i$ does not, the n-gram value for w_i will be 3. The value set for each word hence ranges from 0 to 4. Similarly, we compute the same value for the word aligned to w_i in the source language model. Additionally, we consider also the *backoff behaviour* [17] of the target language model to the word w_i, according to how many times it has to back-off in order to assign a probability to the word sequence.

3.2 Lexical Features

A prominent lexical feature that has been widely explored in WCE researches is word's Part-Of-Speech (POS). We use TreeTagger[2] toolkit for POS annotation task and obtain the following features for each target word:

- Its POS
- Sequence of POS of all source words aligned to it
- Bigram and trigram sequences between its POS and the POS of previous and following words. Bigrams are POS_{i-1}/POS_i , POS_i/POS_{i+1} and trigrams are: $POS_{i-2}/POS_{i-1}/POS_i$; $POS_{i-1}/POS_i/POS_{i+1}$ and $POS_i/POS_{i+1}/POS_{i+2}$

In addition, we also build four other binary features that indicate whether the word is a: *stop word* (based on the stop word list for target language), *punctuation* symbol, *proper name* or *numerical*.

3.3 Syntactic Features

The syntactic information about a word is a potential hint for predicting its correctness. If a word has grammatical relations with the others, it will be more likely to be correct than those which has no relation. In order to obtain the links between words, we select the Link Grammar Parser[3] as our syntactic parser, allowing us to

[2] http://www.ims.uni-stuttgart.de/projekte/corplex/TreeTagger/
[3] http://www.link.cs.cmu.edu/link/

build a syntactic structure for each sentence in which each pair of grammar-related words is connected by a labeled link. Based on this structure, we get a binary feature called *"Null Link"*: 0 in case of word has at least one link with the others, and 1 if otherwise. Another benefit yielded by this parser is the "constituent" tree, representing the sentence's grammatical structure (showing noun phrases, verb phrases, etc.). This tree helps to produce more word syntactic features, including *its constituent label*Ⓟ and *its depth in the tree*Ⓟ (or the distance between it and the tree root).

Figure 1 represents the syntactic structure as well as the constituent tree for a MT output: *"The government in Serbia has been trying to convince the West to defer the decision until by mid 2007."*. It is intuitive to observe that the words in brackets

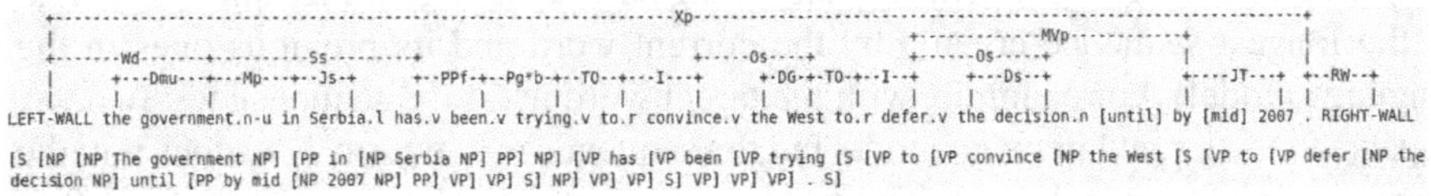

Fig. 1 Example of parsing result generated by Link Grammar

(including "*until*" and "*mid*") have no link with the others, meanwhile the remaining ones have. For instance, the word "*trying*" is connected with "*to*" by the link "TO" and with "*been*" by the link "Pg*b". Hence, the value of "Null Link" feature for "*mid*" is 1 and for "*trying*" is 0. The figure also brings us the constituent label and the distance to the root of each word. In case of the word "*government*", these values are "NP" and "2", respectively.

3.4 Semantic Features

The word semantic characteristic that we study is its polysemy. We hope that the *number of senses* of each target word given its POS can be a reliable indicator for judging if it is the translation of a particular source word. The feature *"Polysemy count"*Ⓟ is built by applying a Perl extension named Lingua::WordNet[4], which provides functions for manipulating the WordNet database.

4 Baseline WCE Experiments

4.1 Experimental Settings

4.1.1 SMT System

Our French - English SMT system is constructed using the Moses toolkit [12]. We keep the Moses's default setting: log-linear model with 14 weighted feature functions. The translation model is trained on the Europarl and News parallel corpora

4 http://search.cpan.org/dist/Lingua-Wordnet/Wordnet.pm

used for WMT10[5] evaluation campaign (1,638,440 sentences). Our target language model is a standard n-gram language model trained by the SRI language modeling toolkit [11] on the news monolingual corpus (48,653,884 sentences).

4.1.2 Corpus Preparation

We used our SMT system to obtain the translation hypothesis for 10,881 source sentences taken from news corpora of the WMT evaluation campaign (from 2006 to 2010). Our post-editions were generated by using a crowdsourcing platform: Amazon Mechanical Turk [13]. We extract 10,000 triples (source, hypothesis and post edition) to form the training set, and keep the remaining 881 triples for the test set.

4.1.3 Word Label Setting

This task is performed by TERp-A toolkit [14]. Table 2 illustrates the labels generated by TERp-A for one hypothesis and reference pair. Each word or phrase in the hypothesis is aligned to a word or phrase in the reference with different types of edit: I (insertions), S (substitutions), T (stem matches), Y (synonym matches), and P (phrasal substitutions). The lack of a symbol indicates an exact match and will be replaced by E thereafter. We do not consider the words marked with D (deletions) since they appear only in the reference. Then, to train a binary classifier, we re-categorize the obtained 6-label set into binary set: The E, T and Y belong to the *Good* (G), whereas the S, P and I belong to the *Bad* (B) category. Finally, we observed that out of total words (train and test sets) are 85% labeled G, 15% labeled B.

Table 2 Example of training label obtained using TERp-A

Reference	The	consequence	of	the	fundamentalist	movement		also	has	its importance	.
		S			S	Y	I		D	P	
Hyp After Shift	The	result	of	the	hard-line	trend	is	also		important	.

4.1.4 Classifier Selection

We apply several conventional models, such as: *Decision Tree*, *Logistic Regression* and *Naive Bayes* using KNIME platform[6]. However, since our intention is to treat WCE as a sequence labeling task, we employ also the *Conditional Random Fields* (CRF) model [15]. Among CRF based toolkits, we selected WAPITI [16] to train our classifier. We also compare our classifier with two naive baselines: in Baseline 1, all words in each MT hypothesis are classified into *G* label. In Baseline 2, we assigned them randomly: 85% into *G* and 15% into *B* label (similar to the percentage of these labels in the corpus).

[5] http://www.statmt.org/wmt10/

[6] http://www.knime.org/knime-desktop

4.2 Preliminary Results and Analysis

We evaluate the performance of our classifier by using three common evaluation metrics: Precision (Pr), Recall (Rc) and F-score (F). We perform the preliminary experiments by training a CRF classifier with the combination of all 25 features, and another one with only "conventional" features (not suggested by us). The classification task is then conducted multiple times, corresponding to a threshold increase from 0.300 to 0.975 (step = 0.025). When threshold = α, all words in the test set which the probability of G class exceeds α will be labelled as "G", and otherwise, "B". The values of Pr and Rc of G and B label are tracked along this threshold variation, and then are averaged and shown in Table 3, for "all-feature", "conventional feature" and baseline systems. These values imply that in our systems: (1) Good

Table 3 Average Pr, Rc and F for labels of systems and two baselines

System	Label	Pr(%)	Rc(%)	F(%)
All features	Good	85.99	88.18	87.07
	Bad	40.48	35.39	37.76
Baseline 1	Good	81.78	100.00	89.98
	Bad	-	0	-
Baseline 2	Good	81.77	85.20	83.45
	Bad	18.14	14.73	16.26
Conventional features	Good	85.12	87.84	86.45
	Bad	38.67	34.91	36.69

label is much better predicted than Bad label, (2) The combination of all features helped to detect the translation errors significantly above the "naive" baselines as well as that with only conventional features.

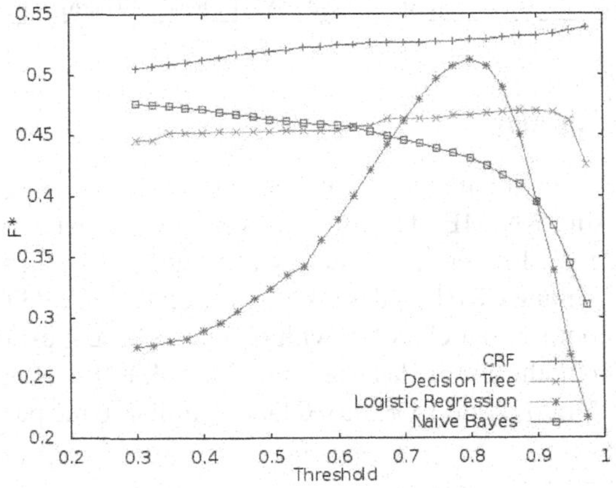

Fig. 2 Performance (F^*) of different "all feature" classifiers (by different models)

In an attempt of investigating the performance of CRF model, we compare the "all feature" system with those built by several other models, as stated in Section 4.1. The pivotal problem is how to define an appropriate metric to compare them efficiently? Due to the fact that in our training corpus, the number of G words sharply outperforms the B ones, so it is fair to say that with our classifiers, detecting a translation error should be more appreciated than identifying a good translated word. Therefore, we propose a "composite" score called F^* putting more weight on the system capability of detecting B words: $F^* = 0.70 * Fscore(B) + 0.30 * Fscore(G)$. We track all scores along the threshold variation and then plot them in Figure 2. The topmost position of CRF curve shown in the figure reveals that the CRF model performs better than all the remaining ones, and it is more suitable to deal with our features and corpus. In the next sections, which propose ideas to improve the prediction capability, we work only with the CRF classifier.

5 Feature Selection for WCE

In Section 4, the all-feature system yielded promising F scores for G label, but not very convincing F scores for B label. That can be originated from the risk that not all of features are really useful, or in other words, some are poor predictors and might be the obstacles weakening the other ones. In order to prevent this drawback, we propose a method to filter the best features based on the "Sequential Backward Selection" algorithm[7]. We start from the full set of N features, and in each step sequentially remove the most useless one. To do that, all subsets of (N-1) features are considered and the subset that leads to the best performance gives us the weakest feature (not included in the considered set). Obviously, the discarded feature is not considered in the following steps. We iterate the process until there is only one remaining feature in the set, and use the following score for comparing systems: $F_{avg}(all) = 0.30 * F_{avg}(G) + 0.70 * F_{avg}(B)$, where $F_{avg}(G)$ and $F_{avg}(B)$ are the averaged F scores for G and B label, respectively, when threshold varies from 0.300 to 0.975. This strategy enables us to sort the features in descending order of importance, as displayed in Table 4. Figure 3 shows the evolution of the WCE performance as more and more features are removed, and the details of 3 best feature subsets yielding the highest $F_{avg}(all)$.

Table 4 reveals that the system-based and lexical features seemingly outperform the other types in terms of usefulness, since in top 10, they contribute 8 (5 system-based + 3 lexical). However, 2 out of 3 syntactic features appear in top 10, indicating that their role cannot be disdained. Observation in 10-best and 10-worst performing features suggests that features belonging to word origin (the word itself, POS) perform well, meanwhile those from word statistical knowledge sources (target and source language models) are less beneficial. In addition, in Figure 3, when the size of feature set is small (from 1 to 7), we can observe sharply the growth of the system performance ($F_{avg}(all)$). Nevertheless the scores seem to saturate as the feature set increases from 8 up to 25. This phenomenon raises a hypothesis about our

[7] `http://research.cs.tamu.edu/prism/lectures/pr/pr_l11.pdf`

Table 4 The rank of each feature (in term of usefulness) in the set. The letter represents category: "S" for system-based , "L" for lexical, "T" for syntactic, and "M" for semantic feature; and the symbol "*" indicates our proposed features.

Rank	Feature name	Rank	Feature name
1L	Source POS	14L	Punctuation
2S	Source word	15M*	Polysemy count
3S	Target word	16S*	Longest source gram length
4S	Backoff behaviour	17S	Number of occurrences
5S	WPP *any*	18L	Numeric
6L	Target POS	19L	Proper name
7T*	Constituent label	20S	Left target context
8S	Left source context	21S*	Min
9T	Null link	22S*	Longest target gram length
10L	Stop word	23S	Right source context
11S*	Max	24T*	Distance to root
12S	Right target context	25S	WPP *exact*
13S*	Nodes		

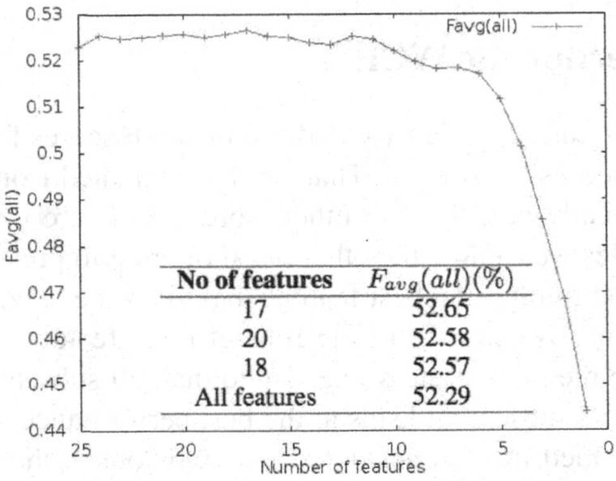

Fig. 3 Evolution of system performance ($F_{avg}(all)$) during Feature Selection process

classifier's learning capability when coping with a large number of features, hence drives us to an idea for improving the classification scores, which is detailed in the next section.

6 Classifier Performance Improvement Using Boosting

If we build a number of "weak" (or "basic") classifiers by using subsets of our features and a machine learning algorithm (such as *Boosting*), should we get a single "strong" classifier? When deploying this idea, our hope is that multiple models can complement each other as one feature set might be specialized in a part of the data where the others do not perform very well.

First, we prepare 23 feature subsets ($F_1, F_2, ..., F_{23}$) to train 23 basic classifiers, in which: F_1 contains all features, F_2 is the Top 17 in Table 4 and F_i $(i = \overline{3..23})$ contains 9 randomly chosen features. Next, the 10-fold cross validation is applied on our usual 10K training set. We divide it into 10 equal subsets (S_1, S_2, …, S_{10}). In the loop i $(i = \overline{1..10})$, S_i is used as the test set and the remaining data is trained with 23 feature subsets. After each loop, we obtain the results from 23 classifiers for each word in S_i. Finally, the concatenation of these results after 10 loops gives us the training data for Boosting. The detail of this algorithm is described below:

Algorithm to build Boosting training data

```
for i :=1 to 10 do
begin
    TrainSet(i) := ∪S_k (k = 1..10, k ≠ i)
    TestSet(i) := S_i
    for j := 1 to 23 do
    begin
        Classifier C_j := Train TrainSet(i) with F_j
        Result R_j := Use C_j to test S_i
        Column P_j := Extract the "probability of word to be G label" in R_j
    end
    Subset D_i (23 columns) := {P_j} (j = 1..23)
end
Boosting training set D := ∪D_i (i = 1..10)
```

Next, Bonzaiboost toolkit[8] (based on decision trees and implements Boosting algorithm) is used for building Boosting model. In the training command, we invoked: algorithm = "AdaBoost", and number of iterations = 300. The Boosting test set is prepared as follows: we train 23 feature sets with the usual 10K training set to obtain 23 classifiers, then use them to test the CRF test set, finally extract the 23 probability columns (like in the above pseudo code). In the testing phase, similar to what we did in Section 5, the *averaged* Pr, Rc and F scores against threshold variation for *G* and *B* labels are tracked as seen in Table 5.

Table 5 Comparison of the average Pr, Rc and F between CRF and Boosting systems

System	Pr(G)	Rc(G)	F(G)	Pr(B)	Rc(B)	F(B)
Boosting	90.10	84.13	**87.02**	34.33	49.83	**40.65**
CRF (all)	85.99	88.18	**87.07**	40.48	35.39	**37.76**

The scores suggest that using Boosting algorithm on our CRF classifiers' output is an efficient way to make them predict better: on the one side, we maintain the already good achievement on *G* class (only 0.05% lost), on the other side we augment 2.89% the performance in *B* class. It is likely that Boosting enables different models to better complement one another, in terms of the later model becomes experts for instances handled wrongly by the previous ones. Another advantage is that Boosting algorithm weights each model by its performance (rather than treating them equally), so the strong models (come from all features, top 17, etc.) can make more dominant impacts than the others.

[8] `http://bonzaiboost.gforge.inria.fr/#x1-20001`

7 Using WCE in Sentence Confidence Estimation (SCE)

WCE helps not only in detecting translation errors, but also in improving the sentence level prediction when combined with other sentence features. To verify this, firstly we build a SCE system (called SYS1) based on our WCE outputs (prediction labels). The seven features used to train SYS1 are:

- The ratio of number of good words to total number of words. (1 feature)
- The ratio of number of good nouns to total number of nouns. The similar ones are also computed for other POS: verb, adjective and adverb. (4 features)
- The ratio of number of n consecutive good word sequences to total number of consecutive word sequences. Here, n=2 and n=3 are applied. (2 features)

Then, we inherit the script used in WMT12[9] for extracting 17 sentence features, to build an another SCE system (SYS2). In both SYS1 and SYS2, each sentence training label is an integer score from 1 to 5, based on its TER score, as following:

$$score(s) = \begin{cases} 5 & \text{if } TER(s) \leq 0.1 \\ 4 & \text{if } 0.1 < TER(s) \leq 0.3 \\ 3 & \text{if } 0.3 < TER(s) \leq 0.5 \\ 2 & \text{if } 0.5 < TER(s) \leq 0.7 \\ 1 & \text{if } TER(s) > 0.7 \end{cases} \quad (1)$$

Two conventional metrics are used to measure the SCE system's performance: Mean Absolute Error (MAE) and Root of Mean Square Error (RMSE)[10]. To observe the impact of WCE on SCE, we design a third system (called SYS1+SYS2), which takes the results yielded by SYS1 and SYS2, post-processes them and makes the final decision. For each sentence, SYS1 and SYS2 generate five probabilities for five integer labels it can be assigned, then select the label which highest probability as the official result. Meanwhile, SYS1+SYS2 collects probabilities come from both systems, then updates the probability for each label by the sum of two appropriate values in SYS1 and SYS2. Similarly, the label with highest likelihood is assigned to this sentence. The results are shown in Table 6.

Table 6 Scores of 3 different SCE systems

System	MAE	RMSE
SYS1	0.5584	0.9065
SYS2	0.5198	0.8707
SYS1+SYS2	0.4835	0.8415

9 https://github.com/lspecia/QualityEstimation/blob/master/baseline_system

10 http://www.52nlp.com/mean-absolute-error-mae-and-mean-square-error-mse/

Scores observed reveal that when WMT12 baseline features and those based on our WCE are separately exploited, they yield acceptable performance. More interesting, the contribution of WCE is definitively proven when it is combined with a SCE system: The combination system SYS1+SYS2 sharply reduces MAE and RMSE of both single systems. It demonstrates that in order to judge effectively a sentence, besides global and general indicators, the information synthesized from the quality of each word is also very useful.

8 Conclusions and Perspectives

We proposed some ideas to deal with WCE for MT, starting with the integration of our proposed features into the existing features to build the classifier. The first experiment's results show that precision and recall obtained in *G* label are very promising, and *B* label reaches acceptable performance. A feature selection strategy is then deployed to identify the valuable features, find out the best performing subset. One more contribution we made is the protocol of applying Boosting algorithm, training multiple "weak" classifiers, taking advantage of their complementarity to get a "stronger" one. Especially, the integration with SCE enlightens the WCE contribution in judging the sentence quality.

In the future, we will take a deeper look into linguistic features of word, such as the grammar checker, dependency tree, semantic similarity, etc. Besides, we would like to investigate the segment-level confidence estimation, which exploits the context relation between surrounding words to make the prediction more accurate. Moreover, a methodology to conclude the sentence confidence relied on the word- and segment- level confidence will be also deeply considered.

References

1. Blatz, J., Fitzgerald, E., Foster, G., Gandrabur, S., Goutte, C., Kulesza, A., Sanchis, A., Ueffing, N.: Confidence Estimation for Machine Translation. Technical report, JHU/-CLSP Summer Workshop (2003)
2. Gandrabur, S., Foster, G.: Confidence Estimation for Text Prediction. In: Conference on Natural Language Learning (CoNLL), Edmonton, pp. 315–321 (May 2003)
3. Ueffing, N., Macherey, K., Ney, H.: Confidence Measures for Statistical Machine Translation. In: MT Summit IX, New Orleans, LA, pp. 394–401 (September 2003)
4. Blatz, J., Fitzgerald, E., Foster, G., Gandrabur, S., Goutte, C., Kulesza, A., Sanchis, A., Ueffing, N.: Confidence Estimation for Machine Translation. In: Proceedings of COLING 2004, Geneva, pp. 315–321 (April 2004)
5. Ueffing, N., Ney, H.: Word-level Confidence Estimation for Machine Translation Using Phrased-based Translation Models. In: Human Language Technology Conference and Conference on Empirical Methods in NLP, Vancouver, pp. 763–770 (2005)
6. Xiong, D., Zhang, M., Li, H.: Error Detection for Statistical Machine Translation Using Linguistic Features. In: 48th ACL, Uppsala, Sweden, pp. 604–611 (July 2010)
7. Soricut, R., Echihabi, A.: Trustrank: Inducing Trust in Automatic Translations via Ranking. In: 48th ACL (Association for Computational Linguistics), Uppsala, Sweden, pp. 612–621 (July 2010)

8. Nguyen, B., Huang, F., Al-Onaizan, Y.: Goodness: A Method for Measuring Machine Translation Confidence. In: 49th ACL, Portland, Oregon, pp. 211–219 (June 2011)
9. Felice, M., Specia, L.: Linguistic Features for Quality Estimation. In: 7th Workshop on Statistical Machine Translation, Montreal, Canada, June 7-8, pp. 96–103 (2012)
10. Ueffing, N., Och, F.J., Ney, H.: Generation of Word Graphs in Statistical Machine Translation. In: Conference on Empirical Methods for Natural Language Processing (EMNLP 2002), Philadelphia, PA, pp. 156–163 (2002)
11. Stolcke, A.: Srilm - an Extensible Language Modeling Toolkit. In: 7th International Conference on Spoken Language Processing, Denver, USA, pp. 901–904 (2002)
12. Koehn, P., Hoang, H., Birch, A., Callison-Burch, C., Federico, M., Bertoldi, N., Cowan, B., Shen, W., Moran, C., Zens, R., Dyer, C., Bojar, O., Constantin, A., Herbst, E.: Moses: Open source toolkit for statistical machine translation. In: 45th Annual Meeting of the Association for Computational Linguistics, Prague, Czech Republic, pp. 177–180 (June 2007)
13. Potet, M., Rodier, E.E., Besacier, L., Blanchon, H.: Collection of a Large Database of French-English SMT Output Corrections. In: 8th International Conference on Language Resources and Evaluation, Istanbul, Turkey, May 23-25 (2012)
14. Snover, M., Madnani, N., Dorr, B., Schwartz, R.: Terp System Description. In: MetricsMATR workshop at AMTA (2008)
15. Lafferty, J., McCallum, A., Pereira, F.: Conditional Random Fields: Probabilistic Models for Segmenting and Labeling Sequence Data. In: CML 2001, pp. 282–289 (2001)
16. Lavergne, T., Cappé, O., Yvon, F.: Practical Very Large Scale CRFs. In: 48th Annual Meeting of the Association for Computational Linguistics, pp. 504–513 (2010)
17. Raybaud, S., Langlois, D., Smaïli, K.: This sentence is wrong. Detecting errors in machine - translated sentences. Machine Translation 25(1), 1–34 (2011)
18. Luong, N.Q.: Integrating Lexical, Syntactic and System-based Features to Improve Word Confidence Estimation in SMT. In: JEP-TALN-RECITAL, Grenoble, France, June 4-8, pp. 43–56 (2012)

An Improvement of Prosodic Characteristics in Vietnamese Text to Speech System

Thanh Son Phan, Anh Tuan Dinh, Tat Thang Vu, and Chi Mai Luong

Abstract. One important goal of TTS system is to generate natural-sounding synthesized voice. To meet the goal, a variety of tasks are performed to model the prosodic aspects of TTS voice. The task being discussed here is POS and Intonation tagging. The paper examines the effects of POS and Intonation information on the naturalness of a hidden Markov model (HMM) based speech when other resources are not available. It is discovered that, when a limited feature set is used for HMM context labels, the POS and Intonation tags improve the naturalness of the synthesized voice.

1 Introduction

In HMM-based speech synthesis systems, the prominent attribute is the ability to generate speech with arbitrary speaker's voice characteristics and various speaking styles without large amount of speech data [1].

The development of text-to-speech (TTS) voices for limited speech corpus has been a challenge today. The speech corpus suffers from the problem of lacking available data such as texts and recorded speech, and linguistic expertise. In recent years, researches in Vietnamese TTS have made considerable results. There are high quality systems such as 'Voice of Southern (VOS)', 'Sao Mai', ... However, the concatenation system needs a large amount of data to train and generate voice. The disadvantage of concatenation approach is unavoidable. Another approach emerg-

Thanh Son Phan
Faculty of Information Technology, Le Qui Don Technical University, 100 Hoang Quoc Viet Street, Cau Giay Dist., Hanoi City, Vietnam
e-mail: sonphan.hts@gmail.com

Anh Tuan Dinh · Tat Thang Vu · Chi Mai Luong
Institute of Information Technology, Vietnam Academy of Science and Technology, 18 Hoang Quoc Viet Street, Cau Giay Dist., Hanoi City, Vietnam
e-mail: {anhtuan,vtthang,lcmai}@ioit.ac.vn

V.-N. Huynh et al. (eds.), *Knowledge and Systems Engineering, Volume 1*, Advances in Intelligent Systems and Computing 244,
DOI: 10.1007/978-3-319-02741-8_10,

ing and becoming a promising solution to make more natural voice with limited sound resource is Hidden Markov Model (HMM) based Statistical Speech Synthesis (HTS) [2].

The naturalness of a Vietnamese TTS system is mainly affected by prosody. Prosody consists of accent, intonation and Vietnamese tones (6 tones). The features with part-of-speech (POS) tagging have close relationships and determine the naturalness and the intelligibility of synthesized voice.

Phonemically, Vietnamese syllable system consists 2376 base syllables and 6492 full (tonal) syllables in Vietnamese [4]. Therefore, if these tones are correctly identified, the number of candidate words is greatly reduced such that the performance of the automatic speech recognition can be improved in terms of accuracy and efficiency. In writing, the tones in the Vietnamese language are represented by a diacritic mark. Vietnamese tones consist of level, falling, broken, curve, rising, drop. One syllable can change its meaning when it goes with different tones. So the tonal feature has a strong impact on the intelligibility of synthetic voice. The following Table 1 shows an example about the name, the mark of tone in Vietnamese.

Phonemically, Vietnamese syllable system consists 2376 base syllables and 6492 full (tonal) syllables in Vietnamese [4]. Therefore, if these tones are correctly identified, the number of candidate words is greatly reduced such that the performance of the automatic speech recognition can be improved in terms of accuracy and efficiency. In writing, the tones in the Vietnamese language are represented by a diacritic mark.

Vietnamese tones consist of level, falling, broken, curve, rising, drop. One syllable can change its meaning when it goes with different tones. So the tonal feature has a strong impact on the intelligibility of synthetic voice. The following Table 1 shows an example about the name, the mark of tone in Vietnamese.

Table 1 Six tones in Vietnamese

Name	Tone mark	Example
LEVEL (ngang)	Unmarked	ta - me
FALLING (huyền)	Grave	tà - bad
BROKEN (ngã)	Tilde	tã - napkin
CURVE (hoi)	Hook above	ta? - describe
RISING (sac)	Acute	tá - dozen
DROP (nang)	Dot below	ta. - quintal

Accent is how one phoneme is emphasized in a syllable. In European language such as Russian, English and French, accent plays an important role. But, in Vietnamese and other tonal languages, the accent's role is less important. In Vietnamese accent, a phoneme is underlined with a long duration. Some Vietnamese words don't have accents such as 'cái' (Vietnamese classifier); the others have a strong accent such as 'hao huc'. Most of the Vietnamese nouns, verbs and adjectives have accent. Identifying the stress in one sentence needs POS tagging.

Intonation is the variation of spoken pitch which is not used to distinguish words; it is used for a variety of functions such as indicating the attitudes and emotions of the speaker, emphasizing the difference between statement and question, focusing the attention to important parts of a sentence. The above aspects of intonation may benefit from identifying question words in Vietnamese through POS tagging [8].

Solve the POS problem, Accent indication and Intonation tagging are steps to reach the goal of natural TTS voices. One question arises is that whether it can be possible to get around the traditional approaches to prosodic modeling by learning directly from speech data using POS, accent and intonation information? In other words, does the improvement of POS, accent and intonation features to the full context labels of an HMM-based Vietnamese synthesizer increase the naturalness and intelligibility of a synthesized voice? [5]

This is the question we aim to answer in our work, context-dependent HMMs were trained from Vietnamese prosodic rich corpus. The voices are compared with and without POS and Prosody (tone, accent and intonation) tagging. The paper is structured as follows: in Section 2, a concept of HMM-based TTS is described. In Section 3, 4 and 5, the related works of Tones, Intonation and Part of Speech are discussed. In Section 6, an experiment was implemented and results were recorded. Finally, section 6 draws some conclusions about the results.

2 Text to Speech System

A TTS system, showed in Fig. 1, is the production of speech from text. It includes the following stages [5]:

- Text tokenization splits the input text stream into smaller units named sentences and tokens. In the phase, written forms of Vietnamese syllables are discovered and tagged. The process is called tokenization.
- Text normalization decodes non-standard word into one or more pronounceable words. Non-standard tokens including numbers, dates, time, abbreviations... are normalized in the phase. The process is also called homograph disambiguation.
- Text parsing investigates lexical, syntactic structures from words which are used for pronunciation and prosodic modeling stages. The stage consists of POS tagging and chunking.
- Pronunciation modeling maps each word to its phonemes. It looks up the words in a lexicon or use grapheme to phoneme (G2P) rules to finish the task. Accents and tones are assigned.
- Prosodic modeling predicts the prosody of sentences. Sentence-level stress is identified, the intonation is assigned to sentences which make melody or tune of entire sentences.
- Speech synthesizer generates the speech waveforms from the above information. In Hidden Markov Model-based synthesis, a source filter paradigm is used

to model the speech acoustics; information from previous stages are used to make full-context label file of each phoneme in the input sentence. Excitation (fundamental frequency F_0), spectrum and duration parameters are estimated from recorded speech and modeled by context-dependent HMMs.

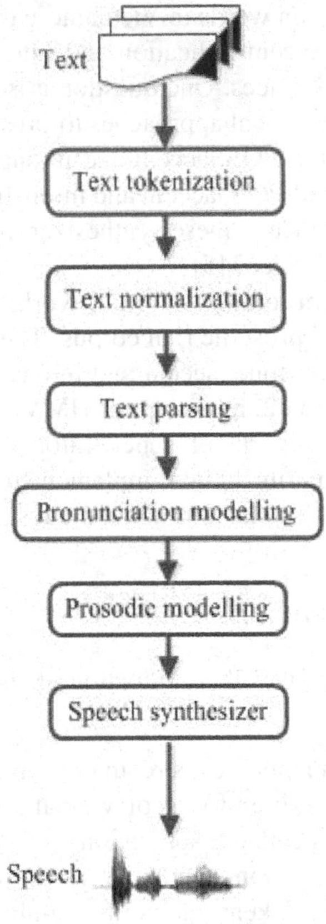

Fig. 1 A TTS system

Fig. 2 is an example of full context model used in HMM-based Speech Synthesis: Based on the full context model, HMM-based TTS is very flexible easy to add more prosodic information.

3 The Improvement of Tonal and Accentual Features

It is thought that tone lies on vowel; however, tone plays an important role on all over a syllable in Vietnamese. However, tonal features are not as explicit as other features in speech signal. According to Doan Thien Thuat [10], a syllable's structure can be described in Table 2.

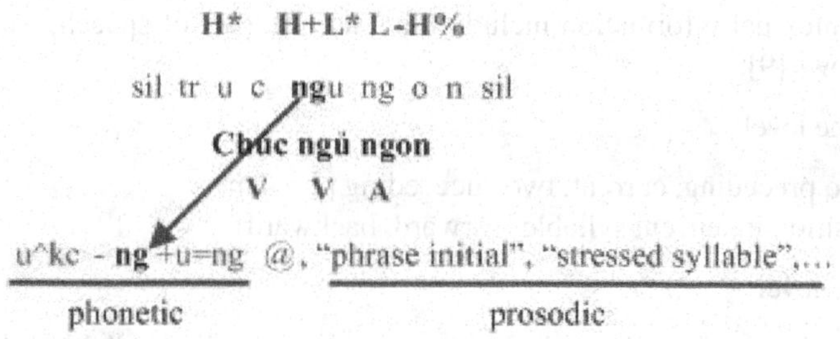

Fig. 2 A HMM full context model

Table 2 Structure of Vietnamese syllable

Tone			
[Initial]	Final		
	[Onset]	Nucleus	[Coda]

In the first consonant, we can hear a little of the tone. Tone becomes clearer in rhyme and finished completely at the end of the syllable. The pervading phenomenon determines the non-linear nature of tone. So, with mono syllabic language like Vietnamese, a syllable can't easily be separated into small acoustic parts like European languages.

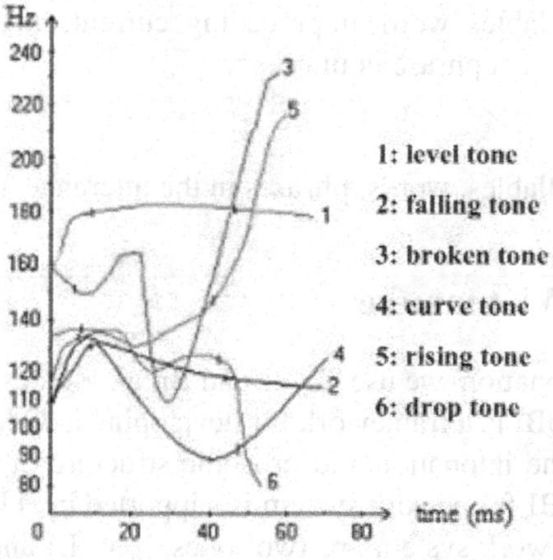

Fig. 3 The Vietnamese tone system

In syllable tagging process, contextual features must be considered. There are many contextual factors (ex, phonetic, stress, dialect, tone) affecting the signal spectral, fundamental frequency and duration. In additional, constructing a decision tree to classify the phonemes based on contextual information. The construction of the decision is very important in HMM-based Vietnamese TTS system [6].

Some contextual information include tone, accent, part-of-speech, was considered as follows [9]:

- Phoneme level:
  - Two preceding, current, two succeeding phonemes
  - Position in current syllable (forward, backward)
- Syllable level:
  - Tone types of two preceding, current, two succeeding syllables
  - Number of phonemes in preceding, current, succeeding syllables
  - Position in current word (forward, backward)
  - Stress-level
  - Distance to previous, succeeding stressed syllable
- Word level:
  - Part-of-speech of preceding, current, succeeding words
  - Number of syllables in preceding, current, succeeding words
  - Position in current phrase
  - Number of content words in current phrase before, after current word
  - Distance to previous, succeeding content words
  - Interrogative flag for the word
- Phrase level:
  - Number of syllables, words in preceding, current, succeeding phrases
  - Position of current phrase in utterance
- Utterance level:
  - Number of syllables, words, phrases in the utterance

4 Intonation in Vietnamese

In order to present intonation, we use Tones and Break Indices (ToBI) in intonation transcription phase. ToBI is a framework for developing a widely accepted convention for transcribing the intonation and prosodic structure of spoken sentences in various languages. ToBI framework system is supported in HTS engine. The primitives in a ToBI framework system are two tones, low (L) and high (H). The distinction between the tones is paradigmatic. That is L is lower than H in the same context. Utterances can consist of one or more intonation phrases. The melody of an intonation phrase is separated into a sequence of elements, each made up of either one or two tones. In our works, the elements can be classified into 2 main classes [3]:

4.1 Phrase-Final Intonation

Intonation tones, mainly phrase-final tones, were analyzed in our work. Boundary tones are associated with the right edge of the prosodic phrase and mark the end of a phrase. It can be established in Vietnamese that, a high boundary tone can change a declarative into an interrogative. To present the boundary tone, 'L-L%', 'L-H%' tags are used. 'L-L%' refers to a low tone; and 'L-H%' describes a high tone. This is a common declarative phrase. The 'L-L%' boundary tone causes the intonation to be low at the end of the prosodic phrase. On the other hand, the effect of 'L-H%' is that first it will drop to a low value and then it will rise towards the end of the prosodic phrase.

4.2 Pitch Accent

Pitch Accent is the falling or rising trends in the top line or baseline of pitch contour. Most noun, verb and adjective in Vietnamese are accented words. An 'H*' (high-asterisk) tends to produce a pitch peak while an 'L*' (low-asterisk) pitch accent pro-duces a pitch trough. In addition, the two other tag 'L+H*' and 'H+L*' are also used. 'L+H*' rises steeply from a much lower preceding F0 value while 'H+L*' falls from a much higher preceding F0 value.

It was showed in the experiment that: the intonation tags add valuable contextual information to Vietnamese syllables in training process. Spoken sentences can be distinguished easily between declarative and interrogative utterances. Import information in a speech is strongly highlighted.

5 Part of Speech

A POS tag is a linguistic category assigned to a word in a sentence based upon its morphological behavior. Words are classified into POS categories such as noun (N), verb (V), adjective (A), pronoun (P), determine (D), adverb (R), apposition (S), conjunction (C), numeral (M), interjection (I) and residual (X). Words can be ambiguous in their POS categories. The ambiguity normally solved by looking at the context of the word in the sentence.

Automatic POS tagging is processed with Conditional Random Fields (CRFs). The training of CRFs model is basically to maximize the likelihood between model distribution and empirical distribution. So, CRFs model training is to find the maximum of a log - likelihood function.

Suppose that training data consists of a set of N pairs, each pair includes an observation sequence and a status sequence, $D = (x(i), y(i))\ \forall i = 1 \ldots N$. Log-likelihood function:

$$l(\theta) = \sum_{x,y} \tilde{p}(x,y) \log(p(y|x,\theta)) \tag{1}$$

Here, $\theta(\lambda_1, \lambda_2, ..., \mu_1, \mu_2, , ...)$ is the parameter of the model and $\tilde{p}(x,y)$ is concurrent empirical experiment of x, y in training set. Replace $p(y|x)$ of (1), we have:

$$l(\theta) = \sum_{x,y} \tilde{p}(x,y) \left[\sum_{i=1}^{n+1} \lambda f + \sum_{i=1}^{n} \mu g \right] - \sum_{x} \tilde{p}(x) \log Z, \qquad (2)$$

Here, $\lambda\,(\lambda_1, \lambda_2, ..., \lambda_n)$ and $\mu\,(\mu_1, \mu_2, ..., \mu_m)$ are parameter vectors of the model, f is a vector of transition attributes, and g is a vector of status attributes.

6 Experiment and Evaluation

In the experiment, we used phonetically balanced 400 in 510 sentences (recorded female and male voices, Northern dialect) from Vietnamese speech database for training. Speech signals were sampled at 48 kHz, mono channel and coded in PCM 16 bit then the signal is downgraded to 16 kHz in waveform format and windowed by a 40-ms Hamming window with an 8-ms shift. All sentences were segmented at the phonetic level. The phonetic labeling procedure was performed as text-to-phoneme conversion through a forced alignment using a Vietnamese speech recognition engine [11]. During the text processing, the short pause model indicates punctuation marks and the silence model indicates the beginning and the end of the input text.

For the evaluation, we used remain 110 sentences in the speech database, these sentences are used as synthesize data for testing and evaluating. MFCCs and fundamental frequency F_0 was calculated for each utterance using the SPTK tool [12]. Feature vector consists of spectral, tone, duration and pitch parameter vectors: spectral parameter vector consists of 39 Mel-frequency cepstral coefficients including the zero-th coefficient, their delta and delta-delta coefficients; pitch feature vector consists of $logF_0$, its delta and delta-delta [7].

A couple of comparisons of synthesized speech qualities, include male and female speech models with only tone and with additional POS, stress and intonation. The information is added to full context model of each phoneme in a semi automatic way.

6.1 Objective Test

The objective measurement is described through comparing of waveform, pitch and spectrogram between natural speech and synthesized testing sentences in both cases:

6.2 MOS test

As a further subjective evaluation, MOS tests were used to measure the quality of synthesized speech signals in comparison with natural ones. The rated levels were: bad (1), poor (2), fair (3), good (4), and excellent (5). In this test, fifty sentences were randomly selected. With three types of audio, (1) natural speech signals, (2) the synthetic speech signals without POS, accent and intonation, and (3) the synthetic speech signals with POS, accent and intonation, the number of listeners were 50 people. The speech signals were played in random order in the tests.

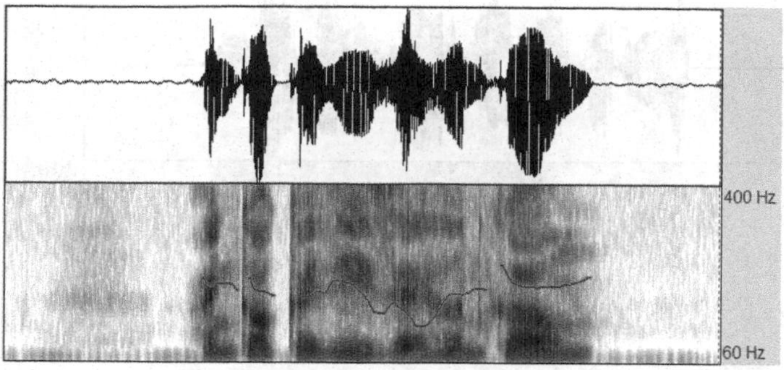

Fig. 4 Examples of wave form, F_0 and spectrogram extracted from utterance "Anh co cai gi re hon khong?" (In English "Do you have anything cheaper?") in natural speech, male voice

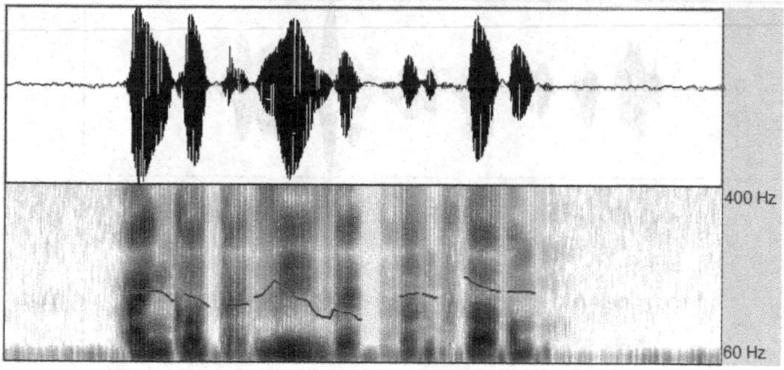

Fig. 5 Examples of wave form, F_0 and spectrogram extracted from utterance "Anh co cai gi re hon khong?" (In English "Do you have anything cheaper?") in synthesized speech without POS and Prosody tagging, male voice

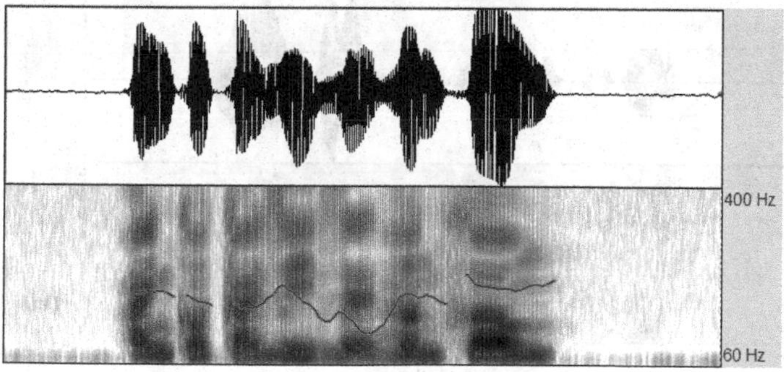

Fig. 6 Examples of wave form, F_0 and spectrogram extracted from utterance "Anh co cai gi re hon khong?" (In English "Do you have anything cheaper?") in synthesized speech with POS and Prosody tagging, male voice

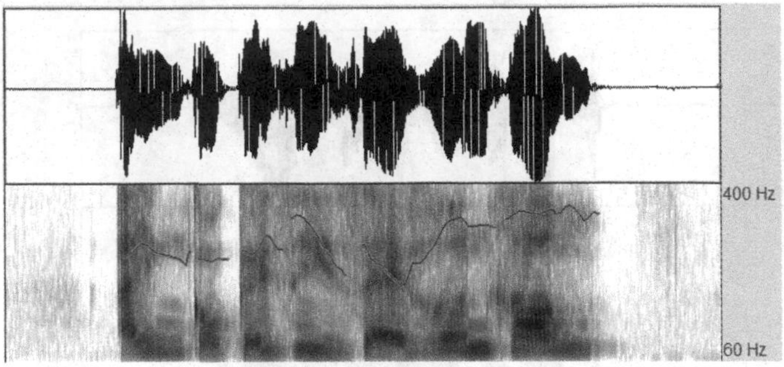

Fig. 7 Examples of wave form, F_0 and spectrogram extracted from utterance "Anh co cai gi re hon khong?" (In English "Do you have anything cheaper?") in natural speech, female voice

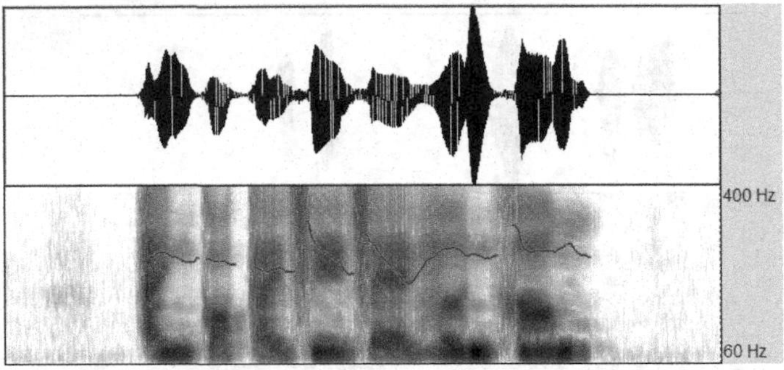

Fig. 8 Examples of wave form, F_0 and spectrogram extracted from utterance "Anh co cai gi re hon khong?" (In English, "Do you have anything cheaper?") in synthesized speech without POS and Prosody tagging, female voice

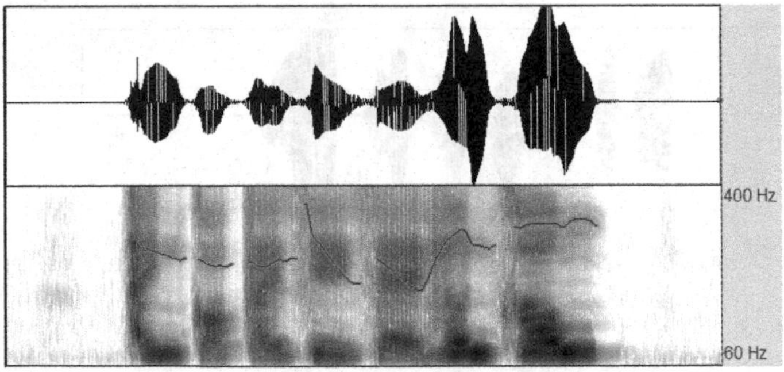

Fig. 9 Examples of wave form, F_0 and spectrogram extracted from utterance "Anh co cai gi re hon khong?" (In English "Do you have anything cheaper?") in synthesized speech with POS and Prosody tagging, female voice

Table 3 shows the mean of opinion scores which were given by all the subjects. The MOS result implied that the quality of natural speech is from good to excellence, and the quality of synthesis speech is from fair to good.

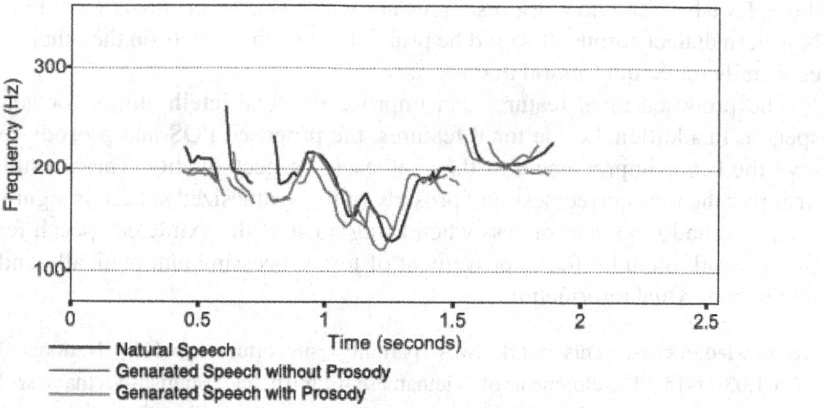

Fig. 10 Comparison F_0 contour of Natural Speech and Generated Speeches, male voice

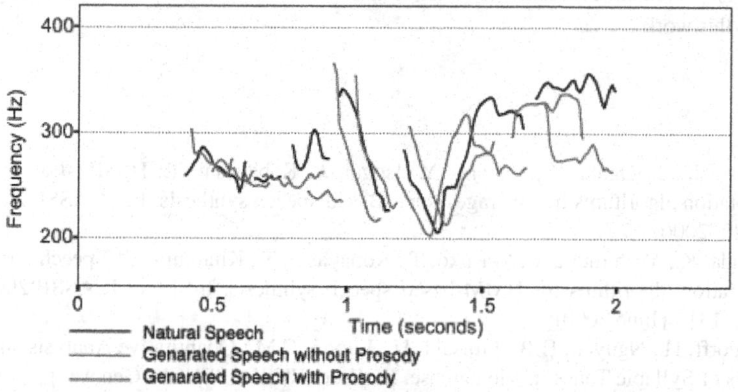

Fig. 11 Comparison F_0 contour of Natural Speech and Generated Speeches, male voice

Table 3 Results of the MOS test

Speech	Mean Opinion Score
Natural	**4.57**
Without POS, Accent, Intonation	**3.26**
With POS, Accent, Intonation	**3.92**

7 Conclusion

The experimental results, shown that POS and prosody information do contribute to the naturalness (specifically in terms of pitch) of a TTS voice when it forms part of a small phoneme identity-based feature set in the full context HTS labels. However, the same effect, even an improvement, can be accomplished by including segmental counting (phrase) and positional information in segment instead of the POS tags in the HTS labels-and no extra resources are used. The experiments were limited by Northern dialect corpus. It would be prudent to test the effects on the other dialects, especially the South Central dialect.

The proposed tonal features can improve the tone intelligibility for generated speech. In addition, beside tonal features, the proposed POS and prosody features give the better improvement of the synthesized speech quality. These results confirm that the tone correctness and prosody of the synthesized speech is significantly improved and more naturalness when using most of the extracted speech features. Future work includes the improvement of text processing automatically and work on the contextual information.

Acknowledgments. This work was partially supported by ICT National Project KC.01.03/11-15 "Development of Vietnamese–English and English–Vietnamese Speech Translation on specific domain". Authors would like to thank all staff members of Department of Pattern Recognition and Knowledge Engineering, Institute of Information Technology (IOIT) - Vietnam Academy of Science and Technology (VAST) for their support to complete this work.

References

1. Yamagishi, J., Ogata, K., Nakano, Y., Isogai, J., Kobayashi, T.: HSMM-Based Model adaptation algorithms for Average-Voice-Based speech synthesis. In: ICASSP 2006, pp. 77–80 (2006)
2. Tokuda, K., Yoshimura, T., Masuko, T., Kobayashi, T., Kitamura, T.: Speech parameter generation algorithms for HMM-based speech synthesis. In: Proc. ICASSP 2000, pp. 1315–1318 (June 2000)
3. Mixdorff, H., Nguyen, H.B., Fujisaki, H., Luong, C.M.: Quantitative Analysis and Synthesis of Syllabic Tones in Vietnamese. In: Proc. EUROSPEECH, Geneva, pp. 177–180 (2003)
4. Le, P.N., Ambikairajah, E., Choi, E.H.C.: Improvement of Vietnamese Tone Classification using FM and MFCC Features. In: Computing and Communication Technologies RIVF 2009, pp. 01–04 (2009)
5. Schlunz, G.I., Barnard, E., Van Huyssteen, G.B.: Part-of-speech effects on text-to-speech synthesis. In: 21st Annual Symposium of the Pattern Recognition Association of South Africa (PRASA), Stellenbosch, South Africa, November 22-23, pp. 257–262 (2010)
6. Phan, S.T., Vu, T.T., Duong, C.T., Luong, M.C.: A study in Vietnam-ese statistical parametric speech synthesis base on HMM. IJACST 2(1), 01–06 (2013)
7. Phan, S.T., Vu, T.T., Luong, M.C.: Extracting MFCC, F0 feature in Vietnamese HMM-based speech synthesis. International Journal of Electronics and Computer Science Engineering 2(1), 46–52 (2013)
8. Lê, T.-H., Nguyen, A.-V., Truong, H.V., Van Bui, H., Lê, D.: A Study on Vietnamese Prosody. In: Nguyen, N.T., Trawiński, B., Jung, J.J. (eds.) New Challenges for Intelligent Information and Database Systems. SCI, vol. 351, pp. 63–73. Springer, Heidelberg (2011)
9. Vu, T.T., Luong, M.C., Nakamura, S.: An HMM-based Vietnamese Speech Synthesis System. In: Proc. Oriental COCOSDA, pp. 116–121 (2009)

10. Doan, T.T.: Vietnamese Acoustic, Vietnamese National Editions, 2nd edn. (2003)
11. Vu, T.T., Nguyen, D.T., Luong, M.C., Hosom, J.P.: Vietnamese large vocabulary continuous speech recognition. In: Proc. INTERSPEECH, pp. 1689–1692 (2005)
12. Department of Computer Science, Nagoya Institute of Technology: Speech Signal Processing Toolkit, SPTK 3.6. Reference manual, Japan (December 2003), http://sourceforge.net/projects/sp-tk/ (updated December 25, 2012)

Text-Independent Phone Segmentation Method Using Gaussian Function

Dac-Thang Hoang and Hsiao-Chuan Wang

Abstract. In this paper, an effective method is proposed for the automatic phone segmentation of speech signal without using prior information about the transcript of utterance. The spectral change is used as the criterion for hypothesizing the phone boundary. Gaussian function can be used to measure the similarity of two vectors. Then a dissimilarity function is derived from the Gaussian function to measure the variation of speech spectra between mean feature vectors before and after the considered location. The peaks in the dissimilarity curve indicate locations of phone boundaries. Experiments on the TIMIT corpus show that the proposed method is more accurate than previous methods.

1 Introduction

Phone segmentation is an important stage in some areas of speech processing, such as acoustic-phonetic analysis, speech/speaker recognition, speech synthesis, and annotations of speech corpus. It can be performed manually and often get reliable results. However, this approach is time consuming, tedious, and subjective. Therefore, many researchers have been interested in automatic phone segmentation. A typical approach is to align the speech signal to its phone transcripts in an utterance. The forced alignment based on hidden Markov model (HMM) is a way to locate phone boundaries when the phone transcripts of the target utterance are available. This text-dependent phone segmentation method usually obtains high accuracy.

Dac-Thang Hoang
Department of Electrical Engineering, National Tsing Hua University,
No. 101, Kuang-Fu Road, Hsinchu, Taiwan 30013, and Department of Network System, Institute of Information Technology, No. 18 Hoang Quoc Viet Road, Hanoi, Vietnam
e-mail: d947924@oz.nthu.edu.tw

Hsiao-Chuan Wang
Department of Electrical Engineering, National Tsing Hua University,
No. 101, Kuang-Fu Road, Hsinchu, Taiwan 30013
e-mail: hcwang@ee.nthu.edu.tw

V.-N. Huynh et al. (eds.), *Knowledge and Systems Engineering, Volume 1*, Advances in Intelligent Systems and Computing 244,
DOI: 10.1007/978-3-319-02741-8_11, © Springer International Publishing Switzerland 2014

However, the training speech and its transcript are not always available. Therefore, phone segmentation must be conducted without prior knowledge of text contents in an utterance. This results in text-independent phone segmentation, which usually relies on the technique of spectral change detection. This approach is difficult to obtain a high accuracy. The experimental results are quite similar in those text-independent methods [1, 2, 3]. How to obtain a high level of accuracy using the text-independent method is a challenge.

The text-independent methods often consider spectral changes as potential transition points that correspond to phone boundaries. Some methods have been proposed. For example, Aversano et al. [4] used a jump function that represented the absolute difference between the mean values of spectra calculated in the previous frames and subsequent frames. Dusan and Rabiner [5] presented the relationship between maximum spectral transition positions and phone boundaries. Bosch and Cranen [6] used the angle between two smoothed feature vectors before and after the considered location as the spectral variation. Räsänen et al. [3] used cross correlation between two feature vectors to measure spectral variation and subsequently applied a 2D-filter and a minmax-filter to find the phone boundaries. Estevan et al. [2] examined the use of maximum margin clustering to detect the points of spectral changes.

Some statistical methods had been proposed to detect phone boundaries. Almpanidis et al. [7] used model selection criteria for the robust detection of phone boundaries, while Qiao et al. [8] introduced a time-constrained agglomerative clustering algorithm to find optimal segmentation. Lee and Glass [9] investigated the problem of unsupervised acoustic modeling. An iterative process was proposed to learn the phone models and then guide its hypotheses on phone boundaries. Instead of using Fourier transform, some researchers used wavelet transform for the parameterization of speech signal. Cherniz et al. [10] proposed a new speech feature based on continuous wavelet transform, continuous multi-resolution divergence and principal component analysis. The authors then modified the algorithm of Aversano et al. [4] and obtained an improvement. Khanagha et al. [11] proposed the use of microcanonical multiscale formalism to develop a new automatic phonetic segmentation algorithm. The authors also proposed a two-step technique to improve the segmentation accuracy [12].

In this paper, a method based on Gaussian function for automatic phone segmentation is studied. The Gaussian function is used to measure the variation of speech spectra. Experiments on the TIMIT corpus show that the proposed method is more accurate than previous methods. An F-value of 78.18% and R-value of 81.38% can be obtained under the condition of zero over-segmentation.

2 Proposed Method

2.1 Speech Feature

The speech signal is first pre-emphasized using a filter with a transfer function: $H(z) = 1 - 0.97z^{-1}$. The signal is subsequently divided into overlapping frames.

The speech signal in each frame is weighted using a Hamming window [13] and transformed to the frequency domain through fast Fourier transform (FFT). The speech energy of each frame is computed for discriminating silent regions. A Mel-scaled filter bank [14], which is composed of twenty overlapped triangular filters, is used to separate the spectral energy into twenty bands. For each frame, the outputs of filter bank are twenty band energies (BEs). At frame n, these twenty BEs form a feature vector denoted as s[n]. Fig. 1(a) shows the waveform of a speech signal extracted from TIMIT corpus. The vertical lines indicate the manually-labeled phone boundaries. The content of the utterance is "Don't ask me to carry an oily rag like that". Fig. 1(b) shows the spectral features expressed by twenty BEs.

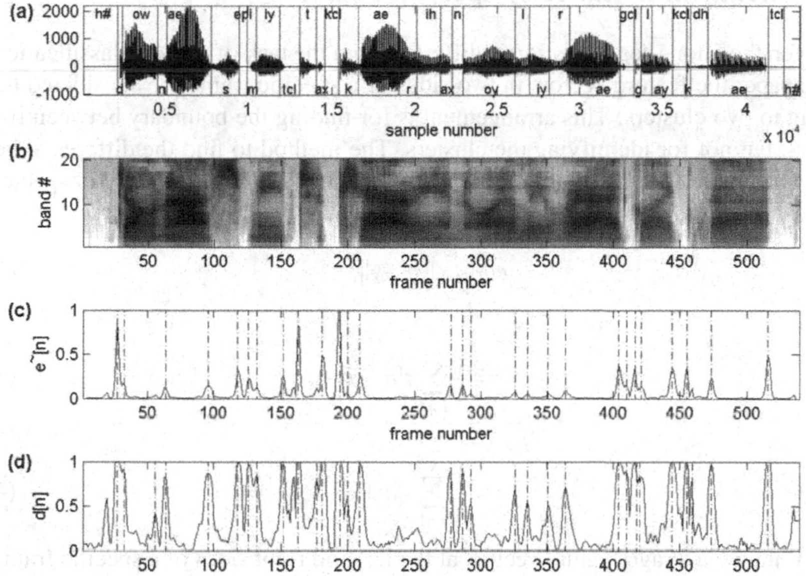

Fig. 1 (a) Speech waveform and manually labeled phone boundaries, (b) Spectrograph represented by BE of 20 bands, (c) Euclidean distance, (d) Dissimilarity function and phone boundaries detected

2.2 *Gaussian Function*

Gaussian function is widely used in the realm of pattern recognition, usually employed as similarity measurement or kernel function, and kernel density function [15]. Given two patterns x and y, the Gaussian function can be defined as followed:

$$g(x,y) = \exp\left(-\frac{\|x-y\|^2}{2\sigma^2}\right) \tag{1}$$

where σ is Gaussian function parameter. Gaussian function presents the similarity between x and y. Two patterns are quite similar when the value of Gaussian function is close to one.

The Gaussian parameter σ in (1) performs a role of resolution between x and y. A large σ will make patterns hard to be distinguished. On the contrary, the small σ will make patterns distinguishable even the difference is quite small. To obtain a good classification, a proper σ is chosen according to the distribution of patterns. There are some methods to select a proper value for parameter σ. In this paper, the method proposed by Peng et al. [15] is used. In their paper, authors found the parameter optimization via similarity comparison within class and between classes.

2.3 Euclidean Distance of Spectral Features

However, the phone segmentation is different from the task of pattern classification. In this paper the feature vectors in two sides of a specific frame are considered belonging to two clusters. This arrangement is for finding the boundary between two clusters, but not for identifying the clusters. The method to find the difference between two clusters is based on a rough measure using Euclidean distance of spectral features. At frame n, Euclidean distance is defined as:

$$e[n] = \|x - y\|^2 \tag{2}$$

where

$$x = \frac{1}{k}\sum_{i=1}^{k} s[n-i] \tag{3}$$

and

$$y = \frac{1}{k}\sum_{i=1}^{k} s[n+i] \tag{4}$$

x and y are the average feature vectors at the left and right sides of a specific frame, respectively. Parameter k is a positive integer defining the length of the average region. This length must be sufficiently short to describe the phone transition (about 20 ms). For convenience, the Euclidean distance is normalized to limit its values within the range of [0, 1],

$$\tilde{e}[n] = \frac{e[n] - e_{min}}{e_{max} - e_{min}} \tag{5}$$

where e_{min} and e_{max} are the minimum and maximum of $e[n]$ in the whole utterance. Fig. 1(c) shows the curve of normalized Euclidean distance.

A threshold e^* is set to select the significant peaks of $\tilde{e}[n]$ curve. The peaks in the silent regions must be removed. Silent regions are detected by assessing the frame energy of the speech signal. If two peaks are close to each other (occur at smaller than 20 ms apart), they will be replaced by an average value at the middle. The dash-dot vertical lines in Fig. 1(c) mark the selected peaks. That is, Fig. 1(c) shows the roughly segmented utterance. The Gaussian parameter σ at each selected peak is computed as in Peng et al.[15]. All the Gaussian parameters in the utterance are averaged. The averaged parameter is then used in Gaussian function for finding possible boundaries in this utterance. Because of using the average Gaussian parameters, the threshold e^* can be roughly chosen in the range of [0.03, 0.05].

2.4 Phone Segmentation

Notice that Gaussian function is non-negative, and its value is in the range of [0, 1]. At frame n, a dissimilarity function $d[n]$ is defined as:

$$d[n] = 1 - g(x,y) \tag{6}$$

where x and y are computed as in (3) and (4), respectively. $d[n]$ is a non-negative value which presents the dissimilarity between x and y. For an utterance, $d[n]$ shows the spectral variation with respect to time. Fig. 1(d) shows the curve of $d[n]$. Therefore, a peak of $d[n]$ curve indicates the possibility of a phone boundary. A threshold h is set to select those most possible boundaries. Delacourt and Wellekens [16] proposed the threshold in the form $h = \beta \cdot \gamma$, where γ is the standard derivation of $\{d[n], n = 1, 2, \cdots\}$ in the utterance, and β is an adjustable factor. The spurious peaks must be removed. The peaks in the silent regions must be removed. Silent regions are detected by assessing the frame energy of the speech signal. If two peaks are close to each other (occur at smaller than 20 ms apart), they will be replaced by one boundary at the middle. As shown in Fig. 1(d), the detected phone boundaries are indicated by dash-dot vertical lines.

3 Experiments

3.1 Quality Measures

The quality measures to evaluate the method for automatic phone segmentation may be different by researchers. The methods are typically evaluated by measuring their hit rate. This requires a reliable reference in which the true phone boundaries are manually produced. The hit rate of the correctly detected boundaries can be calculated by comparing the detected phone boundaries with the manually-labeled phone boundaries. When a detected phone boundary falls in a preset tolerance around the manually-labeled phone boundary, it is marked as a hit or match; otherwise, it is a false alarm or insertion. If a manually-labeled boundary is not detected, it is a missed detection error or deletion. Let N_{hit} be the number of hits and N_{ref} be the total number of manually-labeled boundaries in the reference. The hit rate (HR) or recall rate (RCL) is defined as:

$$HR = \frac{N_{hit}}{N_{ref}} \times 100\% \tag{7}$$

However, the hit rate is not the only measure for evaluating a phone segmentation method. The precision rate (*PRC*), which defines the ratio of hit number to the number of found phone boundaries (N_{find}), must also be considered.

$$PRC = \frac{N_{hit}}{N_{find}} \times 100\% \tag{8}$$

The recall rate and precision rate can be combined to obtain a single scalar value, the F-value [17]

$$F = \frac{2 \times PRC \times RCL}{PRC + RCL} \times 100\% \tag{9}$$

The F-value is often used to measure the overall performance. A high F-value implies good performance. To balance the false alarms and missed detection errors, a measure called over-segmentation (*OS*) is defined as:

$$OS = \left(\frac{N_{find}}{N_{ref}} - 1\right) \times 100\% \tag{10}$$

OS measures the exceeded number of detected boundaries. A negative *OS* implies under-segmentation. A higher value of *OS* can result in higher *RCL*. The performance at zero *OS* indicates that two error types, false alarm and missed detection, are equal. Under this condition, $F = RCL = PRC$.

Räsänen et al [18] proposed another measure, the R-value, which indicates the closeness of the resulting performance to an ideal point (HR=100%, OS=0) and the zero insertion limit ($100 - HR = -OS$). The formula used to calculate R-value is

$$R = \left(1 - \frac{|r_1| + |r_2|}{200}\right) \times 100\% \tag{11}$$

where

$$r_1 = \sqrt{(100 - HR)^2 + OS^2} \tag{12}$$

$$r_2 = \frac{-OS + HR - 100}{\sqrt{2}} \tag{13}$$

A good performance implies that R-value is close to one.

In this study, the F-value and R-value were calculated to evaluate the performance of the proposed method under the condition of $OS = 0$.

3.2 Experiment Setup and Results

The TIMIT corpus was used for evaluating the proposed method. All 4,620 utterances provided by 462 speakers in the training set were used for the performance evaluation. Each speech frame length was 10 ms with a frame shift of 5 ms. The parameter k for x and y in (3) and (4) was set to 3. A tolerance of 20 ms was used as in many studies. By adjusting factor β, the performance was plotted in Fig. 2.

When β increased, *OS* decreased and crossed the zero line; *RCL* decreased; *PRC* increased; *F*-value and *R*-value increased then decreased. These trends were resulted by the increase of missed rate and decrease of false alarm rate. *F*-value and *R*-value were relatively stable when β was in the range of $[1.0, 1.2]$. That was, the method performed effectively when β was set in this range. The *OS* is 0.00% when β is set to 1.055. Under this condition, we got *F*=78.18% and *R*=81.38%. A higher recall rate can be obtained if β decreased. For example, *RCL* can reach to 81.5% if *OS*=10.68% was acceptable.

In the condition of *OS* = 0.00%, the missed error rate was 21.82%. There were 37,635 missed detection errors in 172,460 manually-labeled boundaries. The highest missed detection error rate came from the concatenation of vowels. This resulted from the difficulty of detecting the transition from vowel to vowel. The error rate of this transition was 84.37% (3438/4075).

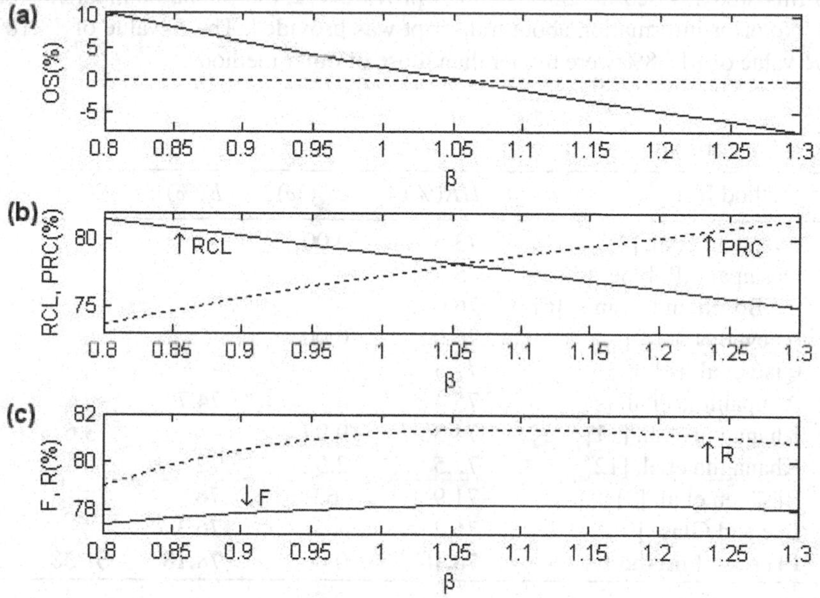

Fig. 2 (a) Over-segmentation, (b) Recall rate and Precision rate, (c) *F*-value and *R*-value

3.3 *Comparison with Other Methods*

It might not be fair to directly compare the results of this study with the results of other text-independent methods because the evaluation may not be performed on the same basis. The size of the database and the quality measures might differ. Table 1 shows the reported results in chronological order. All results of phone segmentation experiments were based on a tolerance of 20 ms. The notation "–" means that the data were not available.

All studies shown in Table I, except Almpanidis et al. [7], used the TIMIT corpus to evaluate their methods. Aversano et al. [4] used 480 utterances, which provided 17,930 boundaries. Dusan and Rabiner [5] used 4,620 utterances, which provide 172,460 phone boundaries. Their detected rate must be 75.2% as explained in [8]. Estevan et al. [2] used a predefined test set consisting of 1,344 utterances (the SA sentences are excluded). Qiao et al. [8] also used 4,620 utterances, which provided more than 170,000 phone boundaries. The number of phone boundaries in each sentence was assumed to be known. A high accuracy was obtained because of the use of prior information. Almpanidis et al. [7] performed their experiments on 256 utterances in NTIMIT corpus. Khanagha et al. [11] tested their algorithm over whole TIMIT database. In another paper, Khanagha et al. [12] evaluated their method using the full train set, which contain 4,620 sentences. Räsänen et al. [3] used a full test set including 560 female and 1,120 male sentences. Lee and Glass [9] used 3,696 utterances.

In this study, 4,620 utterances which provide 172,460 phone boundaries were used. No prior information about transcript was provided. The F-value of 78.18% and R-value of 81.38% were higher than those of other methods.

Table 1 Comparisons

Method [ref]	HR(%)	OS(%)	F(%)	R(%)
Aversano et al. [4]	73.58	0.00	–	–
Dusan and Rabiner [5]	75.2	–	–	–
ten Bosch and Cranen [6]	76.0	–	–	–
Estevan et al. [2]	76.0	0.00	–	–
Qiao et al. [8]	77.5	–	–	–
Almpanidis et al. [7]	75.4	–	74.7	–
Khanagha et al. [11]	74.38	9.97	–	73.6
Khanagha et al. [12]	72.5	2.5	72	76
Räsänen et al. [3]	71.9	-6.90	76	78
Lee and Glass [9]	76.2	–	76.3	–
Proposed method	**78.18**	0.00	**78.18**	**81.38**

4 Conclusion

This paper proposes the use of Gaussian function for the text-independent phone segmentation. A method is suggested to properly determine the Gaussian function parameter. The dissimilarity function, which is derived from Gaussian function, is used to effectively detect phone boundaries. Experiments conducted on TIMIT corpus showed that the proposed method is more accurate than previous methods. An F-value of 78.18% and R-value of 81.38% were obtained under the condition of zero over-segmentation. The method is computational efficiency and easy to be implemented. The future work will pay more attention to the concatenation of voiced phones.

Acknowledgment. This research was partially sponsored by the National Science Council, Taiwan, under contract number NSC-97-2221-E-007-075-MY3. The authors would like to thank Dr. Chi-Yueh Lin, ITRI Taiwan, for his enthusiastic discussions and valuable suggestions in this research work.

References

1. Scharenborg, O., Wan, V., Ernestus, M.: Unsupervised speech segmentation: an analysis of the hypothesized phone boundaries. J. Acoust. Soc. Amer. 172(2), 1084–1095 (2010)
2. Estevan, Y.P., Wan, V., Scharenborg, O.: Finding Maximum Margin Segments in Speech. In: Proc. IEEE Int. Conf. Acoust., Speech, Signal Process 2007, ICASSP 2007, pp. 937–940 (2007)
3. Räsänen, O., Laine, U.K., Altosaar, T.: Blind segmentation of speech using non-linear filtering methods. In: Ipsic, I. (ed.) Speech Technologies, pp. 105–124. InTech Publishing (2011)
4. Aversano, G., Esposito, A., Esposito, A., Marinaro, M.: A New Text-Independent Method for Phoneme Segmentation. In: Proc. the 44th IEEE Midwest Symposium on Circuit and System 2001, vol. 2, pp. 516–519 (2001)
5. Dusan, S., Rabiner, L.: On the Relation between Maximum Spectral Transition Position and Phone Boundaries. In: Proc. INTERSPEECH 2006, pp. 17–21 (2006)
6. ten Bosch, L., Cranen, B.: A computational model for unsupervised word discovery. In: Proc. INTERSPEECH 2007, pp. 1481–1484 (2007)
7. Almpanidis, G., Kotti, M., Kotropoulos, C.: Robust Detection of Phone Boundaries Using Model Selection Criteria with Few Observation. IEEE Trans. on Audio, Speech, and Lang. Process. 17(2), 287–298 (2009)
8. Qiao, Y., Shimomura, N., Minematsu, N.: Unsupervised Optimal Phoneme Segmentation: Objective, Algorithm, and Comparisons. In: Proc. IEEE Int. Conf. Acoust., Speech, Signal Process. 2008, ICASSP 2008, pp. 3989–3992 (2008)
9. Lee, C.Y., Glass, J.: A nonparametric Bayesian Approach to Acoustic Model Discovery. In: Proc. 50th Annual Meeting of the Association for Computational Linguistics, pp. 40–49 (2012)
10. Cherniz, A.S., Torres, M.E., Rufiner, H.L.: Dynamic Speech Parameterization for Text-Independent Phone Segmentation. In: Proc. 32nd Annual International Conference of the IEEE EMBS, pp. 4044–4047 (2010)
11. Khanagha, V., Daoudi, K., Pont, O., Yahia, H.: A novel text-independent phonetic segmentation algorithm based on the microcanonical multiscal formalism. In: Proc. INTERSPEECH 2010, pp. 1393–1396 (2010)
12. Khanagha, V., Daoudi, K., Pont, O., Yahia, H.: Improving Text-Independent Phonetic Segmentation based on the microcanonical multiscal formalism. In: Proc. IEEE Int. Conf. Acoust., Speech, Signal Process. 2011, ICASSP 2011, pp. 4484–4487 (2011)
13. Huang, X., Acero, A., Hon, H.W.: Section 5.4 Digital Filters and Windows. In: Spoken Language Processing. Prentice Hall PTR (2001)
14. Deller Jr., J.R., Hansen, J.H.L., Proakis, J.G.: Section 6.2.4 Other Forms and Variations on the stRC Parameters. In: Discrete-Time Processing of Speech Signals. IEEE Press (2000)

15. Peng, H., Luo, L., Lin, C.: The parameter optimization of Gaussian function via the similarity comparison within class and between classes. In: Proc. Third Pacific-Asia Conference on Circuits, Communications and System 2011, PACCS 2011, pp. 1–4 (2011)
16. Delacourt, P., Wellekens, C.J.: DISTBIC: A Speaker-based segmentation for audio data indexing. Speech Commun. 32(1-2), 111–126 (2000)
17. Ajmera, J., McCowan, I., Bourlard, H.: Robust Speaker Change Detection. IEEE Signal Processing Letters 11(8), 649–651 (2004)
18. Räsänen, O.J., Laine, U.K., Altosaar: An Improved Speech Segmentation Quality Measure: the R-value. In: Proc. INTERSPEECH 2009, pp. 1851–1854 (1854)

New Composition of Intuitionistic Fuzzy Relations*

Bui Cong Cuong and Pham Hong Phong

Abstract. Fuzzy relations have applications in fields such as psychology, medicine, economics, and sociology. Burillo and Bustince introduced the concepts of intuitionistic fuzzy relation and a composition of intuitionistic fuzzy relations using four triangular norms or conorms $\alpha, \beta, \lambda, \rho$ (we abbreviate to $\alpha, \beta, \lambda, \rho$-composition). In this paper, we define a new composition of intuitionistic fuzzy relations using two intuitionistic fuzzy triangular norms or intuitionistic fuzzy triangular conorms (Φ, Ψ-composition for short). It is shown that $\alpha, \beta, \lambda, \rho$-composition is special case of Φ, Ψ-composition. Many properties of Φ, Ψ-composition are stated and proved.

Keywords: intuitionistic fuzzy relation, composition of intuitionistic fuzzy relations, intuitionistic triangular norm, *t*- representability, medical diagnosis.

1 Introduction

Relations are a suitable tool for describing correspondences between objects. The use of fuzzy relations originated from the observation that object x can be related to object y to a certain degree. However, there may be a hesitation or uncertainty about the degree that is assigned to the relationship between x and y. In fuzzy set theory there is no means to incorporate that hesitation in the membership degrees. A possible solution is to use intuitionistic fuzzy sets defined by Atanassov in 1983 [1, 2].

Bui Cong Cuong
Institute of Mathematics, Vietnam Academy of Science and Technology, Vietnam
e-mail: bccuong@gmail.com

Pham Hong Phong
Faculty of Information Technology, National University of Civil Engineering, Vietnam
e-mail: phphong84@yahoo.com

* This research is funded by Vietnam National Foundation for Science and Technology Development (NAFOSTED) under grant number 102.01-2012.14.

V.-N. Huynh et al. (eds.), *Knowledge and Systems Engineering, Volume 1*, Advances in Intelligent Systems and Computing 244,
DOI: 10.1007/978-3-319-02741-8_12, © Springer International Publishing Switzerland 2014

One of the main concepts in relational calculus is the composition of relations.This makes a new relation using two relations. For example, relation between patients and illnesses can be obtained from relation between patients and symptoms and relation between symptoms and illnesses (see medical diagnosis [8, 11, 12]).

In 1995 [3], Burillo and Bustince introduced the concepts of intuitionistic fuzzy relation and a composition of intuitionistic fuzzy relations using four triangular norms or conorms. It is seen that two triangular norms or conorms can be represented via an intuitionistic norm or conorm (triangular norms or conorms [5] are defined similarly to triangular norms or conorms, when the interval $[0,1]$ is substituted by the lattice L^*). However, it is not possible to represent any intuitionistic triangular norm or conorm as two triangular norms or conorms. So, the composition using two intuitionistic triangular norms or conorms is more general than the one defined by Burillo and Bustince. Moreover, using new composition, if we make a change in non-membership components of two relations then the membership components of the result may change (see Example 0.1). This is more realistic.

2 Preliminaries

We give here some basic definitions which are used in our next sections.

2.1 *Intuitionistic Fuzzy Set*

The notions of intuitionistic fuzzy set (IFS) and intuitionistic L-fuzzy set ($ILFS$) were introduced in [1] and [2] respectively, as a generalization of the notion of fuzzy set. Let a set E be fixed, an $ILFS$ A in E is an object having the form

$$A = \{ \langle x, \mu_A(x), \nu_A(x) \rangle \mid x \in E \} , \tag{1}$$

where $\mu_A : E \to L$ and $\nu_A : E \to L$ respectively define the degree of membership and the degree of non-membership of the element $x \in E$ to $A \subset E$, μ_A and ν_A satisfy

$$\mu_A(x) \leq \mathcal{N}(\nu_A(x)) \; \forall x \in E, \tag{2}$$

where $\mathcal{N} : L \to L$ is an involutive order reversing operation in the lattice $\langle L, \leq \rangle$. When $L = [0,1]$, A is an IFS and the following condition holds:

$$0 \leq \mu_A(x) + \nu_A(x) \leq 1 \; \forall x \in E . \tag{3}$$

2.2 *Triangular Norms and Triangular Conorms*

We now review the notions of triangular norm (t-norm) and triangular conorm (t-conorm) used in the framework of probabilistic metric spaces and in multi-valued logic, specifically in fuzzy logic.

Definition 0.1. 1. A triangular norm is a commutative, associative, increasing $[0,1]^2 \to [0,1]$ mapping T satisfying $T(x,1) = x$, for all $x \in [0,1]$.
2. A triangular conorm is a commutative, associative, increasing $[0,1]^2 \to [0,1]$ mapping S satisfying $S(x,0) = x$, for all $x \in [0,1]$.

Deschrijver and Kerre [6] shown that intuitionistic fuzzy sets can also be seen as L-fuzzy sets in the sense of Goguen [8]. Consider the set L^* and operation $\leq_{L^*}$ defined by:

$$L^* = \left\{ (x_1,x_2) | \, (x_1,x_2) \in [0,1]^2 \wedge x_1 + x_2 \leq 1 \right\} , \tag{4}$$

$$(x_1,x_2) \leq_{L^*} (y_1,y_2) \Leftrightarrow x_1 \leq y_1 \wedge x_2 \geq y_2 \; \forall (x_1,x_2), (y_1,y_2) \in L^* . \tag{5}$$

Then, $\langle L^*, \leq_{L^*} \rangle$ is a complete lattice [6]. Its units are denoted by $0_{L^*} = (0,1)$ and $1_{L^*} = (1,0)$. The first and second projection mapping pr_1 and pr_2 on L^* are defined as $pr_1(x_1,x_2) = x_1$ and $pr_2(x_1,x_2) = x_2$ for all $(x_1,x_2) \in L^*$. From now on, it is assumed that if $x \in L^*$, then x_1 and x_2 denote respective the first and the second component of x, i.e. $x = (x_1,x_2)$.

Definition 0.2. [5]

1. An intuitionistic fuzzy triangular norm (*it*-norm for short) is a commutative, associative, increasing $(L^*)^2 \to L^*$ mapping $\mathscr{T}$ satisfying $\mathscr{T}(x, 1_{L^*}) = x$, for all $x \in L^*$.
2. An intuitionistic fuzzy triangular conorm (*it*-conorm for short) is a commutative, associative, increasing $[0,1]^2 \to [0,1]$ mapping $\mathscr{S}$ satisfying $\mathscr{S}(x, 0_{L^*}) = x$, for all $x \in L^*$.

It-norm and *it*-conorm can be constructed using *t*-norms and *t*-conorms in the following way. Let T be a *t*-norm and S be a *t*-conorm. If $T(a,b) \leq 1 - S(1-a, 1-b)$ for all a, $b \in [0,1]$, then a mapping $\mathscr{T}$ defined by $\mathscr{T}(x,y) = (T(x_1,y_1), S(x_2,y_2))$, for all x, $y \in L^*$ is an *it*-norm and a mapping $\mathscr{S}$ defined by $\mathscr{S}(x,y) = (S(x_1,y_1), T(x_2,y_2))$, for all x, $y \in L^*$ is an *it*-conorm. Note that the condition $T(a,b) \leq 1 - S(1-a, 1-b)$ for all a, $b \in [0,1]$ is necessary and sufficient for $\mathscr{T}(x,y)$ and $\mathscr{S}(x,y)$ to be elements of L^* for all x, $y \in L^*$. We write $\mathscr{T} = (T,S)$ and $\mathscr{S} = (S,T)$. Unfortunately, the converse is not always true. It is not possible to find for any *it*-norm $\mathscr{T}$ a *t*-norm T and a *t*-conorm S such that $\mathscr{T} = (T,S)$.

Example 0.1. Consider the *it*-norm $\mathscr{T}_W$ (Lukasiewicz *it*-norm) given by

$$\mathscr{T}_W(x,y) = (\max(0, x_1 + y_1 - 1), \min(1, x_2 + 1 - y_1, y_2 + 1 - x_1)) \, \forall x, y \in L^* . \tag{6}$$

Let $x = (0.5, 0.3)$, $x' = (0.3, 0.3)$ and $y = (0.2, 0)$. Then $pr_2(\mathscr{T}_W(x,y)) = 0.5 \neq pr_2(\mathscr{T}_W(x',y)) = 0.7$. Hence, there is no T and S such that $\mathscr{T} = (T,S)$, since this would imply that $pr_2(\mathscr{T}_W(x,y))$ is independent from x_1.

To distinguish between these two kinds of *it*-norm, the notion of *t*-representability was introduced [4]. An *it*-norm $\mathscr{T}$ is *t*-representable if there exist a *t*-norm T and *a* *t*-conorm S such that $\mathscr{T} = (T,S)$. An *it*-conorm $\mathscr{S}$ is *t*-representable if there exist a *t*-norm T and *a* *t*-conorm S such that $\mathscr{S} = (S,T)$.

2.3 Intuitionistic Fuzzy Relations

Let X and Y be non-empty sets.

Definition 0.3. [3] An intuitionistic fuzzy relation (IFR for short) R between X and Y is an intuitionistic fuzzy subset of $X \times Y$, that is, R is given by

$$R = \{ \langle (x,y), \mu_R(x,y), \nu_R(x,y) \rangle | \ (x,y) \in X \times Y \} \tag{7}$$

satisfying the condition $0 \leq \mu_R(x,y) + \nu_R(x,y) \leq 1$ for all $(x,y) \in X \times Y$. The set of all intuitionistic fuzzy relation R between X and Y is denoted by $IFR(X \times Y)$.

Definition 0.4. [3] Inverse relation of $R \in IFR(X \times Y)$ is relation $R^{-1} \in IFR(Y \times X)$ defined as

$$\mu_{R^{-1}}(y,x) = \mu_R(x,y) \ , \nu_{R^{-1}}(y,x) = \nu_R(x,y) \ , \forall (x,y) \in X \times Y \, . \tag{8}$$

Definition 0.5. [3] Let $P, R \in IFR(X \times Y)$, we can define

1. $P \leq R$ if $\mu_P(x,y) \leq \mu_R(x,y)$ and $\nu_P(x,y) \geq \nu_R(x,y)$ for all $(x,y) \in X \times Y$;
2. $P \preccurlyeq R$ if $\mu_P(x,y) \leq \mu_R(x,y)$ and $\nu_P(x,y) \leq \nu_R(x,y)$ for all $(x,y) \in X \times Y$;
3. $P \vee R = \{ \langle (x,y), \mu_P(x,y) \vee \mu_R(x,y), \nu_P(x,y) \wedge \nu_R(x,y) \rangle | \ (x,y) \in X \times Y \}$;
4. $P \wedge R = \{ \langle (x,y), \mu_P(x,y) \wedge \mu_R(x,y), \nu_P(x,y) \vee \nu_R(x,y) \rangle | \ (x,y) \in X \times Y \}$;
5. $R_c = \{ \langle (x,y), \nu_R(x,y), \mu_R(x,y) \rangle | \ (x,y) \in X \times Y \}$.

Theorem 0.1. *[3] Let R, P, $Q \in IFR(X \times Y)$.*

1. *$R \leq P$ then $R^{-1} \leq P^{-1}$;*
2. *$(R \vee P)^{-1} = R^{-1} \vee P^{-1}$;*
3. *$(R \wedge P)^{-1} = R^{-1} \wedge P^{-1}$;*
4. *$\left(R^{-1}\right)^{-1} = R$;*
5. *$R \wedge (P \vee Q) = (R \wedge P) \vee (R \wedge Q)$, $R \vee (P \wedge Q) = (R \vee P) \wedge (R \vee Q)$;*
6. *$(R \vee P) \geq R$, $(R \vee P) \geq P$, $(R \wedge P) \leq R$, $(R \wedge P) \leq P$;*
7. *If $R \geq P$ and $R \geq Q$ then $R \geq P \vee Q$, if $R \leq P$ and $R \leq Q$ then $R \leq P \wedge Q$.*

Burillo and Bustince defined a composition of IFR using four t-norms or t-conorms α, β, λ, ρ (we write α, β, λ, ρ- composition for short).

Definition 0.6. [3] Let α, β, λ, ρ be t-norms or t-conorms, $R \in IFR(X \times Y)$ and $P \in IFR(Y \times Z)$. Relation $P \underset{\lambda,\rho}{\overset{\alpha,\beta}{\circ}} R \in (X \times Z)$ is the one defined by

$$P \underset{\lambda,\rho}{\overset{\alpha,\beta}{\circ}} R = \left\{ \left\langle (x,z), \mu_{P \underset{\lambda,\rho}{\overset{\alpha,\beta}{\circ}} R}(x,z), \nu_{P \underset{\lambda,\rho}{\overset{\alpha,\beta}{\circ}} R}(x,z) \right\rangle \middle| (x,z) \in X \times Z \right\}, \tag{9}$$

where

$$\mu_{P \underset{\lambda,\rho}{\overset{\alpha,\beta}{\circ}} R}(x,z) = \underset{y}{\alpha}\left\{\beta\left[\mu_R(x,y), \mu_P(y,z)\right]\right\}, \tag{10}$$

$$\nu_{P \underset{\lambda,\rho}{\overset{\alpha,\beta}{\circ}} R}(x,z) = \underset{y}{\lambda}\left\{\rho\left[\nu_R(x,y), \nu_P(y,z)\right]\right\}, \tag{11}$$

whenever

$$0 \le \mu_{P \underset{\lambda,\rho}{\overset{\alpha,\beta}{\circ}} R}(x,z) + \nu_{P \underset{\lambda,\rho}{\overset{\alpha,\beta}{\circ}} R}(x,z) \le 1 \;\forall (x,z) \in X \times Z. \tag{12}$$

Various properties of α, β, λ, ρ- composition are presented [3]. In the remain of this paper, we define a new composition, and show that almost of above properties are still correct.

3 New Composition of Intuitionistic Fuzzy Relations

3.1 Definition

In the following, we define a new composition of intuitionistic fuzzy relations.

Definition 0.7. Let Φ, Ψ be *it*-norms or *it*-conorms, $R \in IFR(X \times Y)$ and $P \in IFR(Y \times Z)$. Relation $P \underset{\Psi}{\overset{\Phi}{\circ}} R \in IFR(X \times Z)$ is defined by

$$P \underset{\Psi}{\overset{\Phi}{\circ}} R = \left\{ \left\langle (x,z), \mu_{P \underset{\Psi}{\overset{\Phi}{\circ}} R}(x,z), \nu_{P \underset{\Psi}{\overset{\Phi}{\circ}} R}(x,z) \right\rangle \middle| (x,z) \in X \times Z \right\}, \tag{13}$$

where

$$\left(\mu_{P \underset{\Psi}{\overset{\Phi}{\circ}} R}(x,z), \nu_{P \underset{\Psi}{\overset{\Phi}{\circ}} R}(x,z) \right) = \underset{y}{\Phi}\left\{\Psi\left[(\mu_R(x,y), \nu_R(x,y)), (\mu_P(y,z), \nu_P(y,z))\right]\right\}. \tag{14}$$

Remark 0.1. The composition $\underset{\Psi}{\overset{\Phi}{\circ}}$ uses two *it*-norms or *it*-conorms Φ, Ψ. So, we call Φ, Ψ-composition. In the following, it will be seen that α, β, λ, ρ- composition is a special case of Φ, Ψ-composition.

Proposition 0.1. *If Φ and Ψ are t-representable, then Φ, Ψ-composition is reduced to α, β, λ, ρ- composition.*

Proof. There exist t-norms or t-conorms α, ρ, λ, β so that $\Phi = (\alpha,\lambda)$, $\Psi = (\beta,\rho)$ and

$$\alpha(a,b) \leq 1 - \lambda(1-a, 1-b)\,, \tag{15}$$

$$\rho(a,b) \leq 1 - \beta(1-a, 1-b)\,. \tag{16}$$

We have

$$\begin{pmatrix} \mu_{P \underset{\Psi}{\overset{\Phi}{\circ}} R}(x,z), \nu_{P \underset{\Psi}{\overset{\Phi}{\circ}} R}(x,z) \end{pmatrix}$$

$$= \underset{y}{\Phi}\left\{\Psi\left[\left(\mu_R(x,y), \nu_R(x,y)\right), \left(\mu_P(y,z), \nu_P(y,z)\right)\right]\right\}$$

$$= \underset{y}{\Phi}\left\{\left(\beta\left[\mu_R(x,y), \mu_P(y,z)\right], \rho\left[\nu_R(x,y), \nu_P(y,z)\right]\right)\right\}$$

$$= \left(\underset{y}{\alpha}\left\{\beta\left[\mu_R(x,y), \mu_P(y,z)\right]\right\}, \underset{y}{\lambda}\left\{\rho\left[\nu_R(x,y), \nu_P(y,z)\right]\right\}\right).$$

Then

$$\mu_{P \underset{\Psi}{\overset{\Phi}{\circ}} R}(x,z) = \underset{y}{\alpha}\left\{\beta\left[\mu_R(x,y), \mu_P(y,z)\right]\right\} = \mu_{P \underset{\lambda,\rho}{\overset{\alpha,\beta}{\circ}} R}(x,z)\,,$$

$$\nu_{P \underset{\Psi}{\overset{\Phi}{\circ}} R}(x,z) = \underset{y}{\lambda}\left\{\rho\left[\nu_R(x,y), \nu_P(y,z)\right]\right\} = \nu_{P \underset{\lambda,\rho}{\overset{\alpha,\beta}{\circ}} R}(x,z)\,.$$

Conditions (15) and (16) can be rewritten as $\alpha \leq \lambda^*$ and $\beta \leq \rho^*$. Then, the inequality (12) is met (see Proposition 1, [3]).

3.2 Properties

In this Section, we examine various properties of Φ, Ψ-composition fulfilled for α, β, λ, ρ- composition [3].

Theorem 0.2. *Let $R \in IFR(X \times Y)$, $P \in IFR(Y \times Z)$ and Φ, Ψ be it-norms or it-conorms, we have*

$$\begin{pmatrix} \Phi \\ P \circ R \\ \Psi \end{pmatrix}^{-1} = R^{-1} \underset{\Psi}{\overset{\Phi}{\circ}} P^{-1}. \tag{17}$$

Proof.

$$\left(\mu_{\begin{pmatrix} \Phi \\ P \circ R \\ \Psi \end{pmatrix}^{-1}}(z,x), \nu_{\begin{pmatrix} \Phi \\ P \circ R \\ \Psi \end{pmatrix}^{-1}}(z,x) \right) = \left(\mu_{\underset{\Psi}{\overset{\Phi}{P \circ R}}}(x,z), \nu_{\underset{\Psi}{\overset{\Phi}{P \circ R}}}(x,z) \right)$$

$$= \underset{y}{\Phi} \{\Psi[(\mu_R(x,y), \nu_R(x,y)), (\mu_P(y,z), \nu_P(y,z))]\}$$

$$= \underset{y}{\Phi} \{\Psi[(\mu_{P^{-1}}(z,y), \nu_{P^{-1}}(z,y)), (\mu_{R^{-1}}(y,x), \nu_{R^{-1}}(y,x))]\}$$

$$= \left(\mu_{\underset{\Psi}{\overset{\Phi}{R^{-1} \circ P^{-1}}}}(z,x), \nu_{\underset{\Psi}{\overset{\Phi}{R^{-1} \circ P^{-1}}}}(z,x) \right) \forall (z,x) \in Z \times X.$$

Theorem 0.3. *Let $R_1, R_2 \in IFR(X \times Y)$, $P_1, P_2 \in IFR(Y \times Z)$ and Φ, Ψ be it-norms or it-conorms, we have:*

1. *If $P_1 \le P_2$ then $P_1 \underset{\Psi}{\overset{\Phi}{\circ}} R \le P_2 \underset{\Psi}{\overset{\Phi}{\circ}} R$, for every $R \in IFR(X \times Y)$;*
2. *If $R_1 \le R_2$ then $P \underset{\Psi}{\overset{\Phi}{\circ}} R_1 \le P \underset{\Psi}{\overset{\Phi}{\circ}} R_2$, for every $P \in IFR(Y \times Z)$;*
3. *If $P_1 \preccurlyeq P_2$ then $P_1 \underset{\Psi}{\overset{\Phi}{\circ}} R \preccurlyeq P_2 \underset{\Psi}{\overset{\Phi}{\circ}} R$, for every $R \in IFR(X \times Y)$;*
4. *If $R_1 \preccurlyeq R_2$ then $P \underset{\Psi}{\overset{\Phi}{\circ}} R_1 \preccurlyeq P \underset{\Psi}{\overset{\Phi}{\circ}} R_2$, for every $P \in IFR(Y \times Z)$.*

Proof. $P_1 \leq P_2$ then $\mu_{P_1}(y,z) \leq \mu_{P_2}(y,z)$ and $\nu_{P_1}(y,z) \geq \nu_{P_2}(y,z)$. Then

$$
\begin{aligned}
& (\mu_{P_1}(y,z), \nu_{P_1}(y,z)) \leq_{L^*} (\mu_{P_2}(y,z), \nu_{P_2}(y,z)) \\
\Rightarrow \quad & \Psi[(\mu_R(x,y), \nu_R(x,y)), (\mu_{P_1}(y,z), \nu_{P_1}(y,z))] \\
& \leq_{L^*} \Psi[(\mu_R(x,y), \nu_R(x,y)), (\mu_{P_2}(y,z), \nu_{P_2}(y,z))] \\
\Rightarrow \quad & \underset{y}{\Phi} \{\Psi[(\mu_R(x,y), \nu_R(x,y)), (\mu_{P_1}(y,z), \nu_{P_1}(y,z))]\} \\
& \leq_{L^*} \underset{y}{\Phi} \{\Psi[(\mu_R(x,y), \nu_R(x,y)), (\mu_{P_2}(y,z), \nu_{P_1}(y,z))]\} \\
\Rightarrow \quad & P_1 \overset{\Phi}{\underset{\Psi}{\circ}} R(x,z) \leq_{L^*} P_2 \overset{\Phi}{\underset{\Psi}{\circ}} R(x,z) \ \forall (x,z) \in X \times Z \\
\Rightarrow \quad & P_1 \overset{\Phi}{\underset{\Psi}{\circ}} R \leq P_2 \overset{\Phi}{\underset{\Psi}{\circ}} R.
\end{aligned}
$$

The rest of the items are done in a similar way.

Theorem 0.4. *Let $R \in IFR(X \times Y)$, P, $Q \in IFR(Y \times Z)$ and Φ, Ψ be it-norms or it-conorms, we have*

$$
(P \vee Q) \overset{\Phi}{\underset{\Psi}{\circ}} R \geq \left(P \overset{\Phi}{\underset{\Psi}{\circ}} R \right) \vee \left(Q \overset{\Phi}{\underset{\Psi}{\circ}} R \right). \tag{18}
$$

Proof. By (6) -(7), Theorem 0.1 and , (1) Theorem 0.3,

$$
\begin{cases} P \vee Q \geq P \\ P \vee Q \geq Q \end{cases} \Rightarrow \begin{cases} (P \vee Q) \overset{\Phi}{\underset{\Psi}{\circ}} R \geq P \overset{\Phi}{\underset{\Psi}{\circ}} R \\ (P \vee Q) \overset{\Phi}{\underset{\Psi}{\circ}} R \geq Q \overset{\Phi}{\underset{\Psi}{\circ}} R \end{cases} \Rightarrow (P \vee Q) \overset{\Phi}{\underset{\Psi}{\circ}} R \geq \left(P \overset{\Phi}{\underset{\Psi}{\circ}} R \right) \vee \left(P \overset{\Phi}{\underset{\Psi}{\circ}} R \right).
$$

Remark 0.2. Let I be a finite family of indices and $\{a_i\}_{i \in I}$, $\{b_i\}_{i \in I} \subset L^*$. Let Φ, Ψ be *it*-norms or *it*-conorms, we have:

- $\underset{i}{\Phi}(a_i \vee b_i) \geq \underset{i}{\Phi}(a_i) \vee \underset{i}{\Phi}(b_i)$, where $\vee : (L^*)^2 \to L^*$ is the supremum operator;
- $\underset{i}{\Phi}(a_i \wedge b_i) \geq \underset{i}{\Phi}(a_i) \wedge \underset{i}{\Phi}(b_i)$, where $\wedge : (L^*)^2 \to L^*$ is the infimum operator.

In [7], Fung and Ku show that α is an idempotent t-norm (idempotent t-conorm) if and only if $\alpha = \wedge$ ($\alpha = \vee$). Here, similar these results, we have following lemma.

Lemma 0.1. *For every it-norm $\mathscr{T}$ and it-conorm $\mathscr{S}$:*

1. *$\mathscr{T}$ is idempotent if and only if $\mathscr{T}$ is the infimum operator ;*
2. *$\mathscr{S}$ is idempotent if and only if $\mathscr{S}$ is the supremum operator.*

Proof. We prove 1 (2 is proved analogously).

$\Leftarrow$) Implied from definition of the infimum operator.

$\Rightarrow$) For all $x, y \in L^*$

$$\mathscr{T}(x,y) \geq \mathscr{T}(x \wedge y, x \wedge y) = x \wedge y\,, \tag{19}$$

$$\mathscr{T}(x,y) \leq \mathscr{T}(x, 1_{L^*}) = x\,, \tag{20}$$

$$\mathscr{T}(x,y) \leq \mathscr{T}(1_{L^*}, y) = y\,. \tag{21}$$

By (20), (21)

$$\mathscr{T}(x,y) \leq x \wedge y\,. \tag{22}$$

(19), (22) imply $\mathscr{T}(x,y) = x \wedge y$.

Lemma 0.2. *Let $\{a_i\}_{i\in I}$, $\{b_i\}_{i\in I}$ be two finite element families of L^* and Φ, Ψ be not null t-norms or t-conorms. Then*

1. $\underset{i}{\Phi}(a_i \vee b_i) = \underset{i}{\Phi}(a_i) \vee \underset{i}{\Phi}(b_i)$ *if and only if* $\Phi = \vee$ *;*

2. $\underset{i}{\Psi}(a_i \wedge b_i) = \underset{i}{\Psi}(a_i) \wedge \underset{i}{\Psi}(b_i)$ *if and only if* $\Phi = \wedge$.

Proof. 1) $\Leftarrow$) Implied from the associativity of the *it*-conorm $\vee$.

$\Rightarrow$) Supposing $\underset{i}{\Phi}(a_i \vee b_i) = \underset{i}{\Phi}(a_i) \vee \underset{i}{\Phi}(b_i)$, we will prove the idempotency of Φ.

- *Case 1.* Φ is an *it*-conorm. We have

$$\Phi(x,x) = \Phi(x \vee 0_{L^*}, 0_{L^*} \vee x) = \Phi(x, 0_{L^*}) \vee \Phi(0_{L^*}, x) = x \vee x = x\ \forall x \in L^*\,.$$

Then Φ is idempotent. By Lemma 0.1, $\Phi = \vee$.

- *Case 2.* Φ is an *it*-norm. We have

$$\Phi(x,x) = \Phi(x \vee 0_{L^*}, 0_{L^*} \vee x) = \Phi(x, 0_{L^*}) \vee \Phi(0_{L^*}, x) = 0_{L^*} \vee 0_{L^*} = 0_{L^*}\ \forall x \in L^*\,.$$

Let $(x,y) \in (L^*)^2$, we have $\Phi(x,y) \leq \Phi(x \vee y, x \vee y) = 0_{L^*}$. So $\Phi(x,y) \leq \Phi(x \vee y, x \vee y) = 0_{L^*}$ for all $(x,y) \in (L^*)^2$, in opposition to $\Phi \neq 0$.

The rest of items are proved in a way similar to the previous one.

Theorem 0.5. *Let Φ, Ψ be not null it-norms or it-conorms. Then* $(P_1 \vee P_2) \overset{\Phi}{\underset{\Psi}{\circ}} R = \left(P_1 \overset{\Phi}{\underset{\Psi}{\circ}} R\right) \vee \left(P_2 \overset{\Phi}{\underset{\Psi}{\circ}} R\right)$ *for all* $R \in IFR(X \times Y)$, P_1, $P_2 \in IFR(Y \times Z)$ *if and only if* $\Phi = \vee$.

Proof. $\Rightarrow$)

$$\left(\mu_{(P_1 \vee P_2) \overset{\Phi}{\underset{\Psi}{\circ}} R}(x,z), \nu_{(P_1 \vee P_2) \overset{\Phi}{\underset{\Psi}{\circ}} R}(x,z)\right)$$

$$= \underset{y}{\Phi} \{\Psi[(\mu_R(x,y), \nu_R(x,y)), (\mu_{P_1 \vee P_2}(y,z), \nu_{P_1 \vee P_2}(y,z))]\}$$

$$= \underset{y}{\Phi} \{\Psi[(\mu_R(x,y), \nu_R(x,y)), (\mu_{P_1}(y,z), \nu_{P_1}(y,z)) \vee (\mu_{P_2}(y,z), \nu_{P_2}(y,z))]\}$$

$$= \underset{y}{\Phi} \{\Psi[(\mu_R(x,y), \nu_R(x,y)), (\mu_{P_1}(y,z), \nu_{P_1}(y,z))]$$
$$\vee \Psi[(\mu_R(x,y), \nu_R(x,y)), (\mu_{P_2}(y,z), \nu_{P_2}(y,z))]\} \ \forall (x,z) \in X \times Z.$$

Because of hypothesis of the theorem, the last formula equals to

$$\underset{y}{\Phi} \{\Psi[(\mu_R(x,y), \nu_R(x,y)), (\mu_{P_1}(y,z), \nu_{P_1}(y,z))]\}$$

$$\vee \underset{y}{\Phi} \{\Psi[(\mu_R(x,y), \nu_R(x,y)), (\mu_{P_2}(y,z), \nu_{P_2}(y,z))]\} \forall (x,z) \in X \times Z.$$

- If Ψ is an *it*-norm, we choose R such that there exists x_0 satisfying

  $$(\mu_R(x_0,y), \nu_R(x_0,y)) = 1_{L^*} \forall y \in Y,$$

  and define (for z_0 fixed and for every y)

  $$a_y = (\mu_{P_1}(y,z_0), \nu_{P_1}(y,z_0)), b_y = (\mu_{P_2}(y,z_0), \nu_{P_2}(y,z_0)).$$

  Then

$$\underset{y}{\Phi}\{\Psi[(\mu_R(x_0,y),\nu_R(x_0,y)),(\mu_{P_1}(y,z_0),\nu_{P_1}(y,z_0))]$$
$$\vee\Psi[(\mu_R(x_0,y),\nu_R(x_0,y)),(\mu_{P_2}(y,z_0),\nu_{P_2}(y,z_0))]\}$$
$$=\underset{y}{\Phi}[\Psi(1_{L^*},a_y)\vee\Psi(1_{L^*},b_y)]=\underset{y}{\Phi}(a_y,b_y)\ ,$$

and

$$\underset{y}{\Phi}\{\Psi[(\mu_R(x_0,y),\nu_R(x_0,y)),(\mu_{P_1}(y,z_0),\nu_{P_1}(y,z_0))]\}$$

$$\vee\underset{y}{\Phi}\{\Psi[(\mu_R(x_0,y),\nu_R(x_0,y)),(\mu_{P_2}(y,z_0),\nu_{P_2}(y,z_0))]\}$$

$$=\underset{y}{\Phi}[\Psi(1_{L^*},a_y)]\vee\underset{y}{\Phi}[\Psi(1_{L^*},b_y)]=\underset{y}{\Phi}(a_y)\vee\underset{y}{\Phi}(b_y)\ .$$

So, $\underset{y}{\Phi}(a_y,b_y)=\underset{y}{\Phi}(a_y)\vee\underset{y}{\Phi}(b_y)$. Using lemma 0.2, $\Phi=\vee$.

- If Ψ is *it*-conorm, we choose R such that there exist x_0 satisfying

$$(\mu_R(x_0,y),\nu_R(x_0,y))=0_{L^*}\ \forall y\in Y\ ,$$

and define (for z_0 fixed and for every y)

$$a_y=(\mu_{P_1}(y,z_0),\nu_{P_1}(y,z_0))\ ,b_y=(\mu_{P_2}(y,z_0),\nu_{P_2}(y,z_0))\ .$$

We conclude $\underset{y}{\Phi}(a_y,b_y)=\underset{y}{\Phi}(a_y)\vee\underset{y}{\Phi}(b_y)$, and then $\Phi=\vee$.

$\Leftarrow$) Let us take $\Phi=\vee$, and Ψ any *it*-norm or *it*-conorm, we have

$$\left(\mu_{(P_1\vee P_2)\ \underset{\Psi}{\overset{\Phi}{\circ}}\ R}(x,z),\nu_{(P_1\vee P_2)\ \underset{\Psi}{\overset{\Phi}{\circ}}\ R}(x,z)\right)$$

$$=\underset{y}{\vee}\{\Psi[(\mu_R(x,y),\nu_R(x,y)),(\mu_{P_1}(y,z),\nu_{P_1}(y,z))]$$
$$\vee\Psi[(\mu_R(x,y),\nu_R(x,y)),(\mu_{P_2}(y,z),\nu_{P_2}(y,z))]\}\ .$$

Using the associative property of the *it*-conorm $\vee$, we have

$$\left(\mu_{(P_1 \vee P_2) \overset{\Phi}{\underset{\Psi}{\circ}} R}(x,z), \nu_{(P_1 \vee P_2) \overset{\Phi}{\underset{\Psi}{\circ}} R}(x,z)\right)$$

$$= \bigvee_y \{\Psi[(\mu_R(x,y), \nu_R(x,y)), (\mu_{P_1}(y,z), \nu_{P_1}(y,z))]\}$$

$$= \vee \bigvee_y \{\Psi[(\mu_R(x,y), \nu_R(x,y)), (\mu_{P_2}(y,z), \nu_{P_2}(y,z))]\}$$

$$= \left(\mu_{P_1 \overset{\Phi}{\underset{\Psi}{\circ}} R}(x,z), \nu_{P_1 \overset{\Phi}{\underset{\Psi}{\circ}} R}(x,z)\right) \vee \left(\mu_{P_2 \overset{\Phi}{\underset{\Psi}{\circ}} R}(x,z), \nu_{P_2 \overset{\Phi}{\underset{\Psi}{\circ}} R}(x,z)\right)$$

$$\forall (x,z) \in X \times Z .$$

Analogous to Theorems 0.4, 0.5, we have Theorems 0.6, 0.7.

Theorem 0.6. *Let $R \in IFR(X \times Y)$, P, $Q \in IFR(Y \times Z)$ and Φ, Ψ be it-norms or it-conorms, we have*

$$(P \wedge Q) \overset{\Phi}{\underset{\Psi}{\circ}} R \leq \left(P \overset{\Phi}{\underset{\Psi}{\circ}} R\right) \wedge \left(Q \overset{\Phi}{\underset{\Psi}{\circ}} R\right) . \tag{23}$$

Theorem 0.7. *Let Φ, Ψ be not null it-norms or it-conorms. Then $(P_1 \wedge P_2) \overset{\Phi}{\underset{\Psi}{\circ}} R =$ $\left(P_1 \overset{\Phi}{\underset{\Psi}{\circ}} R\right) \wedge \left(P_2 \overset{\Phi}{\underset{\Psi}{\circ}} R\right)$ for all $R \in IFR(X \times Y)$, P_1, $P_2 \in IFR(Y \times Z)$ if and only if $\Phi = \wedge$.*

Theorem 0.8. *For all $P \in IFR(X \times Y)$, $Q \in IFR(Y \times Z)$, $R \in IFR(Z \times U)$, we have*

$$\left(R \overset{\vee}{\underset{\Psi}{\circ}} Q\right) \overset{\vee}{\underset{\Psi}{\circ}} P = R \overset{\vee}{\underset{\Psi}{\circ}} \left(Q \overset{\vee}{\underset{\Psi}{\circ}} P\right) , \tag{24}$$

$$\left(R \overset{\wedge}{\underset{\Psi}{\circ}} Q\right) \overset{\wedge}{\underset{\Psi}{\circ}} P = R \overset{\wedge}{\underset{\Psi}{\circ}} \left(Q \overset{\wedge}{\underset{\Psi}{\circ}} P\right) . \tag{25}$$

Proof.

For convenience, in this proof, for each relation R, we denote $(\mu_R(x,y), \mu_R(x,y))$ by $R(x,y)$. We will use the facts that:

- For each Ψ *it*-norm or *it*-conorm:

$$\Psi\left(a, \underset{i}{\vee} b_i\right) = \underset{i}{\vee}\left[\Psi(a,b_i)\right], \Psi\left(\underset{i}{\vee} a_i, b\right) = \underset{i}{\vee}\left[\Psi(a_i,b)\right] \; \forall a, b_i \in L^* \; ; \quad (26)$$

- Ψ is associative.

Then

$$\begin{aligned}
\left(R \overset{\vee}{\underset{\Psi}{\circ}} Q\right) \overset{\vee}{\underset{\Psi}{\circ}} P(x,u) &= \underset{y}{\vee}\left\{\Psi\left[P(x,y), R \overset{\vee}{\underset{\Psi}{\circ}} Q(y,u)\right]\right\} \\
&= \underset{y}{\vee}\left\{\Psi\left[P(x,y), \underset{z}{\vee}\left\{\Psi\left[Q(y,z), R(z,u)\right]\right\}\right]\right\} \\
&= \underset{y}{\vee}\left\{\underset{z}{\vee}\left[\Psi\left\{P(x,y), \Psi\left[Q(y,z), R(z,u)\right]\right\}\right]\right\} \\
&= \underset{y}{\vee}\left\{\underset{z}{\vee}\left[\Psi\left\{\Psi\left[P(x,y), Q(y,z)\right], R(z,u)\right\}\right]\right\} \\
&= \underset{z}{\vee}\left\{\Psi\left[\underset{y}{\vee}\left\{\Psi\left[P(x,y), Q(y,z)\right], R(z,u)\right\}\right]\right\} \\
&= \underset{z}{\vee}\left\{\Psi\left[P \overset{\vee}{\underset{\Psi}{\circ}} Q(y,z), R(z,u)\right]\right\} \\
&= R \overset{\vee}{\underset{\Psi}{\circ}} \left(Q \overset{\vee}{\underset{\Psi}{\circ}} P\right)(x,u) \forall (x,u) \in (X \times U) \, .
\end{aligned}$$

The remainder is also proved in a analogous way.

4 Conclusion

In this research, we define a new composition of intuitionistic fuzzy relations. It is also proved that the Burrilo and Bustinces notion is a special case of our notion when *it*-norms and *it*-conorms are *t*-representable. Various properties are stated and proved.

References

1. Atanassov, K.T.: Intuitionistic fuzzy sets. Fuzzy Sets and Systems 20, 87–96 (1986)
2. Atanassov, K.T., Stoeva, S.: Intuitionistic Lfuzzy sets. Cybernetics and Systems Research 2, 539–540 (1984)
3. Baets, B., De, K.E.E.: Fuzzy relations and applications. Adv. Electron. Electron. Phys. 89, 255–324 (1994)
4. Burillo, P., Bustince, H.: Intuitionistic Fuzzy Relations. Part I, Mathware and Soft Computing 2(1), 5–38 (1995)
5. Chen, S., Chen, H., Duro, R., Honaver, V., Kerre, E., Lu, M., Romay, M.G., Shih, T.K., Ventura, D., Wang, P.P., Yang, Y.: Proceedings of the 6th Joint Conference on Information Sciences, pp. 105–108 (2002)
6. Deschrijver, G., Cornelis, C., Kerre, E.E.: On the representation of intuitionistic fuzzy t-norms and t-conorms. IEEE Trans. Fuzzy Systems 12, 45–61 (2004)
7. Deschrijver, G., Kerre, E.E.: On the relationship between some extensions of fuzzy set theory. Fuzzy Sets and Systems 133(2), 227–235 (2003)
8. De, S.K., Biswas, R., Roy, A.R.: An application of intuitionistic fuzzy sets in medical diagnosis. Fuzzy Sets and Systems 117, 209–213 (2001)
9. Fung, L.W., Fu, K.S.: An axiomatic approach to rational decision making in fuzzy environment. In: Zadeh, L.A., Fu, K.S., Tanaka, K., Shimura, M. (eds.) Fuzzy Sets and their Applications to Cognitive and Decision Processes, pp. 227–256. Academic Press (1975)
10. Goguen, J.: L-fuzzy sets. J. Math. Anal. Appl. 18, 145–174 (1967)
11. Sanchez, E.: Solutions in composite fuzzy relation equation. Application to Medical diagnosis in Brouwerian Logic. In: Gupta, M.M., Saridis, G.N., Gaines, B.R. (eds.) Fuzzy Automata and Decision Process. Elsevier, North-Holland (1977)
12. Sanchez, E.: Resolution of composition fuzzy relation equations. Inform. Control 30, 38–48 (1976)

Using Unicode in Encoding the Vietnamese Ethnic Minority Languages, Applying for the Ede Language

Le Hoang Thi My, Khanh Phan Huy, and Souksan Vilavong

Abstract. The problem of the displaying characters on a computer is one of the issues to be considered early. Encoding characters, designing fonts, building typing tool to display the characters on the screen is indispensable in the text typing. With the problems relating to text typing and the restrictions of ethnic typing tools so far, the paper proposes a solution using Unicode in encoding the Vietnamese minority languages, applying for the Ede Language from that we can type the ethnic language text in multilingual environment using Unicode fonts and the common Vietnamese typing tool such as Vietkey, Unikey, etc. Based on the results of using Unicode in encoding Ede language, we continue to expand for solving problems of text typing on computers with any ethnic language that has the Latin letters. This paper will also be the basis for our proposal to the competent authorities who unity the using Unicode fonts in the ethnic language text in general and Ede text in particular.

1 Introduction

Currently, Vietnam people have 55 ethnic groups, Kinh people is called Viet people and account for about 86% of the country's population [13]. Since the Independence Day in 1945, Vietnamese - Kinh language has officially become the national language in Vietnam. In the educational system from kindergarten to university, Vietnamese-Kinh language is the most common language, a tool to impart knowledge, a tool to communication and the State management of all ethnic groups in Vietnam. Each ethnic group has its own language and traditional culture, in which there are 24 ethnic groups having their own letters such as Thai, H'mong, Tay, Nung, Khmer and Gia Rai, Ede, Chinese, Cham, etc. The script of some ethnic minorities such as Thai, Chinese, Khmer, Cham, Ede, Tay-Nung, Ho and Lao is used in schools [10].

Le Hoang Thi My · Khanh Phan Huy · Souksan Vilavong
Danang University of Education, The University of Danang, Vietnam

V.-N. Huynh et al. (eds.), *Knowledge and Systems Engineering, Volume 1*, Advances in Intelligent Systems and Computing 244,
DOI: 10.1007/978-3-319-02741-8_13,

Ede is an ethnic group in the national community of Vietnam. Currently, there are over 330 thousand people, residing in the provinces of Vietnam, such as Dak Lak, southern Gia Lai, western Phu Yen and Khanh Hoa. Comparing with other ethnic minorities in Vietnam, The Ede ethnic group has used the script based on the Latin letters early since the 1920s. The Ede language teaching in schools have been carried out over 30 years. According to statistics reported by Dak Lak Department of Education and Training there were 88 schools, 487 classes with more than 12 thousand pupils learning the Ede language subjects in the 2011 - 2012 school years. The primary school textbooks in Ede language have been used in the Ethnic boarding schools. The province has also organized teaching Ede language for the native staffs, movement staffs.

The goal of Ede language teaching at primary level is to form and to develop skills in listening, speaking, reading, writing Ede language for pupils. That contributes to develop in thinking and learning skills in the Vietnamese language subjects. The Ede language teaching provides the necessary knowledge of the phonetics, script, vocabulary, grammar of Ede language; that is very necessary for communication skills training, expanding the understanding about human life, culture and literature of Ede people as well as other ethnic groups. Moreover, Ede language also fosters the love of the native language for pupils, contributes to the preservation and development of the national cultural character.

In Vietnam, processing the Vietnamese - Kinh language problems has deployed fairly soon, had many results [14], and has been continued. However, the script problem on the computer for the ethnic minority languages has not been interested much. Especially, in the explosive development of information and communication technologies as well as internet, the services on the internet have been relatively familiar to the people in almost all regions of the country. However, there is not any website in ethnic minority languages. Even in the website of the Committee for Ethnic Minorities Vietnamese CEMA, there is not any ethnic minority language [9], the websites of the locals where the ethnic people live are only in Vietnamese - Kinh language, or accompanied by English.

Currently, the typing tools of the ethnic minority languages such as Cham, Ede, Jarai, Katu, MNong, KHo, SanChi... have been built, but the authors have created the own fonts for their typing tool. Therefore, these typing tools have many restrictions such as the texts of the ethnic languages do not integrate into the Unicode fonts as Vietnamese - Kinh language and the display of the ethnic language characters on websites depends on own fonts.

So far, the script processing of the ethnic minority on the computer has only been solved locally for each ethnic language, have not been a national unity and have not satisfied the needs of the culture development and integration of ethnic minority communities in Vietnam. On the other hand, the problem of the ethnic language processing is been rarely interested by the scientists.

Therefore, the paper presents a solution that use Unicode in encoding the Vietnam minority languages, applying for the Ede Language from that we can type ethnic language text in multilingual environment with Unicode fonts and use the common Vietnamese typing tool such as Vietkey, Unikey, etc. Furthermore, the paper will present a summary of character encoding issues; as well as the solution using Unicode in encoding the Vietnamese ethnic minority languages, applying for the Ede language.

2 Character Encoding

Character encoding for each language to store and display information on the computer is the first problem of natural language processing (NLP). The using of many different encodings in the same country is a major obstacle in the development of large information systems. The optimal solution for the incompatibility between encodings is using Unicode. This solution has been accepted by many countries and has put to the standard for information exchanging. Unicode was designed in order to overcome the above disadvantages and to develop a universal code for all languages in the world.

Unicode encodings: Unicode (UCS-2 ISO 10646) is a 16-bit character encoding that contains all characters (2^{16} = 65,536 different characters) of major languages in the world, including Vietnamese [15]. The Unicode character set contains all Vietnamese characters, including uppercase and lowercase letters that do not conflict with the control characters. It allows us to use Vietnamese language in multilingual environment. Using Unicode should have:

- The software allows to type and display characters in the Unicode standard on the screen.
- Unicode fonts are installed on computer system.

2.1 Vietnamese in Unicode

Vietnamese language belongs to the extended Latin letters. Vietnamese characters are not distributed continuously in Unicode encoding and have two forms, including pre-compound and compound characters. For example, the word "dân" is in Unicode as follows:

- Pre-compound: d + â + n
- Compound: d + a +^+ n

All the characters of the Vietnamese language are present in Unicode fonts. For displaying Unicode Vietnamese language, therefore, it is necessary to install Windows or Internet Explorer or MS Office, from that the Vietnamese Unicode fonts will be installed automatically into computer system. The basic fonts of Microsoft combined with the above application programs supporting Unicode Vietnamese fonts include Times New Roman, Courier New, Arial, Tahoma, Verdana, Palatino Linotype. On the other hand, we can also download the Unicode fonts (supporting Vietnamese) on the Internet, such as Verdana, Arial Narrow, Arial Black, Bookman Old Style, Garamond, Impact, Lucida Sans, Comic Sans, etc ...

Unikey typing tool developed by Pham Kim Long author and Vietkey typing tool developed by Dang Minh Tuan author are the success of using Unicode in encoding Vietnamese and building Vietnamese typing tools with Unicode fonts.

The three typing modes that are commonly used in Vietnamese typing tool include Telex, VNI, and VIQR as seen in table 1.

Table 1 Three methods typing accent of Vietnamese letters

	Typing TELEX	Typing VNI	Typing VIQR
Acute accent	s	1	'
Grave accent	f	2	`
Hook above	r	3	?
Tilde	x	4	~
Dot bellow	j	5	.
Letter â, ê, ô	aa, ee, oo	a6, e6, o6	a^, e^, o^
Letter ơ, ư	ow or [, uw or]	o7, u7	o*, u*
Letter ă	aw	a8	a(
Letter đ	dd	d9	dd

2.2 Ede Language in Unicode

The Ede letters also uses the Latin characters, with 76 Ede characters (uppercase and lowercase letters), including 66 characters of Unicode and 10 characters of non-Unicode as in group 3 of table 4. So far, there is still no solution using Unicode in encoding the Ede language. The following sections of this paper will display the solution of this problem.

2.3 Assessment of Unicode Using in Vietnamese and Ede Languages

The using of Unicode in Vietnamese and Ede languages is compared, commented and assessed as seen in table 2.

3 Using Unicode in Encoding the Ethnic Minority Languages

3.1 Ethnic Character Encoding

Most of the scripts of 24 ethnic groups in Vietnam derive from the Latin letters with accents and phonetics being similar to Vietnamese-Kinh language. Only some letters, accents and phonetics of ethnic minority languages are variation. Our solution is using Unicode in encoding the ethnic characters that inherits the common method of typing accented Vietnamese of Unikey, Vietkey typing tools. The main ideas of the solution are as follows:

Table 2 Evaluating the using Unicode of Vietnamese - Ede Languages

The problem of NLP	Vietnamese language	Ede language
Unicode encoding	Coded	Do not code
	Evaluating: Concentrating on researching the use of Unicode in encoding the Vietnam's minority languages.	
Unicode fonts	Used	Do not use
Displaying Unicode fonts	There are Unikey,Vietkey typing tools	There is not
	Evaluating: Concentrating on researching the displaying the letters of minority language with Unicode fonts..	

- Based on Unicode encodings, the alphabet of the ethnic minority languages is divided into three groups as follows:
  - Group 1: The characters are in Vietnamese-Kinh language.
  - Group 2: The characters are in Unicode but not in the script of Vietnamese language.
  - Group 3: The characters are neither in Unicode nor in the script of Vietnamese language. For this case, combining diacritical marks from the range 0300036F in table of Unicode code are used [8]. For example as figure 1:

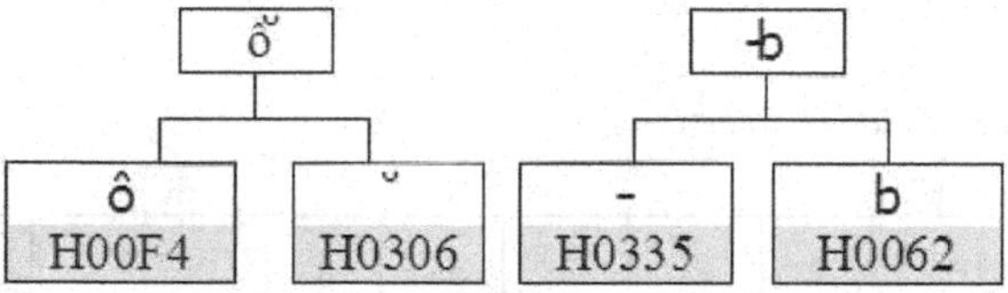

Fig. 1 Example of Combining diacritical marks for the characters of group 3

- For group 1: Typing and displaying the characters are similar to those of the Vietnamese language using the common typing tools such as Unikey, Vietkey, etc ...
- For group 2, 3: There are typing rules and then using the text conversion tool that is integrated into menu bar of application programs displays ethnic characters.

Figure 2 shows the solution using Unicode in encoding the ethnic minority languages, applying to editing ethnic language text in multilingual environment with Unicode fonts (Arial, Times New Roman, Tahoma, Verdana, etc), not using own typing tool, but inheriting Vietnamese typing tools (Unikey, Vietkey, etc ...).

3.2 Ede Character Encoding

The Ede letters also derive from the Latin characters, with 76 Ede characters (including uppercase and lowercase letters) as seen in table 3.

Table 3 Ede alphabet

Consonant				Vowel			
Uppercase		*Lowercase*		*Uppercase*		*Lowercase*	
B	M	b	m	A	Ŏ	a	ŏ
Ƀ	N	ƀ	n	Ă	Ô	ă	ô
Č	Ñ	č	ñ	Â	Ô̆	â	ô̆
D	P	d	p	E	Ơ	e	ơ
Đ	R	đ	r	Ĕ	Ơ̆	ĕ	ơ̆
G	S	g	s	Ê	U	ê	u
H	T	h	t	Ê̆	Ŭ	ê̆	ŭ
J	W	j	w	I	Ư	i	ư
K	Y	k	y	Ĭ	Ư̆	ĭ	ư̆
L		l		O		o	

Table 4 The groups of Ede characters

Group 1	A	a	Ă	ă	Â	â	E	e	Ê	ê
	I	i	O	o	Ô	ô	Ơ	ơ	U	u
	Ư	ư	B	b	D	d	Đ	đ	G	g
	H	h	J	j	K	k	L	l	M	m
	N	n	P	p	R	r	S	s	T	t
	W	w	Y	y						
Group 2	Č	č	Ñ	ñ	Ĕ	ĕ	Ĭ	ĭ	Ŏ	ŏ
	Ŭ	ŭ								
Group 3	Ƀ	ƀ	Ê̆	ê̆	Ô̆	ô̆	Ơ̆	ơ̆	Ư̆	ư̆

Based on the ideas of the solution using Unicode in encoding the ethnic minority languages, using Unicode in the encoding the Ede characters was implemented as follows:

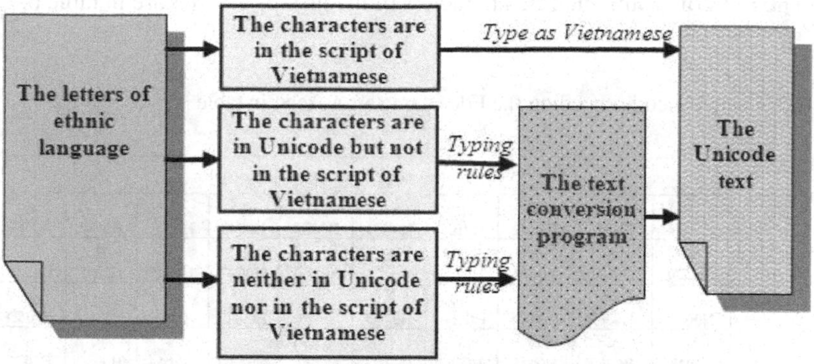

Fig. 2 Scheme of using Unicode in editing ethnic language text in multilingual environment

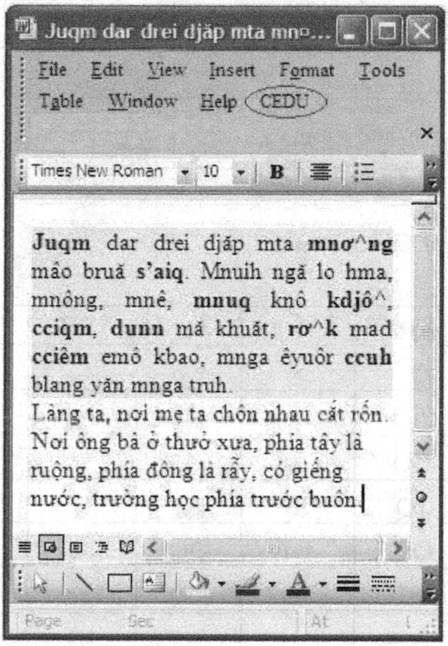

Fig. 3 CEDU command on the menu bar of Microsoft Word

- The Ede Alphabet is divided into three group:
  - Group 1: The characters are in Vietnamese-Kinh language.
  - Group 2: The characters are in Unicode, but not in the script of Vietnamese language.
  - Group 3: The characters are neither in Unicode nor in the script of Vietnamese language. For this case, diacritical combining marks in table of Unicode code are used. Table 4 displays three groups of Ede characters.

- Using Unicode encoding the Ede characters as seen in table 5

- The rules of typing the Ede characters of the groups 2 and 3 are in table 6.

Table 5 Using Unicode encoding the Ede characters as seen in table

	Value Hecxa of Ede Characters in Uniocode									
Group 1	A H0041	a H0061	Ă H0102	ă H0103	Â H00C2	â H00E2	B H0042	b H0062	D H0044	d H0064
	Đ H00D0	đ H0111	E H0045	e H0065	Ê H00CA	ê H00EA	G H0047	g H0067	H H0048	h H0068
	I H0049	i H0069	J H004A	j H006A	K H004B	k H006B	L H004C	l H006C	M H004D	m H006D
	N H004E	n H006E	O H004F	o H0067	Ô H00D4	ô H00F4	Ơ H01A0	ơ H01A1	P H0050	p H0070
	R H0052	r H0072	S H0053	s H0073	Y H0059	t H0074	U H0055	u H0075	Ư H01AF	ư H01B0
Group 2	Č H010C	č H010D	Ĕ H0114	ĕ H0115	Ĭ H012C	ĭ H012D	Ñ H00D1	ñ H00F1	Ŏ H014E	ŏ H014F
	Ŭ H016C	ŭ H016D								
Group 3	B̵ H0335 H0042	b̵ H0335 H0062	Ê̆ H00CA H0306	ê̆ H00EA H0306	Ô̆ H00D4 H0306	ô̆ H00F4 H0306	Ơ̆ H01A0 H0306	ơ̆ H01A1 H0306	Ư̆ H016C H0306	ư̆ H016D H0306

Table 6 The typing rules of Ede characters

Characters		Typing Telex		Typing VNI		Typing VIQR	
Č	č	CC	cc	C6	c6	C^	c^
Ñ	ñ	NN	nn	N4	n4	N~	n~
Ĕ	ĕ	EQ	eq	EQ	eq	EQ	eq
Ĭ	ĭ	IQ	iq	IQ	iq	IQ	iq
Ŏ	ŏ	OQ	oq	OQ	oq	OQ	oq
Ŭ	ŭ	UQ	uq	UQ	uq	UQ	uq
B̵	b̵	BB	bb	B9	b9	BB	bb
Ê̆	ê̆	Ê^	ê^	Ê^	ê^	Ê[	ê[
Ô̆	ô̆	Ô^	ô^	Ô^	ô^	Ô[	ô[
Ơ̆	ơ̆	Ơ^	ơ^	Ơ^	ơ^	Ơ[	ơ[
Ư̆	ư̆	Ư^	ư^	Ư^	ư^	Ư[	ư[

Table 7 The formation of .RTF file

Type methods Telex, VNI, VIQR	Ede characters	Hex	Type methods Telex, VNI, VIQR	Ede characters	Hex
cc, c6, c^	č	H010D	bb, b9, bb	H0335 H0062	b̵
CC, C6, C^	Č	H010C	BB, B9, BB	H0335 H0042	B̵
nn, n6, n~	ñ	H00F1	ê^,ê^, ê[	H00EA H0306	ê̆
NN, N6, N~	Ñ	H00D1	Ê^, Ê^, Ê[	H00CA H0306	Ê̆
eq, eq, eq	ĕ	H0115	ô^, ô^, ô[	H00F4 H0306	ô̆
EQ, EQ, EQ	Ĕ	H0114	Ô^, Ô^, Ô[	H00D4 H0306	Ô̆
iq, iq, iq	ĭ	H012D	ơ^, ơ^, ơ[	H01A1 H0306	ơ̆
IQ, IQ, IQ	Ĭ	H012C	Ơ^, Ơ^, Ơ[	H01A0 H0306	Ơ̆
oq, oq, oq	ŏ	H014F	ư^, ư^, ư[	H016D H0306	ư̆
OQ, OQ, OQ	Ŏ	H014E	Ư^, Ư^, Ư[	H016C H0306	Ư̆
uq, uq, uq	ŭ	H016D			
UQ, UQ, UQ	Ŭ	H016C			

Where:

[shaded box] The characters of the group 2 [white box] The characters of the group 3

3.3 *Building CEDU Program for Converting Ede Characters in Text Document into Unicode*

In the text document, the Ede characters of the group 2 and 3 are written following table 6. Those characters are converted into the Ede characters by CEDU (Convert Ethnic Document into Unicode) program. It is built as follows:

- Creating the data source for CEDU program with .RTF file, which is formed as seen in table 7.
- Writing coding for program bases on the data source of program.
- After installing CEDU program, CEDU command is displayed on the menu bar of Microsoft Word (figure 3)

After choosing CEDU command, CEDU function finds the characters group written following the Ede character typing rules in documents and replaces with Ede character based on data source of CEDU program as seen in figure 4.

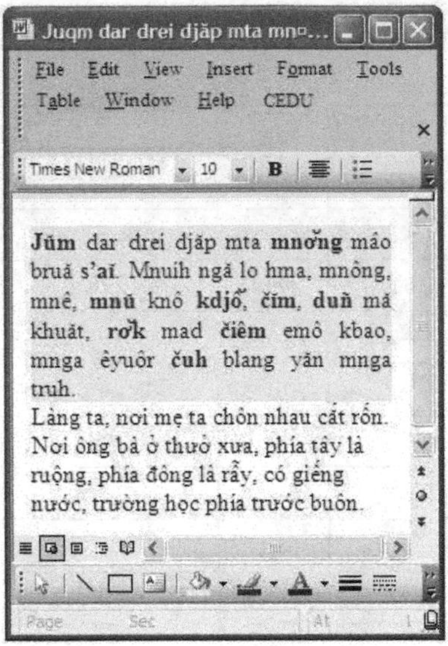

Fig. 4 The result of selecting the CEDU command

Table 8 The example Ede document using Unicode font and typing telex

Ede document typing follow the Telex typing rules	Seleting CEDU command	Vietnamese
MĂ BRUĂ M'KA ÊĐI Ju^m dar drei djăp mta **mnơ^ng** mâo bruă **s'ai^**. Mnuih ngă lo hma, mnông, mnê, **mnu^** knô **kdjô^, cci^m, dunn** mă khuăt, **rơ^k** mad **cciêm** emô kbao, mnga êyuôr **c^uh** blang yăn mnga truh. Hmei ăt mâo bruă **mse^** mơh, mnei hriăm hră. Ngă klei hriăm. Dôk ti sang hmei dru **ami^ amâ^. Djâ^p** bruă anei amă hong klei m'ak, mâo klei **tu^** dưn **s'ai^.**	**MĂ BRUĂ M'KA ÊĐI** Jŭm dar drei djăp mta **mnơ̆ng** mâo bruă **s'aĭ**. Mnuih ngă lo hma, mnông, mnê, **mnŭ** knô **kdjô̆, čĭm, duñ** mă khuăt, **rơ̆k** mad **čiêm** emô kbao, mnga êyuôr **čuh** blang yăn mnga truh. Hmei ăt mâo bruă **msĕ** mơh, mnei hriăm hră. Ngă klei hriăm. Dôk ti sang hmei dru **amĭ amâ̆. Djâ̆p** bruă anei amă hong klei m'ak, mâo klei **tŭ** dưn **s'aĭ**.	**YÊU LÀNG TA** Làng ta, nơi mẹ ta chôn nhau cắt rốn. Nơi ông bà ở thưở xưa, phía tây là ruộng, phía đông là rẫy, có giếng nước, trường học phía trước buôn. Dọc đường cái lớn dấu chân trâu bò, heo gà đạp chi chít. Cơm ngon, canh ngọt quanh năm có nhiều, dấu chân trâu bò, heo gà đầy.

4 Testing

With solution using Unicode in encoding the Vietnam minority languages, applying for the Ede Language and building CEDU program, the problem of text typing in multilingual environment with Unicode fonts (Arial, Times New Roman, Tahoma, Verdana, etc.) has been solved, in which the Ede fonts or the own tying tool is not used as seen in the previous research.

Some application results:

- Ede text editor in multilingual environment (English, Vietnamese, Ede) is carried out using only Unicode font with Unikey typing tool.
- Ede text using Times New Roman font, Unikey typing tool, CEDU program is exampled in table 8.

5 Conclusion

The text editor of Ethnic minority language does not use the own fonts and own typing tool, inherits Unikey, Vietkey typing tools. We offer the solutions using Unicode in encoding the Vietnamese ethnic minority languages in general and Ede language in particular, building CEDU program to display Ede characters with Unicode fonts.

Moreover, CEDU program can expand with the other ethnic languages using the Latin letters that can use Unicode in encoding.

This solution is applicable because it solves the issues of the Ede text editor in multilingual environment with Unicode fonts (Arial, Times New Roman, Tahoma, Verdana, etc...), does not use the own fonts or own typing tool and it also solves the display issues of the Ede text on the Internet.

This paper will also be the basis for our proposal to the competent authorities who unity the using Unicode fonts (Arial, Times New Roman, Tahoma, Verdana, etc...) in the ethnic language text in general and Ede text in particular.

References

1. Huy, K.P.: Using VBA macro programming tool to build text processing utilities. The Journal of Control and Computing Sciences (10) (2005)
2. Trung, V.N.: Vietnamese letters problem on the computer. The Journal of Control and Computing Sciences (3) (1987)
3. My, L.H.T.: Building information processing system for text editor with Ede language. Computer Science Master Thesis, Courses 1, the University of DaNang (2002)
4. Cang Nił Sieng, Y., Swain, K.: The lessons of Ra-de Language. The Vietnamese minority languages bookshelves of Saigon Ministry of Education (1971)
5. Cang Nił Siłng, Y.: Łł - Rade Vocabulary. The Vietnamese minority languages bookshelves of the Summer Language Institute, XIV (1979)
6. Huy, K.P.: Building the text processing grammars, applying for Vietnamese ethnic minority languages. In: Proceedings of the 30th Anniversary of the Establishment Institute of Information Technology, Vietnam Academy of Science and Technology (2006)

7. Huy, K.P., Xuan, D.T.: Building the Cham- Vietnamese-Cham dictionary for Cham language processing. In: Proceedings of the Third National Conference on Computer Sciences, FAIR 2007, Nhatrang Cit (2007)
8. Combining Diacritical Marks, http://www.alanwood.net/unicode/combining_diacritical_marks.html
9. Committee for Ethnic Minorities Vietnamese CEMA CEMA, http://cema.gov.vn/modules.php?name=Content&op=details&mid=498
10. PeoplesLanguage, http://www.vietnamemb.se/vi/index.php?option=com_content&view=article&id=51&Itemid=36
11. Unicode, http://vi.wikipedia.org/wiki/Unicode
12. Vietnam fonts, http://fontchu.com
13. Vietnam General Statistics Office, http://www.gso.gov.vn/default.aspx?tabid=407&idmid=4&ItemID=1346
14. Vietnamese-Kinh language processing in information technology, http://www.jaist.ac.jp/~ bao/Writings/VLSPwhitepaper%20-%20Final.pdf
15. Vietnamese Unicode, http://vietunicode.sourceforge.net/main.html

Improving Moore's Sentence Alignment Method Using Bilingual Word Clustering

Hai-Long Trieu, Phuong-Thai Nguyen, and Kim-Anh Nguyen

Abstract. Sentence alignment plays an extremely important role in machine translation. Most of the hybrid approaches get either a bad recall or low precision. We tackle disadvantages of several novel sentence alignment approaches, which combine length-based and word correspondences. Word clustering is applied in our method in order to improve the quality of the sentence aligner, especially when dealing with the sparse data problem. Our approach overcomes the limits of previous hybrid methods and obtains both highly recall and reasonable precision rates.

1 Introduction

Nowadays, the online parallel texts have a considerable growth, and these are an extremely ample and substantial resource. Nevertheless, in order to apply these abundant materials into useful applications like machine translation, they have to be processed through some stages. Parallel texts are first collected before learning word correspondences. Since the size of the translated segments forming the parallel corpus is usually huge, this task of learning is very ambiguous. Thus, it is necessary to reduce the ambiguity, and the solution is that using sentence alignment to resolve this issue. This process maps sentences in the text of the source language to their corresponding units in the text of the target language. After processed by sentence alignment, the bilingual corpora are greatly useful in many important applications

Hai-Long Trieu
Faculty of Mathematics, Thai Nguyen University of Education, Thainguyen, Vietnam
e-mail: `longspt41@gmail.com`

Phuong-Thai Nguyen
University of Engineering and Technology, Vietnam National University, Hanoi, Vietnam
e-mail: `thainp@vnu.edu.vn`

Kim-Anh Nguyen
Faculty of Mathematics and Technology, Hung Vuong University, Phutho, Vietnam
e-mail: `anhnk@hvu.edu.vn`

V.-N. Huynh et al. (eds.), *Knowledge and Systems Engineering, Volume 1*, Advances in Intelligent Systems and Computing 244,
DOI: 10.1007/978-3-319-02741-8_14,

that machine translation is an apparent instance. Efficient and powerful sentence alignment algorithms, therefore, become increasingly important, especially with the rapid climb of online parallel texts today.

A sentence alignment algorithm should be efficient both in accuracy and speed. With regard to the quality of alignment result, both precision and recall rates should be high. The approach of Moore [9] obtains a high precision and computational efficiency. Nonetheless, the recall rate is at most equal to the proportion of 1-to-1 correspondences contained in the parallel text to align. This evidence is especially problematic when aligning parallel corpora with much noise or sparse data. Meanwhile, Varga, et al. [7] having the same idea as Moore [9] gains a very high recall ratio but lower precision. In this paper, we have proposed a method which overcomes weaknesses of those approaches and uses a new feature in sentence alignment, word clustering, to deal with the sparse data issue. The lack of necessary items in the dictionary used in the lexical stage is supplemented by applying word clustering data sets. This approach obtains a high recall rate while the accuracy still gets a reasonable ratio, which gains a considerably better overall performance than above-mentioned methods.

The rest of this paper is structured as follows. Previous works are described in section 2. In section 3, we present our approach as well as the sentence alignment framework we use and focus on description of applying word clustering feature in our algorithm. Section 4 indicates the experiment results and evaluations on some aligners. Finally, Section 5 gives several conclusions and future works.

2 Related Works

In various sentence alignment algorithms which have been proposed, there are three widespread approaches which are based on a comparison of sentence length (Brown, et al. [3], Gale and Church [6]), lexical correspondences (Chen [4], Melamed [10]), and the combination of these two methods (Moore [9], Varga, et al. [7]).

Length-based algorithms align sentences according to their length (measured by character or word). The algorithms of this type have first been proposed in Brown, et al. [3], and Gale and Church [6]. The primary idea is that long sentences will be translated into long sentences and short ones into short ones. A probabilistic score is assigned to each proposed correspondence of sentences, based on the scaled difference of lengths of the two sentences and the variance of this difference. This can produce a good alignment on language pairs with high length correlation such as French and English and perform in a high speed. Nevertheless, it is not robust because of using only the sentence length information. When there is too much noise in the input bilingual texts, sentence length information will be no longer reliable.

The second approach tries to overcome the disadvantages of length-based approaches by using lexical information from translation lexicons, and/or through the recognition of cognates. This approach could be illustrated in Kay and Röscheisen [4]. Chen [4], meanwhile, constructs a word-to-word translation model during alignment to assess the probability of an alignment. These approaches are usually more robust than the length-based algorithms because they use the lexical information from source and translation lexicons rather than only sentence length to determine the translation relationship between sentences in the source text and the target text. Nevertheless, algorithms based on a lexicon are slower than those based on length sentence since they require considerably more expensive computation. Other algorithms which also use this idea are indicated in Ma [8], Melamed [10], and Wu [10].

Sentence length and lexical information are also combined to achieve more efficient algorithms. Moore [9] describes an algorithm with two phases. Firstly, a length-based approach is used for an initial alignment. This first alignment plays then as training data for a translation model, which is then used within a combination framework of length-based and dictionary. Varga, et al. [7] uses a dictionary-based translation model with a dictionary that can be manually expanded while the approach of Moore [9] works with a IBM-1 translation model. Braune and Fraser [2] also propose an algorithm that has some modifications with Moore [9]. Nonetheless, this approach built more 1-to-many and many-to-1 alignments rather than only making 1-to-1 alignments as Moore [9]. The hybrid approaches achieve a relatively high performance that overcome the limits of the first two approaches and combine the advantages of them. Nonetheless, there are still weaknesses which should be handled in order to obtain a more efficient sentence alignment algorithm.

In addition to these above-mentioned algorithms, there have also been some new methods proposed lately, which based on other approaches like Sennrich and Volk [11] and Fattah [5]. While Sennrich and Volk [11] uses a variant of BLEU in measuring similarity between all sentence pairs, the approach of Fattah [5] is based on classifiers: Multi-Class Support Vector Machine and Hidden Markov Model. In general, the approach based on combination of length sentences and word correspondences is quite effective in overall like the proposal of Moore [9] whose idea has been referred and expanded by some researches as Varga, et al. [7] and Braune and Fraser [2]. We also apply this framework of sentence alignment in order to develop our idea, which is presented in the next section.

3 Our Proposal

In this section, we describe aspects concerning our algorithm. The method we use based on the framework of Moore [9] will be presented in the section 3.1. The section 3.2 illustrates our analyses and evaluations about the impacts of dictionary quality to the performance of sentence aligner, which is the basis in order that we give our algorithm using word clustering feature into sentence alignment described in the section 3.3. This section also briefly introduces about word clustering, a new feature we use in sentence alignment.

3.1 *Sentence Alignment Framework*

We use the framework of Moore [9] with some modifications based on hybrid of length-based and word correspondences, which consists of two phases. Firstly, the corpus is aligned by length of sentences. After extracting sentence pairs with the highest probabilities, a lexical model is trained with sentences just chosen. In the second phase, the corpus is aligned based on the combination of the first model with lexical information, which word clustering is used in this phase. Visually, our approach is illustrated in the Fig. 1.

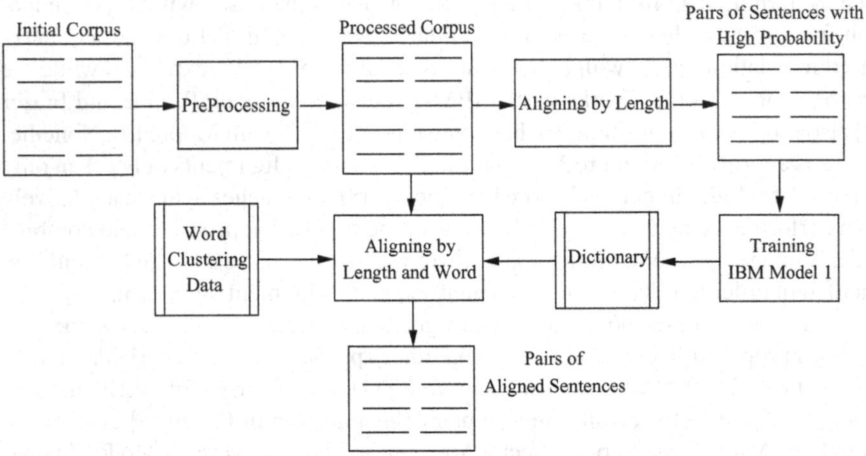

Fig. 1 Framework of sentence alignment in our algorithm

3.2 *The Effect of Dictionary*

Using a dictionary in sentence alignment contributes to create a better result. Varga, et al. [7] uses an extra dictionary in their framework or train IBM Model 1 to make a dictionary when without such a dictionary. Moore [9] uses IBM Model 1 to make a bilingual dictionary. Suppose that there is a pair of sentences (*s*, *t*) where *s* is one sentence in the text of source language, *t* is one sentence in the text of target language:

$s = (s_1, s_2, ..., s_l)$, where s_i is a word of sentence *s*.

$t = (t_1, t_2, ..., t_m)$, where t_j is a word of sentence *t*.

To count alignment probability of this sentence pair, there has to look up all word pairs (s_i, t_j) in the dictionary. However, a dictionary do not contain all word pairs, and this is increasingly evident when processing the sparse data. Moore [9] deals this issue by assigning every word not found in the dictionary to a common word "*other*". This is not really smooth and could lead to declining the quality of sentence aligner. We have dealt this issue by using word clustering introduced in the next sections.

3.3 *Word Clustering*

3.3.1 Brown's Algorithm

Word clustering Brown, et al. [5] is considered as a method for estimating the probabilities of low frequency events that are likely unobserved in an unlabeled data. One of the aims of word clustering is the problem of predicting a word from previous words in a sample of text. This algorithm counts the similarity of a word based on its relations with words on its left and on its right. The input to the algorithm is a corpus of unlabeled data which consist of a vocabulary of words to be clustered. Initially, each word in the corpus is considered to be in its own distinct cluster. The algorithm then repeatedly merges the pair of clusters that maximizes the quality of the clustering result and each word belongs to exactly one cluster until the number of clusters is reduced to the predefined number of clusters. The output of the word cluster algorithm is a binary tree in Fig. 2, in which the leaves of the tree are the words in the vocabulary. A word cluster contains a main word and subordinate words, each subordinate word has the same bit string and corresponding frequency.

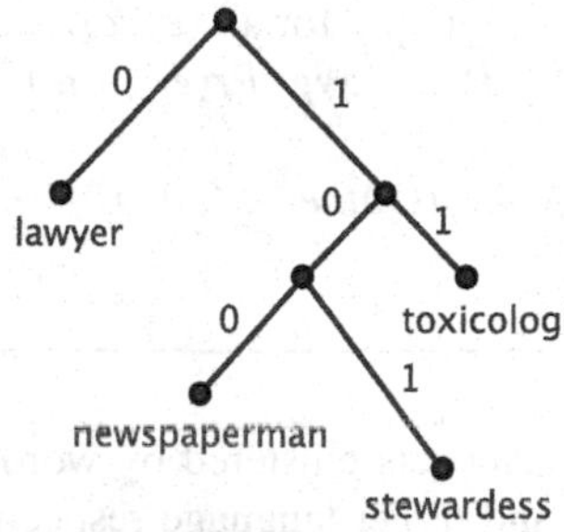

Fig. 2 An example of Brown's cluster algorithm

3.3.2 Algorithm Using Word Cluster Feature

When aligning sentences based on a dictionary, word pairs forming corresponding sentences are looked for in the bilingual word dictionary. All of them, however, do not always appear in the dictionary. Thus, we apply the idea using word clustering data sets. Words in the same cluster have a specific correlation, and in some cases they are able to be replaced to each other. Words that disappear in the dictionary would be replaced by ones in the same cluster rather than assigning all of those words to a common word as the method of Moore [9]. We use two word clustering data sets corresponding to the two languages in the corpus. The idea we use could be indicated as the Algorithm. 1.

In this algorithm, *D* is the dictionary which is created by training IBM Model 1. The dictionary *D* contains word pairs (*e*, *v*), which *e* is the word of the text of source language and *v* is the word of the text of target language, and their word translation probability *Pr*(*e*, *v*).

Algorithm 1: Using word clustering in sentence alignment

Input: A pair of word (e, v);
 Dictionary D, two clusters C_e and C_v
Output: P, word translation probability of (e, v)

 1. **if** (e, v) contained in D **then**
 2. $P \longleftarrow Pr(e, v)$
 3. **else**
 4. **if** (e contained in D) and (v contained in D) **then**
 5. looking for all $(e_1, e_2, ..., e_n)$ in C_e
 6. looking for all $(v_1, v_2, ..., v_m)$ in C_v
 7. $P \longleftarrow$ **avg**$(Pr(e_i, v), Pr(e, v_j)), 1 \leq i \leq n, 1 \leq j \leq m$
 8. **else**
 9. **if** (e contained in D) or (v contained in D) **then**
10. **if** (e contained in D) **then**
11. looking for all $(v_1, v_2, ..., v_m)$ in C_v
12. $P \longleftarrow$ **avg**$(Pr(e, v_j)), 1 \leq j \leq m$
13. **else**
14. looking for all $(e_1, e_2, ..., e_n)$ in C_e
15. $P \longleftarrow$ **avg**$(Pr(e_i, v)), 1 \leq i \leq n$
16. **else**
17. $P \longleftarrow Pr(\text{"other"}, \text{"other"})$
18. **return** P

In addition, there are two data sets clustered by word, which contain words of the texts of source language and target language respectively. Words in these two data sets have been divided into clusters in which C_e is a cluster containing word e, and C_v is a cluster containing word v. When (e, v) is not contained in the dictionary, each word of this pair is replaced by all words in its cluster before looking up these new word pairs in the dictionary. This is also the main idea of our algorithm. The probability of (e, v) is counted by the function **avg** which calculates the average value of probabilities of all word pairs looked up according to the approach in the above-mentioned algorithm.

Considering an English-Vietnamese sentence pair:

```
damodaran ' s solution is gelatin hydrolysate , a protein
known to act as a natural antifreeze .
```

```
gii_php ca damodaran l cht thy_phn gelatin , mt loi protein
c chc_nng nh cht chng ng t_nhiłn .
```

Several word pairs in the Dictionary created by training IBM Model 1 can be listed as follows:

```
damodaran    damodaran    0.2162747606740l324
's           ca           0.11691850681612706
solution     gii_php      0.030837942375885663
is           l            0.5459029836894168
a            mt           0.7344357464467972
as           nh           0.4577810602382339
natural      t_nhiłn      0.43630299483622276
...
```

However, there are not all word pairs in the Dictionary such as the word pair (*act*, *chc_nng*). Thus, first of all, the algorithm finds the cluster of each word in this word pair. The clusters which contain words "*act*" and "*chc_nng*" are as follows:

```
0110001111  act
0110001111  society
0110001111  show
0110001111  departments
0110001111  helps

11111110    chc_nng
11111110    hnh_vi
11111110    pht
11111110    hot_ng
...
```

In these clusters, the bit strings "0110001111" and "11111110" indicate the names of the clusters. The algorithm then looks up word pairs in the Dictionary and achieves following results:

```
departments      chc_nng        9.146747911957206E-4

act              hnh_vi         0.4258088124678187
act              pht            7.407735728457372E-4
act              hot_ng       0.009579429801707052
```

The next step of the algorithm is to calculate the average value of these probabilities, and the probability of word pair (*act*, *chc_nng*) would be:

```
Pr(act, chc_nng) = avg( 9.146747911957206E-4,
                        0.4258088124678187,
                        7.407735728457372E-4,
                        0.009579429801707052 )
                 = 0.1092609226583920
```

This word pair with the probability just calculated, now, can be used as a new item in the Dictionary.

4 Experiment

In this section, we assess the performance of our sentence aligner and conduct experiments to compare to some other hybrid approaches. First of all, we describe data sets used in experiments. The metrics to evaluate the performance of sentence aligners then are introduced. Finally, we illustrate results of the experiments and evaluate the performance of methods conducted in these experiments.

4.1 Data

4.1.1 Training Data

We perform experiments on 66 pairs of bilingual files English-Vietnamese extracted from websites World Bank, Science, WHO, and Vietnamtourism, which consist of 1800 English sentences with 39526 words (6309 different words) and 1828 Vietnamese sentences with 40491 words (5696 different words). We align this corpus at the sentence level by hand and gain 846 sentences pairs. Moreover, to achieve a better result in experiments, we use more 100,000 English-Vietnamese sentence pairs with 1743040 English words (36149 different words) and 1681915 Vietnamese words (25523 different words), which is available at[1]. This data set consists of 80,000 sentence pairs in Economics-Social topics and 20,000 sentence pairs in information technology topic.

In order to ensure the aligned result more accurate, we identify the discrimination between lowercase and upper case. This is reasonable since whether a word is lower case or upper case, it basically is similar in the meaning to the other. Thus, we carry out to convert all words in these corpora into their lowercase form. In addition to this, in Vietnamese, there are quite many compound words, which the accuracy of word translation is able to increase if the compound words are recognized rather than keeping all of them by single words. All words in the Vietnamese data set, therefore, are tokenized into compound words. The tool to perform this task is also available at[2].

4.1.2 Word Clustering Data

Related to applying word clustering feature in our approach, we use two word clustering data sets of English and Vietnamese in experiments which are indicated in Table 1. We create those data sets by using Brown's word clustering algorithm conducting on two input data sets. The input English data set is extracted from a part of British National Corpus with 1044285 sentences (approximately 22 million words). The input Vietnamese data set, meanwhile, is the Viettreebank data set consisting of 700,000 sentences with somewhere in the vicinity of 15 million words including Political-Social topics from 70,000 sentences of Vietnamese treebank and the rest from topics of websites laodong, tuoitre, and PC world. With 700 clusters in each

[1] http://vlsp.vietlp.org:8080/demo/?page=resources
[2] http://vlsp.vietlp.org:8080/demo/?page=resources

Table 1 Word clustering data sets

	Number of word items	**Number of clusters**
English data set	601960	700
Vietnamese data set	198634	700

data set, word items of these data sets cover approximately more than 81 percent those of these input corpora.

4.2 Metrics

We use metrics for evaluation: *Precision, Recall* and *F-measure* to evaluate sentence aligners.

$$Precision = \frac{CorrectSents}{AlignedSents}$$

$$Recall = \frac{CorrectSents}{HandSents}$$

$$F\text{-}measure = 2 * \frac{Recall * Precision}{Recall + Precision}$$

Where:

CorrectSents: number of sentence pairs created by the aligner match those aligned by hand.

AlignedSents: number of sentence pairs created by the aligner.

HandSents: number of sentence pairs aligned by hand.

4.3 Evaluations

We conduct experiments and compare our approach which is implemented on Java with two hybrid algorithms: M-Align (Bilingual Sentence Aligner, Moore [9]) and Hunalign (Varga, et al. [7]). As mentioned in the Moore's approach, to ensure reliable training data, a threshold 0.99 probability of correct alignment is chosen in the first stage in order that the training data is the sentence pairs assigned the highest probability of alignment. In order to have a fair comparison, we use thresholds 0.5 and 0.9 probabilities for the final alignment as in Moore experiments. In stead of computing the most probable overall alignment sequence, these probabilities can be estimated. The thresholds to accept an alignment pair are chosen so that we can make the balance between the precision and recall.

The results in Table 2 for the 0.99 probability threshold of the length-based stage and 0.9 probability threshold of the final alignment illustrate that the precision rate of M-Align is pretty high, but the recall rate is quite poor. This is also a limit in the approach of Moore [9]. Hunalign is 2.06% better than M-Align in recall rate, but the precision rate is approximately 4.23% lower. Our approach (EVS) is 4.56% and

Table 2 Comparison of M-Align (Moore [9]), Hunalign (Varga, et al. [7]), and our proposed approach (EVS)

Aligner	Precision(%)	Recall (%)	F-measure(%)
Hunalign	66.38	45.8	54.2
M-Align	70.61	43.74	54.01
EVS	61.82	75.41	67.94

Table 3 Comparision of approaches on the 0.5 probability threshold of the length-based phase

Aligner	Precision(%)	Recall (%)	F-measure(%)
Hunalign	56.52	73.37	63.58
M-Align	69.3	51.77	59.27
EVS	61.69	75.18	67.77

8.79% lower when comparing to Hunalign and M-Align in precision rate, but the recall rate of our approach is considerably higher, which is approximately 31.67% and 29.61% better than M-Align and Hunalign respectively, or 1.6 to 1.7 times higher. Further, our approach not only gets a considerably higher recall rate but also achieves an F-measure which is approximately 13.93% and 13.74% better than M-Align and Hunalign respectively. This is a good example of comparing different approaches when dealing with sparse data. M-Align usually gets a high precision rate; however, the weakness of this method is the quite low recall ratio, particularly when facing a sparseness of data. This kind of data results in a low accuracy of the dictionary, which is the key factor of a poor recall rate in the approach of Moore [9] because of using only word translation model - IBM Model 1. Our approach, meanwhile, deals this issue flexibly. If the quality of the dictionary is good enough, a reference to IBM Model 1 also gains a rather accurate output. Moreover, using word clustering data sets which assist to give more translation word pairs by mapping them through their clusters resolved sparse data problem rather thoroughly.

In aligning sentence using lexical information, the quality of the dictionary significantly impacts to the accuracy of final alignment. If the dictionary quality is poor, the performance of aligner would be quite bad. This is also a limitation of M-Align and Hunalign. When we use the 0.99 probability threshold in length-based phase indicated in Table 2, the recall rates of these two aligners are relatively low. To indicate those disadvantages clearly, we conduct experiments on the 0.5 threshold probability of length-based. Using a low threshold in this initial phase also means that the computation would be higher and the processing of this phase is more complex and difficult.

To evaluate the impact of a dictionary to the quality of alignment, we perform experiment by decreasing the threshold of the length-based phase by 0.5 rather than 0.99 of prior experiment. It is an indisputable fact that when using a lower threshold, the number of word items in the dictionary will increase that lead to a growth of recall rate. The results of this experiment are shown in Table 3. Examining these results indicates that when reducing the threshold of length-based phase, the recall rates of Hunalign and M-Align increase immediately. Comparing to the 0.99 probability threshold of the previous experiment, the recall rate of Hunalign increases 27.57% while this ratio of M-Align increases approximately 8%. However, these rates of (EVS) are almost invariable compare to the first experiment. In other words, our approach is quite independent with the size of the dictionary. By observing the results indicated in Table 3, our approach is higher than Hunalign in both precision and recall rates, which is more than 5.1% precision and 1.8% recall higher. Thus, our approach obtains an F-measure 4.19% better than Hunalign. When comparing to M-Align, although the precision of our approach is lower than M-Align at 7.61%, our approach achieves a better recall rate which is significantly higher at 23.41% than M-Align. As a result, the F-measure of our approach is more overall absolute.

In all cases, with regard to M-Align, our approach illustrates a considerable climb in both recall and F-measure. At the threshold of 0.99 for the length-based stage, the recall of EVS soars more than 30% while the F-measure of our aligner grows approximately 14% compare to M-Align. Meanwhile, although reducing the threshold of the length-based phase to 0.5 that leads to the recall of M-Align significantly increased, our approach still has seen a higher performance which not only the recall climbs approximately 23% but F-measure also increases more than 8% in comparison with M-Align. The identification of words not found in the dictionary into a common word as in Moore [9] results in quite a low accuracy in lexical phase that many sentence pairs, therefore, are not found by the aligner. In stead of that, using word clustering feature assists to improve the quality of lexical phase, and thus the performance increases significantly.

5 Conclusions and Future Works

The quality of the dictionary significantly impacts to the performance of sentence aligners which based on lexical information. When aligning corpus with sparse data, the dictionary, which is created by training sentence pairs extracted from length-based phase, would be lacked a great number of word pairs. This leads to a low quality of the dictionary, which declines the performance of the aligners. We have dealt this issue by using a new feature which is word clustering in our algorithm. The lack of many necessary items in the dictionary is effectively handled by referring to clusters in word clustering data sets sensibly. We associate this feature with the sentence alignment framework proposed by Moore [9] in our method. The experiments indicated a better performance of our approach compare to other hybrid approaches. It is the fact that word clustering is a useful application and could be utilized in sentence alignment that is able to improve the performance of the aligner.

It is very useful when sentence aligners effectively carry out and gain a high performance. In near future, we try not only to improve the quality of sentence alignment by using more new features such as assessing the correlation of sentence pairs based on word phrases or using another model in length-based phase rather than Poisson distribution as used by Moore [9] but to tackle the difficult issues like handling the noisy data.

Acknowledgments. This paper has been supported by the VNU project "Exploiting Very Large Monolingual Corpora for Statistical Machine Translation" (code QG.12.49).

References

1. Braune, F., Fraser, A.: Improved unsupervised sentence alignment for symmetrical and asymmetrical parallel corpora. In: Proceedings of the 23rd International Conference on Computational Linguistics: Posters, pp. 81–89 (2010)
2. Brown, P.F., Lai, J.C., Mercer, R.L.: Aligning sentences in parallel corpora. In: Proceedings of the 29th Annual Meeting on Association for Computational Linguistics, Berkeley, California, pp. 169–176 (1991)
3. Brown, P.F., Desouza, P.V., Mercer, R.L., Pietra, V.J.D., Lai, J.C.: Class-based n-gram models of natural language. Computational Linguistics 18(4), 467–479 (1992)
4. Chen, S.F.: Aligning sentences in bilingual corpora using lexical information. In: Proceedings of the 31st Annual Meeting on Association for Computational Linguistics, pp. 9–16 (1993)
5. Gale, W.A., Church, K.W.: A program for aligning sentences in bilingual corpora. Computational Linguistics 19(1), 75–102 (1993)
6. Kay, M., Röscheisen, M.: Text-translation alignment. Computational Linguistics 19(1), 121–142 (1993)
7. Ma, X.: Champollion: a robust parallel text sentence aligner. In: LREC 2006: Fifth International Conference on Language Resources and Evaluation, pp. 489–492 (2006)
8. Melamed, I.D.: A geometric approach to mapping bitext correspondence. In: Proceedings of the Conference on Empirical Methods in Natural Language Processing, pp. 1–12 (1996)
9. Fattah, M.A.: The Use of MSVM and HMM for Sentence Alignment. JIPS 8(2), 301–314 (2012)
10. Moore, R.C.: Fast and Accurate Sentence Alignment of Bilingual Corpora. In: Proceedings of the 5th Conference of the Association for Machine Translation in the Americas on Machine Translation: From Research to Real Users, pp. 135–144 (2002)
11. Sennrich, R., Volk, M.: MT-based sentence alignment for OCR-generated parallel texts. In: The Ninth Conference of the Association for Machine Translation in the Americas (AMTA 2010), Denver, Colorado (2010)
12. Varga, D., Németh, L., Halácsy, P., Kornai, A., Trón, V., Nagy, V.: Parallel corpora for medium density languages. In: Proceedings of the RANLP 2005, pp. 590–596 (2005)
13. Wu, D.: Aligning a parallel English-Chinese corpus statistically with lexical criteria. In: Proceedings of the 32nd Annual Meeting on Association for Computational Linguistics, pp. 80–87 (1994)

Frequent Temporal Inter-object Pattern Mining in Time Series

Nguyen Thanh Vu and Vo Thi Ngoc Chau

Abstract. Nowadays, time series is present in many various domains such as finance, medicine, geology, meteorology, etc. Mining time series for useful hidden knowledge is very significant in those domains to help users get fascinating insights into important temporal relationships of objects/phenomena along the time. Hence, in this paper, we introduce a notion of frequent temporal inter-object pattern and accordingly propose two frequent temporal pattern mining algorithms on a set of different time series. As compared to frequent sequential patterns, frequent temporal inter-object patterns are more informative with explicit and exact temporal information automatically discovered from many various time series. The two proposed algorithms which are brute-force and tree-based are efficiently defined in a level-wise bottom-up approach dealing with the combinatorial explosion problem. As shown in experiments on real financial time series, our work can be further used to efficiently enhance the temporal rule mining process on time series.

1 Introduction

Nowadays, there is the increasing popularity of time series in our lives as we record along the time the changes of objects and phenomena of interest. Mining such data will produce knowledge useful and valuable for users to get more understanding about behavioral activities of the objects and phenomena. Time series mining is the third challenging problem, one of the ten challenging problems in data mining research pointed out in [14]. Indeed, [7] has shown this research area has been very active so far. Among time series mining tasks, rule mining is a meaningful but tough mining task shown in [13]. In this paper, our work is dedicated to frequent temporal pattern mining on time series.

Nguyen Thanh Vu · Vo Thi Ngoc Chau
Faculty of Computer Science & Engineering, HCMC Uni. of Technology, Vietnam
e-mail: thanhvu_1710@yahoo.com, chauvtn@cse.hcmut.edu.vn

V.-N. Huynh et al. (eds.), *Knowledge and Systems Engineering, Volume 1*,
Advances in Intelligent Systems and Computing 244,
DOI: 10.1007/978-3-319-02741-8_15, © Springer International Publishing Switzerland 2014

As of this moment, we are aware of many existing works related to the frequent temporal pattern mining task on time series. Some are [3, 4, 5, 9, 10, 11, 12, 14, 15]. Firstly, in these works, it is realized that patterns are often different from work to work and discovered from many various time series datasets. In a few works, the sizes and shapes of patterns are fixed, and time gaps in patterns are pre-specified by users. In contrast, our work would like to discover patterns of interest that can be of any shapes with any sizes and with any time gaps able to be automatically derived from time series. Secondly, comparison between a proposed work with its previous related works is rare. This situation is understandable because there were different datasets and algorithms with different intended purposes in the related works and thus, data and implementation bias could exist in experimental comparison. Moreover, there is neither data benchmarking nor standardized definition of the frequent temporal pattern mining problem on time series. Indeed, whenever we get a mention of frequent pattern mining, market basket analysis appears to be a marvelous example of the traditional association rule mining problem. Such an example is not available in the time series mining research area for frequent temporal patterns.

Based on the aforementioned motivations, we introduce an efficient frequent temporal pattern mining process on time series with two main contributions. Our first contribution is a new knowledge type in terms of frequent temporal inter-object pattern. A frequent temporal inter-object pattern is a frequent pattern that occurs as often as or more often than expectation from users determined by a minimum support count. These patterns are abstracted from a set of different time series in terms of trends to express some repeating behavioral activities of some different objects. In addition, such patterns are composed of one or many components from one or many various time series showing the temporal relationships between the involved objects, i.e. inter-object relationships. The components of a pattern are ordered along the time line obeying the exact time intervals between the sequences of trends. Thus, our frequent patterns are temporal inter-object patterns. As for the second contribution, we design two frequent temporal pattern mining algorithms in support of the efficiency of the mining process. The first algorithm is brute-force in the manner of directly discovering frequent temporal inter-object patterns in a set of different time series according to a minimum support count using QuickSort and hash tables. Since employing appropriate data structures including tree and hash table, the second one is tree-based with a keen sense of reducing the number of spurious candidates generated and checked for resulting patterns. So, it is more efficient than the brute-force one. Nonetheless, the brute-force algorithm plays an important role of a baseline algorithm in helping us check the correctness of our tree-based algorithm by comparing their results. Besides, these algorithms can tackle the combinatorial explosion in a level-wise bottom-up approach.

2 Related Works

In this section, several works [3, 4, 5, 9, 10, 11, 12, 14, 15] related to frequent temporal pattern mining in time series are examined in comparison with our work.

In the most basic form, motifs can be considered as primitive patterns in time series mining. There exist many approaches to finding motifs in time series named a few as [11, 14, 15]. Our work is different from those because the scope of our algorithms does not include the phase of finding primitive patterns that might be concerned with a motif discovery algorithm. We suppose that those primitive patterns are available to our proposed algorithms.

For more complex patterns, [4] introduced a notion of perception-based pattern in time series mining with a methodology of computing with words and perceptions. Also, [9] presented a duration-based linguistic trend summarization of time series using the slope of the line, the fairness of the approximation of the original data points by line segments and the length of a period of time comprising the trend. Differently, our work concentrates on discovering relationships among primitive patterns. It is worth noting that our proposed algorithms are not constrained by the number of pattern types as well as the meanings and shapes of primitive patterns. Moreover, [3] has recently focused on discovering recent temporal patterns from interval-based sequences of temporal abstractions with two temporal relationships: before and co-occur. Mining recent temporal patterns in [3] is one step in learning a classification model for event detection problems. Different from [3], our work aims to discover more complex frequent temporal patterns in many different time series with more temporal relationships. Not directly proposed for frequent temporal patterns in time series, [12] used of Allen's temporal relationships in their so-called temporal abstractions. It is realized that temporal abstractions discovered from [12] are temporal patterns rather similar to our frequent temporal inter-object patterns. However, our work supports richer trend-based patterns and also provides two new efficient pattern mining algorithms as compared to [12]. Based on the temporal concepts of duration, coincidence, and partial order in interval time series, [10] defined pattern types from multivariate time series as Tone, Chord, and Phrase. For another form of patterns, [5] aimed to capture the similarities among stock market time series such that their sequence-subsequence relationships are preserved using temporal relationships such as begin earlier, end later, and are longer. Compared to [5, 10], our work supports more temporal relationships with explicit time information able to be automatically discovered along with resulting patterns.

3 A Frequent Temporal Inter-object Pattern Mining Process

In this section, a mining process is figured out to elaborate our solution to discovering frequent temporal inter-object patterns from a given set of different time series. Using the aforementioned notion of frequent temporal inter-object pattern, our work aims to capture more temporal aspects of the inter-object relationships in our resulting patterns so that they can be more informative and applicable. In addition, interestingness of discovered patterns is measured by means of the degree to which they are frequent in the lifespan of these objects in regard to a minimum support count.

Depicted in Fig. 1, our mining process includes three phases. Phase 1 is responsible for preprocessing to prepare for trend-based time series, phase 2 for the first step to obtain a set of repeating trend-based subsequences, and phase 3 for the primary step to fully discover frequent temporal inter-object patterns.

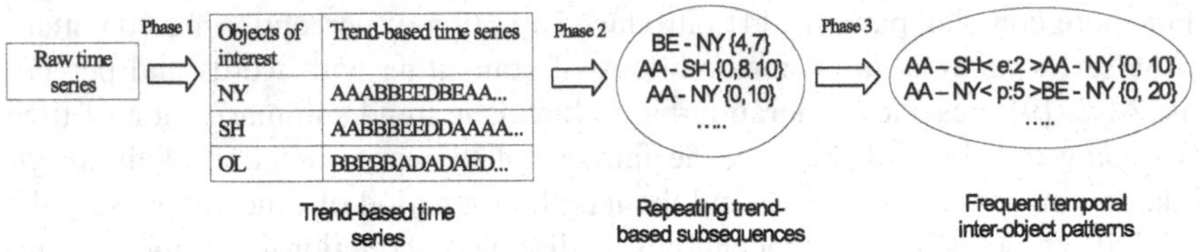

Fig. 1 A frequent temporal inter-object pattern mining process on time series

3.1 Phase 1 for Trend-Based Time Series

The input of this phase consists of a set of raw time series of the same length. Formally, each time series TS is defined as TS = $(v_1, v_2, \ldots, v_n)$. TS is univariate in an n-dimension space. The length of TS is n. $v_1, v_2, \ldots$, and v_n are time-ordered real numbers. Indices 1, 2, ..., n correspond to points in time in our real world on a regular basis. Regarding semantics, time series is understood as the recording of a quantitative characteristic of an object or phenomenon of interest observed regularly over the time. Besides, we pay attention to behavioral changes of objects and the degree to which they change. We focus on the change influences of objects on others to analyze in-depth whether or not the objects change frequently and what has made the objects change. So, we transform the raw time series into trend-based time series. Based on short-term and long-term moving averages defined in [17], a trend-based time series is a symbolic sequence with six trend indicators: A (in a weak increasing trend), B (in a strong increasing trend), C (starting a strong increasing trend), D (starting a weak increasing trend), E (in a strong decreasing trend), and F (in a weak decreasing trend). The output of this phase is a set of trend-based time series each of which is $(s_1, s_2, \ldots, s_n)$ where $s_i \in \{\text{A, B, C, D, E, F}\}$ for i = 1...n.

3.2 Phase 2 for Repeating Trend-Based Subsequences

The input of phase 2 is exactly the output of phase 1, one or many trend-based time series. The main objective of phase 2 is to find repeating trend-based subsequences in the input. Such subsequences are indeed motifs hidden in the trend-based time series. Regarding semantics, motifs themselves are frequent, representing the frequent behavioral changes of the objects being considered. As compared to discrete point-based events, motifs are suitable for the applications where the time spans of an event are significant to user's problems. For example, it is more informative for us to know that a stock keeps strongly increasing three consecutive days denoted by

BBB from Monday to Wednesday in comparison with a simple fact such that a stock increases. Motif discovery is out of the scope of our work. Thus, we implemented a simple brute force algorithm to extract repeating trend-based subsequences which are motifs along with their counts, each of which is the number of occurrences of the subsequence in its corresponding trend-based time series. In short, the output of this phase is a set of repeating trend-based subsequences stemming from different objects.

3.3 Phase 3 for Frequent Temporal Inter-object Patterns

Phase 3 has the input which is the output of phase 2, a set of repeating trend-based subsequences. In addition, phase 3 needs a minimum support count called *min_sup* from users to evaluate the output returned to users. Using *min_sup* and the input, phase 3 first obtains a set of primitive patterns, named L_1, which includes only repeating trend-based subsequences with the counts equal to or greater than *min_-sup*. All elements in L_1 are called frequent temporal inter-object patterns at level 1. Secondly, phase 3 proceeds with a frequent temporal inter-object pattern mining algorithm to discover and return to users a full set of frequent temporal inter-object patterns formally defined below.

In general, a frequent temporal inter-object pattern at level k for k>1 is in the following form: "$m_1 - m_1.ID\langle operatortype_1 : deltatime_1\rangle m_2 - m_2.ID \ldots m_{k-1} - m_{k-1}.ID\langle operatortype_{k-1} : deltatime_{k-1}\rangle m_k - m_k.ID$".

In this form, $m_1, m_2, \ldots, m_{k-1}$, and m_k are primitive patterns in L_1 which might come from different objects whose identifiers are $m_1.ID, m_2.ID, \ldots$, and $m_k.ID$, respectively. Regarding relationships between the components of a pattern at level k, $operatortype_1$, ..., $operatortype_{k-1}$ are well-known Allen's temporal operators [1] to express interval-based relationships along the time, including precedes (p), meets (m), overlaps (o), Finished by (F), contains (D), starts (s), and equals (e). Moreover, we use $deltatime_1$, ..., $deltatime_{k-1}$ to keep time information of the corresponding relationships. Regarding semantics, intuitively speaking, a frequent temporal inter-object pattern at level k for k>1 fully presents the relationships between the trends of changes of different objects of interest over the time. Unlike some in the related works [5, 10], our patterns are in a richer and more understandable form and in addition, our pattern mining algorithms are enabled to automatically discover all such patterns with no limitation on their relationship types and time information.

Example 1: Consider a frequent temporal inter-object pattern on two objects NY and SH: $AA - NY\langle e:2\rangle AA - SH$ $\{0, 10\}$. We state about NY and SH that NY has a two-day weak increasing trend and in the same period of time, SH does too. This fact occurs twice at positions 0 and 10 in their lifetime. It is also worth noting that we absolutely do not know if NY influences SH or vice versa in real life unless their relationships are analyzed in some depth. Nonetheless, such discovered patterns provide us with objective data-driven evidences on the relationships among objects of interest so that we can make further thorough investigations into these objects and their surrounding environment.

4 The Proposed Frequent Temporal Inter-object Pattern Mining Algorithms on Time Series

Since the knowledge type we aim to discover from time series has not yet been considered, we define two novel mining algorithms in a level-wise bottom-up approach: brute-force and tree-based. The brute-force algorithm provides a baseline for our solution and the tree-based one enables us to speed up the pattern mining process by avoiding the combinatorial explosion problem.

4.1 The Brute-Force Algorithm

The main idea of the brute-force algorithm is to find all frequent temporal inter-object patterns hidden in time series in a spirit of the popular Apriori algorithm [2]. Particularly, the brute-force algorithm discovers a set of frequent temporal inter-object patterns in a level-wise bottom-up manner starting from the initial set of primitive patterns, called L_1, which includes motifs with variable lengths in trend-based time series. They are also called repeating trend-based subsequences with the number of occurrences equal to or greater than the user-specified minimum support count *min_sup*. Using L_1, the brute-force algorithm employs QuickSort algorithm to sort instances in L_1 by their starting points in time and then proceeds to combine instances in L_1 to obtain a set C_2 of candidates and select only patterns with support counts satisfying *min_sup* to generate a set of patterns at level 2, called L_2. This routine is repeated till no candidate is found for L_k from L_{k-1} where k $\geq$ 2. Although the idea of the brute-force algorithm is rather simple, combining instances in L_{k-1} to generate candidates for L_k easily leads to combinatorial explosion as we consider interval-based relationships among these instances. To overcome this problem, we utilize a hash table and sorting. The brute-force algorithm is described below.

Step 1 - Construct L_2 from L_1: Let m and n be two instances in L_1 for a combination: m.StartPosition $\leq$ n.StartPosition where m.StartPosition and n.StartPosition are starting points in time of m and n. If any combination has a satisfied support count, it is a frequent pattern at level 2 and added into L_2.

If m and n are from the same object, m must precede n. A combination is in the form of: m-m.ID$\langle$p:delta$\rangle$n-n.ID where m.ID and n.ID are object identifiers, p stands for precedes, and delta (delta>0) for a time interval between m and n.

If m and n are from two different objects, ie. m-m.ID $\neq$ n-n.ID, a combination of m and n might be generated below for the additional Allen's operators where d is a common time interval in m and n.

– *meets* (m): m-m.ID$\langle$m:0$\rangle$n-n.ID
– *overlaps* (o): m-m.ID$\langle$o:d$\rangle$n-n.ID
– *Finished by* (F): m-m.ID$\langle$F:d$\rangle$n-n.ID
-*contains* (D): m-m.ID$\langle$D:d$\rangle$n-n.ID
-*starts* (s): m-m.ID$\langle$s:d$\rangle$n-n.ID
-*equal* (e): m-m.ID$\langle$e:d$\rangle$n-n.ID

Example 2: Consider m-m.ID = EEB-ACB starting at 0 and n-n.ID = ABB-ACB at 7. Their valid combination is EEB-ACB⟨p:4⟩ABB-ACB starting at 0.

Step 2 - Construct L_k from L_{k-1}: In general, we generate a candidate at level k from two instances in L_{k-1}. Let m and n be two instances in L_{k-1} in the following form where $m_1, \ldots, m_{k-1}, n_1, \ldots, n_{k-1}$ are instances in L_1; $operatortype_1$, ..., $operatortype_{k-2}$, $operatortype'_1$, ..., $operatortype'_{k-2}$ are among the Allen's operators {p, m, o, F, D, s, e}; $delta\,time_1$, ..., $delta\,time_{k-2}$, $delta\,time'_1$, ..., $deltatime'_{k-2}$ are time information of the relationships:
m = m_1-m_1.ID $\langle operator\ type_1 : delta\ time_1\rangle$ m_2-m_2.ID $\langle operator\ type_2 : deltatime_2\rangle \ldots \langle operatortype_{k-2} : deltatime_{k-2}\rangle$ m_{k-1}-m_{k-1}.ID
n = n_1-n_1.ID $\langle operatortype'_1 : deltatime'_1\rangle$ n_2-n_2.ID $\langle operatortype'_2 : deltatime'_2\rangle \ldots \langle operatortype'_{k-2} : deltatime'_{k-2}\rangle$ n_{k-1}-n_{k-1}.ID

If m and n have the same starting point in time and their components are also the same except for the last components: m_{k-1}-m_{k-1}.ID≠n_{k-1}-n_{k-1}.ID, a valid combination of m and n is generated in the form: $m_1 - m_1$.ID$\langle operatortype_1 : deltatime_1\rangle$ $m_2 - m_2$.ID$\langle operator\ type_2 : delta\ time_2\rangle \ldots \langle operator\ type_{k-2} : delta\ time_{k-2}\rangle$ $m_{k-1} - m_{k-1}$.ID$\langle operatortype_{k-1} : deltatime_{k-1}\rangle$ $n_{k-1} - n_{k-1}$.ID. Similar to step 1, any candidate with a satisfied support count is a frequent pattern at level k and added into L_k.

Example 3: Consider two instances in L_2: m = AA-NY⟨p:2⟩AAA-NY and n = AA-NY⟨p:10⟩FF-NY. A valid combination for a candidate in C_3 is AA-NY⟨p:2⟩AAA-NY⟨p:5⟩FF-NY.

As soon as valid to be a candidate, a combination is temporarily kept in a hash table. If existed, its support count increases by 1. Using a hash table, we can save search time for finding existing instances of a pattern generated previously as the complexity of an equal search on a hash table is O(1) on average. In addition, we later have the support count of each candidate conveniently to speed up the process. All frequent patterns discovered at level (k-1) are also organized in a hash table to easily support mining frequent patterns at the next level k. In short, our brute-force algorithm is efficiently designed for frequent temporal inter-object pattern mining.

4.2 The Tree-Based Algorithm

Though designed with efficiency by means of sorting and hash tables, the brute-force algorithm needs to create many combinations. The number of combinations is sometimes very large depending on the nature of input time series and the user-specified minimum support count. So, we invent the second algorithm that can as soon as possible select potential candidates and remove any combinations that cannot further form frequent patterns. The second algorithm is based on not only a hash table but also a tree appropriately in the spirit of FP-Growth algorithm [8]. We call the second algorithm the tree-based algorithm and a tree data structure a temporal pattern tree.

A Temporal Pattern Tree (TP-tree):

A temporal pattern tree (TP-tree for short) is a tree with n nodes of the same structure. The structure of a node being considered has the following fields:
ParentNode – a pointer that points to the parent node of the current node.
OperatorType – an operator about the relationship between the current node and its parent node in the form of ⟨p⟩, ⟨m⟩, ⟨e⟩, ⟨s⟩, ⟨F⟩, ⟨D⟩, or ⟨o⟩.
DeltaTime – an exact time interval associated with the temporal relationship in OperatorType field.
ParentLength – the length of the pattern counting up to the current node.
Info – information about the pattern that the current node represents.
ID – an object identifier of the object which the current node stems from.
k – the level of the current node.
List of Instances – a list of all instances corresponding to all positions of the pattern that the current node represents.
List of ChildNodes – a hash table that contains pointers pointing to all children nodes of the current node at level (k+1). Key information of an element in the hash table is: [*OperatorType* associated with a child node + *DeltaTime* + *Info* of a child node + *ID* of a child node].

Each node corresponds to a component of a frequent pattern. In particular, the root is at level 0, all primitive patterns at level 1 are handled by all nodes at level 1, the second components of all frequent patterns at level 2 are associated with all nodes at level 2, and so on. All nodes at level k are created and added into TP-tree from all possible combinations of nodes at level (k-1). Also, only nodes associated with support counts satisfying *min_sup* are inserted into TP-tree.

Building a Temporal Pattern Tree:

In a level-wise approach, a temporal pattern tree is built step by step from level 0 to level k. It is realized that a pattern at level k is only generated from all nodes at level (k-1) which belong to the same parent node. This feature helps us avoid traversing the entire tree to discover and create frequent patterns at higher levels and expand the rest of the tree.

Step 1 - Initialize TP-tree: Create the root of TP-tree labeled 0 at level 0.

Step 2 - Handle L_1: From the input L_1 including m motifs with support counts satisfying *min_sup*, create m nodes and insert them into TP-tree at level 1. Distances between these nodes to the root are 0 and Allen's *OperatorType* of each node is empty. For an illustration, TP-tree after steps 1 and 2 is displayed in Fig. 2 when L_1 has 3 frequent patterns corresponding to nodes 1, 2, and 3.

Step 3 - Handle L_2 from L_1: Generate all possible combinations between the nodes at level 1 as all nodes at level 1 belong to the same parent node which is the root. This step is similar to step 1 in the brute-force algorithm. Nevertheless, a combination in the tree-based algorithm is associated with nodes in TP-tree that help us to early detect if a pattern is frequent. Indeed, if a combination corresponding to an instance of a node that is currently available in TP-tree, we simply update the position of the instance in *List of Instances* field of that node and further ascertain that the combination is associated with a frequent pattern. If a combination corresponds to

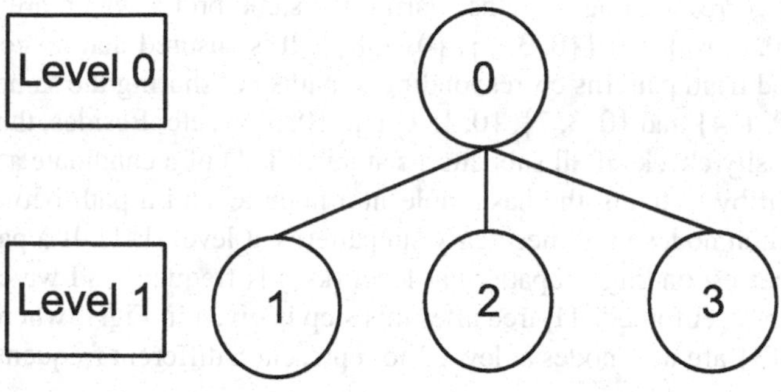

Fig. 2 TP-tree after steps 1 and 2

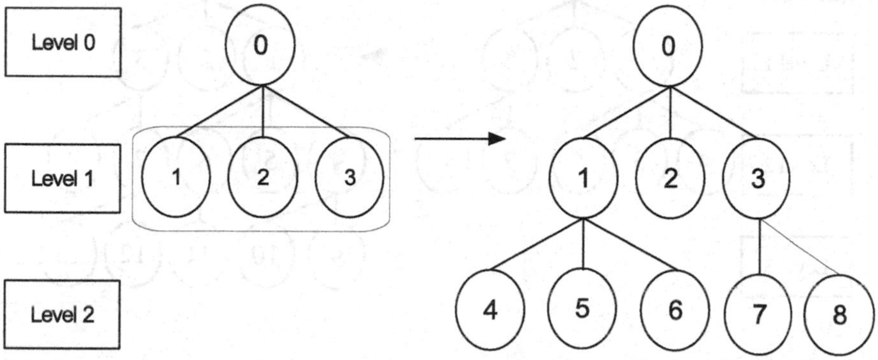

Fig. 3 TP-tree after step 3

a new node not in TP-tree, using a hash table, we easily have the support count of its associated pattern to check if it satisfies *min_sup*. If yes, the new node is inserted into TP-tree by connecting to its parent node. TP-tree after step 3 is illustrated in Fig. 3 with new nodes {4, 5, 6, 7, 8} at level 2 to represent 5 frequent patterns at level 2.

Step 4 - Handle L_3 from L_2: Unlike the brute-force algorithm where we have to generate all possible combinations between patterns at level 2 as candidates for patterns at level 3, we simply traverse TP-tree to generate combinations from branches sharing the same prefix path one level right before the level we are considering. Thus, we can reduce greatly the number of combinations. For instance, consider all patterns at L_2 in Fig. 3. The brute-force algorithm needs to generate combinations from all patterns corresponding to paths {0, 1, 4}, {0, 1, 5}, {0, 1, 6}, {0, 3, 7}, and

{0, 3, 8}. Instead, the tree-based algorithm needs to generate combinations from the patterns corresponding to paths sharing the same prefix which are {{0, 1, 4}, {0, 1, 5}, {0, 1, 6}} and {{0, 3, 7}, {0, 3, 8}}. It is ensured that no combinations are generated from patterns corresponding to paths not sharing the same prefix, for example: {0, 1, 4} and {0, 3, 7}, {0, 1, 4} and {0, 3, 8}, etc. Besides, the tree-based algorithm easily checks if all subpatterns at level (k-1) of a candidate at level k are also frequent by utilizing the hash table in a node to find a path between a node and its children nodes in all necessary subpatterns at level (k-1). If a path exists in TP-tree, a corresponding subpattern at level (k-1) is frequent and we can know if the constraint is enforced. TP-tree after this step is given in Fig. 4 where nodes {9, 10, 11, 12, 13} are new nodes at level 3 to represent 5 different frequent patterns at level 3.

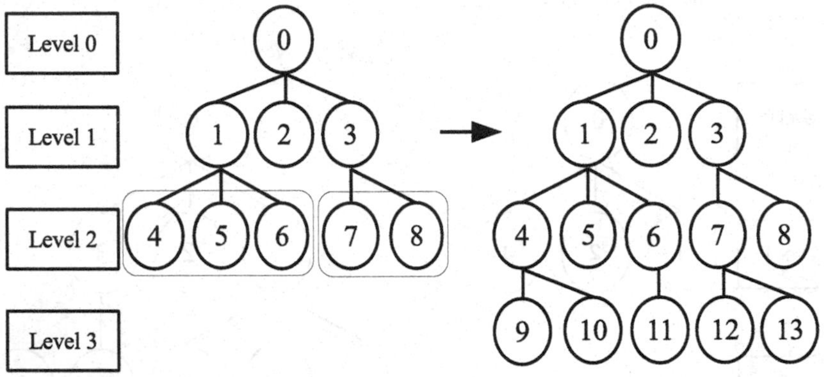

Fig. 4 TP-tree after step 4

Step 5 - Handle L_k from L_{k-1} where k$\geq$ 2: Step 5 is a general case of step 4 to discover frequent patterns at higher levels. Once TP-tree has been expanded up to level (k-1), we generate nodes at level k for frequent patterns in L_k if nodes at level (k-1) at the end of the branches sharing the same prefix path can be combined with support counts satisfying *min_sup*. The routine keeps repeating till no more level is created for TP-tree.

Finding All Frequent Temporal Inter-Object Patterns from TP-tree: As soon as TP-tree is completely constructed, we can traverse TP-tree from the root to derive all frequent temporal inter-object patterns from level 1 to level k. This subprocess recursively forms a frequent pattern represented by each node except the root node in TP-tree with no more checks.

5 Experiments

For an empirical study on the proposed algorithms, we present several experiments and provide discussions about their results. The experiments were done on a 3.0 GHz Intel Core i5 PC with 4.00 GB RAM. There are two groups for the efficiency of the proposed algorithms and how much improvement has been made between them. The first group was done by varying the time series length *Len* and the second one by varying the minimum support count *min_sup*. Each experiment of every algorithm was carried out ten times and then its average processing time in millisecond was reported. In Tables 1 and 2, we recorded the average processing time of the brute-force algorithm represented by BF, the average processing time of the tree-based one by Tree, and the ratio B/T of BF to Tree for comparison. We also captured the number M# of the initial motifs, the number P# of generated patterns excluding ones in L_1.

In the experiments, we used 5 real-life datasets of the daily closing stock prices available at [6]: S&P 500, BA from Boeing company, CSX from CSX Corp., DE from Deere & Company, and CAT from Caterpillar Inc. These time series have been unintentionally collected with variable lengths *Lens* of 20, 40, 60, 80, and 100 days from 01/04/1982. In each group, 5 various sets of time series TS for 1 to 5 objects are processed to obtain frequent temporal inter-object patterns.

Table 1 Results with various lengths *Lens*

TS	*Len*	M#	P#	BF	Tree	B/T
S&P500	20	0	0	≈0.0	≈0.0	1
	40	4	0	1.0	1.0	1
	60	14	9	12.7	7.6	2
	80	26	92	102.7	34.3	3
	100	40	201	316.9	98.0	3
S&P500, Boeing	20	0	0	≈0.0	≈0.0	1
	40	7	4	2.8	2.4	1
	60	22	49	39.2	25.1	2
	80	42	356	474.5	123.1	4
	100	60	964	1735.5	363.4	5
S&P500, Boeing, CAT	20	1	0	≈0.0	≈0.0	1
	40	9	12	8.5	3.6	2
	60	35	184	232.8	67.0	3
	80	72	850	1764.7	399.0	4
	100	100	2646	8203.0	1292.8	6
S&P500, Boeing, CAT, CSX	20	2	0	≈0.0	≈0.0	1
	40	14	14	19.9	10.4	2
	60	48	292	415.0	110.9	4
	80	95	1821	3857.3	545.1	7
	100	130	5394	19419.7	1794.6	11
S&P500, Boeing, CAT, CSX, DE	20	2	0	≈0.0	≈0.0	1
	40	20	16	36.2	14.8	2
	60	62	451	839.1	221.5	2
	80	116	3375	10670.7	1304.3	8
	100	167	8943	69482.7	4659.5	15

Table 2 Results with various *min_sups*

min_sup	M#	P#	BF	Tree	B/T
5	40	201	319.8	97.1	3
6	27	76	169.9	54.9	3
7	18	33	80.2	28.5	3
8	11	17	39.5	14.6	3
9	6	7	14.9	6.5	2
5	60	964	1732.2	382.4	5
6	43	273	698.2	196.3	4
7	31	106	367.1	109.7	3
8	20	47	175.3	56.8	3
9	14	22	95.0	34.6	3
5	100	2646	8248.6	1303.4	6
6	71	591	2222.7	574.2	4
7	51	223	1073.7	294.1	4
8	33	103	530.3	152.4	3
9	24	45	294.0	93.6	3
5	130	5394	19482.2	1976.2	10
6	94	1036	4628.6	1080.7	4
7	68	365	2075.9	546.6	4
8	43	160	972.4	270.7	4
9	30	67	519.9	145.9	4
5	167	8943	69068.7	4600.9	15
6	122	1558	8985.9	1685.1	5
7	89	527	3713.1	880.8	4
8	58	223	1751.0	437.8	4
9	43	92	983.7	256.2	4

Table 1 shows the experimental results in the first group with a minimum support count *min_sup* = 5. In this group, we carried out the experiments by varying the length of every time series from 20 to 100 with a gap of 20. Through these results, the ratio BF/Tree varies from 1 to 15 showing how inefficient the brute-force algorithm is in comparison with the tree-based one. As the length of each time series is bigger, each L_1 has more motifs and the tree-based algorithm works better than the brute-force one.

Table 2 shows the experimental results in the second group with time series length = 100. The experiments in this group are conducted by varying the minimum support count *min_sup* from 5 to 9. Their results let us know that the tree-based algorithm can improve at least 3 up to 15 times the processing time of the brute-force algorithm. Besides, it is worth noting that the larger a minimum support count, the fewer number of combinations that need to be checked for frequent temporal patterns, and thus, the less processing time is required by each algorithm. Once *min_sup* is high, a pattern is required to be more frequent; that is, a pattern needs to repeat more during the length of time series which is in fact the lifespan of each corresponding object. This leads to fewer patterns returned to users. Once *min_sup* is small, many frequent patterns might exist in time series and thus, the number of combinations might be very high. In such a situation, the tree-based algorithm is very useful to filter out combinations in advance and save much more processing time than the brute-force algorithm.

In almost all the cases, no doubt the tree-based algorithm consistently outperformed the brute-force algorithm. Especially, when the number of objects of interest increases, the complexity does too. As a result, the brute-force algorithm requires more processing time while the tree-based algorithm also needs more processing time but much less than the brute-force time. This fact helps us confirm our suitable design of data structures and processing mechanism in the tree-based algorithm. It can discover all frequent temporal inter-object patterns as the brute-force algorithm does. Nevertheless, it is more efficient than the brute-force algorithm to speed up our frequent temporal inter-object pattern mining process on time series.

6 Conclusion

In this paper, we have considered a time series mining task which aimed at frequent temporal inter-object patterns hidden in time series. These so-called frequent temporal inter-object patterns are richer and more informative in comparison with frequent patterns proposed in the existing works. Furthermore, we have developed two novel frequent temporal inter-object pattern mining algorithms. The first algorithm is based on a brute-force approach to provide us with a simple solution while the second one is more efficient using appropriate data structures such as hash table and TP-tree. Indeed, their capabilities of frequent temporal inter-object pattern mining in time series have been confirmed with experiments on real financial time series. However, in the future, we would like to examine the scalability of the proposed algorithms with respect to a very large amount of time series in a much higher dimensional space. Above all, with the resulting knowledge mined from time series, users can have useful information to get fascinating insights into important relationships between objects of interest over the time. Nonetheless, our work needs more investigation for semantically post-processing so that the effect of the surrounding environment on objects or influence of objects on each other can be analyzed in great detail.

References

1. Allen, J.F.: Maintaining Knowledge about Temporal Intervals. Communications of the ACM 26, 832–843 (1983)
2. Agrawal, R., Srikant, R.: Fast Algorithms for Mining Association Rules. In: VLDB, pp. 487–499 (1994)
3. Batal, I., Fradkin, D., Harrison, J., Mörchen, F., Hauskrecht, M.: Mining Recent Temporal Patterns for Event Detection in Multivariate Time Series Data. In: KDD, pp. 280–288 (2012)
4. Batyrshin, I., Sheremetov, L., Herrera-Avelar, R.: Perception Based Patterns in Time Series Data Mining. In: Batyrshin, I., Kacprzyk, J., Sheremetov, L., Zadeh, L.A. (eds.) Perception-based Data Mining and Decision Making in Economics and Finance. SCI, vol. 36, pp. 85–118. Springer, Heidelberg (2007)

5. Dorr, D.H., Denton, A.M.: Establishing Relationships among Patterns in Stock Market Data. Data & Knowledge Engineering 68, 318–337 (2009)
6. Financial Time Series, `http://finance.yahoo.com/` (accessed by May 23, 2013)
7. Fu, T.: A Review on Time Series Data Mining. Engineering Applications of Artificial Intelligence 24, 164–181 (2011)
8. Han, J., Pei, J., Yin, Y.: Mining Frequent Patterns without Candidate Generation. In: Proc. the 2000 ACM SIGMOD, pp. 1–12 (2000)
9. Kacprzyk, J., Wilbik, A., Zadrożny, S.: On Linguistic Summarization of Numerical Time Series Using Fuzzy Logic with Linguistic Quantifiers. In: Chountas, P., Petrounias, I., Kacprzyk, J. (eds.) Intelligent Techniques and Tools for Novel System Architectures. SCI, vol. 109, pp. 169–184. Springer, Heidelberg (2008)
10. Mörchen, F., Ultsch, A.: Efficient Mining of Understandable Patterns from Multivariate Interval Time Series. Data Min. Knowl. Disc. 15, 181–215 (2007)
11. Mueen, A., Keogh, E., Zhu, Q., Cash, S.S., Westover, M.B., BigdelyShamlo, N.: A Disk-Aware Algorithm for Time Series Motif Discovery. Data Min. Knowl. Disc. 22, 73–105 (2011)
12. Sacchi, L., Larizza, C., Combi, C., Bellazzi, R.: Data Mining with Temporal Abstractions: Learning Rules from Time Series. Data Min. Knowl. Disc. 15, 217–247 (2007)
13. Struzik, Z.R.: Time Series Rule Discovery: Tough, not Meaningless. In: Proc. the Int. Symposium on Methodologies for Intelligent Systems, pp. 32–39 (2003)
14. Tanaka, Y., Iwamoto, K., Uehara, K.: Discovery of Time Series Motif from Multi-dimensional Data Based on MDL Principle. Machine Learning 58, 269–300 (2005)
15. Tang, H., Liao, S.S.: Discovering Original Motifs with Different Lengths from Time Series. Knowledge-Based Systems 21, 666–671 (2008)
16. Yang, Q., Wu, X.: 10 Challenging Problems in Data Mining Research. International Journal of Information Technology & Decision Making 5, 597–604 (2006)
17. Yoon, J.P., Luo, Y., Nam, J.: A Bitmap Approach to Trend Clustering for Prediction in Time Series Databases. In: Data Mining and Knowledge Discovery: Theory, Tools, and Technology II (2001)

iSPLOM: Interactive with Scatterplot Matrix for Exploring Multidimensional Data

Tran Van Long

Abstract. The scatterplot matrix is one of the most common methods for multidimensional data visualization. The scatterplot matrix is usually used to display all pairwise of data dimensions, that is organized a matrix. In this paper, we propose an interactive technique with scatterplot matrix to explore multidimensional data. The multidimensional data is projected into all pairwise orthogonal sections and display with scatterplot matrix. A user can select a subset of data set that separates from the rest of data set. A subset of the data set organized as a hierarchy cluster structure. The hierarchy cluster structure is display as a radial tree cluster. The user can select a cluster in the radial tree and all data points in this cluster are display on scatterplot matrix. The user is repeated this process to identify clusters. One of the most useful of our method can be identify the structure of multidimensional data set in an intuition fashion.

1 Introduction

As consequences of the digital revolution, the amount of digital data that analytical systems are facing nowadays goes beyond what humans can fully understand by investigating individual data records. Automatic approaches like clustering need to be couple with interactive visualization methods to extract certain features and display them in an intuitive manner. This is particularly challenging for multidimensional multivariate data, as visual spaces are limited to (at most) three dimensions. Multidimensional multivariate data visualizations need to map the data into a lower dimensional layout that still conveys the important information.

When dealing with large data sets with many records, clustering has proven to be extremely useful. Clustering is a partition method of a data set into subsets of similar observations. Each subset is a cluster, which consists of observations that

Tran Van Long
Department of Basic Sciences, University of Transport and Communications, Hanoi, Vietnam
e-mail: vtran@utc.edu.vn

V.-N. Huynh et al. (eds.), *Knowledge and Systems Engineering, Volume 1*, 175
Advances in Intelligent Systems and Computing 244,
DOI: 10.1007/978-3-319-02741-8_16, © Springer International Publishing Switzerland 2014

are similar within themselves and dissimilar to observations of other clusters. Cluster analysis tasks for multidimensional data have the goal of finding areas where individual records group together to form a cluster.

Cluster analysis divides data into meaningful or useful groups (clusters). Clustering algorithms can be classified into four main approaches: partitioning methods, hierarchical methods, density-based methods, and grid-based methods [13]. In partitioning methods, data sets are divided into *k* disjoint clusters. In hierarchical methods, data sets are display by a hierarchy structure based on their similarity. Clusters are constructed from the hierarchical tree based on similarity threshold. In density-based methods, clusters are a dense region of points separated by low-density regions. In grid-based methods, the data space is divided into a finite number of cells that form a grid structure and all of the clustering operations are performed on the cells.

Unfortunately, all most approaches are not designed to cluster of a high dimensional data, thus the performance of existing algorithms degenerates rapidly with increasing dimensions. To improve efficiency, researchers have proposed optimized clustering techniques. Examples include grid-based clustering; balanced iterative reducing and clustering using hierarchies (BIRCH) [26], which builds on the cluster-feature tree; statistical information grid (STING) [24], which uses a quad-tree-like structure containing additional statistical information; and density based clustering (DENCLUE) [14], which uses a regular grid to improve efficiency. Unfortunately, the curse of dimensionality also severely affects the resulting clusterings effectiveness. A detailed comparison of the existing methods shows problems in effectiveness, especially in the presence of noise. Elsewhere, we showed that the existing clustering methods either suffer from a marked breakdown in efficiency (which proves true for all index based methods) or have notable effectiveness problems (which is basically true for all other methods).

Visual cluster analysis denotes the integration of cluster analysis and information visualization techniques. Seo and Shneiderman [19] introduced the hierarchical clustering explorer (HCE), which couples clustering algorithms (hierarchical algorithm, k-means) with information visualization techniques (dendrograms, scatterplots, parallel coordinates). It supports an interactive dendrogram representation to explore data sets at different levels of similarity. When a cluster is chosen in the dendrogram, all data points of this cluster are displayed in a parallel coordinates layout. The number of clusters varies depending on the level of similarity, such that it becomes difficult for the user to understand the qualitative structure of the clusters. For large data sets, the dendrogram likely exhibits clutter. Long and Lars [18] introduced the MultiClusterTree system that allows both clustering and visualization clusters of multidimensional data set with parallel coordinates and integrated with circular parallel coordinates. Visualization technology can help in performing these tasks. A large number of potential visualization techniques can be using in data mining. Examples include geometric projection techniques such as pro-section matrices and parallel coordinates, icon-based techniques, hierarchical techniques, graph-based techniques, pixel-oriented techniques, and combinations thereof. In general, the visualization techniques work in conjunction with some interaction techniques and some distortion techniques.

Unfortunately, the existing techniques do not effectively support the projection- and separator-finding process needed for an efficient clustering in high-dimensional space. Therefore, we developed a number of new visualization techniques that represent the important features of a large number of projections. These techniques help identify the most interesting projections and select the best separators [15]. For 1D projection, the authors developed a pixel-oriented representation of the point-density projections. Given a large number of interesting projections, the user may also employ a different visualization that represents the most important maxima of the point-density projections and their separation potential by small polygonal icons. The iconic representation reduces the information in the pixel-oriented visualization and allows a quick overview of the data.

Our idea, presented in this article, is to combine an advanced multidimensional data visualization techniques and information visualization techniques for a more effective interactive clustering of the data. We use one of the most common multidimensional data visualization that called a scatterplot matrix (SPLOM). We use the scatterplot on the SPLOM as a two-dimensional projections for identify clusters of the given multidimensional data set. This step performs by an analyzer. Another information visualization technique is radial layout that supports to represent the hierarchical structure. The hierarchy cluster tree and visualization of hierarchical cluster tree are constructed by fully automatically. When new cluster or sub-cluster is created, that is automatic represented on the 2D radial layout. The linked views between 2D radial layout and the SPLOM named iSPLOM. The iSPLOM is a powerful method for exploring multidimensional data set. To show the effectiveness of our new visualization techniques, we used the system for clustering real data. The experiments show the effectiveness of our approach.

The remainder of this paper is organized as follows: Section 2 present and analyze related works. Section 3 present more detail about the scatterplot matrix and the radial tree. Section 4 present a linked view between scatterplot matrix and radial tree of hierarchy clusters and we also implement our method with some data set. Finally, Section 5 present some conclusions and future works.

2 Related Work

2.1 *Clustering Data*

In cluster analysis, clustering algorithms are classified into four main groups: partitioning, hierarchical, density-based, and grid-based algorithms. Our approach is family of the grid-based methods. Typical grid-based algorithms partition the multidimensional space into a set of cells. The clusters are extracted based on the cells in all processing steps. Some well-known grid-based methods are STING [24] and CLIQUE [1]. Ester et al. [11] introduced the DBSCAN algorithm. The first step of the DBSCAN algorithm is to estimate the density using an Eps-neighborhood (like a spherical density estimate). Second, DBSCAN selects a threshold level set MinPts and eliminates all points with density values less than MinPts. Third, a graph is constructed based on the two parameters Eps and MinPts. Finally, the high density clusters are generated by connected components of the graph. Hinneburg and Keim introduced the DENCLUE approach [14], where high density clusters are identified by determining density attraction. Hinneburg et al. [15] further introduced the HD-Eye system that uses visualization to find the best contracting projection into a one- or two-dimensional space. The data is divided based on a sequence of the best projections determined by the high density clusters. The advantage of this method is that it does not divide regions of high density. However, it is difficult for the analyzer to find the contracting projections to generate clusters.

2.2 *Visualization of Multidimensional Data*

Information visualization has been demonstrated the great intuition and benefit in multidimensional data analysis. Scatterplot matrix, star coordinates, parallel coordinates, radviz etc. are most common techniques for visualization multidimensional data set. Here we discuss the scatterplot matrix techniques.

Scatterplot matrix is the most common techniques for display multidimensional data set [6]. The ScatterPLOt Matrix (SPLOM) is also integrated in many statistical software as R programming[1], GGOBI[2]. The scatterplot matrix have been improved by researchers that support as an efficient to display for a large scale multidimensional data set. Carr et al. [4] used a hexagonal shaped symbol whose size increases monotonically as the number of observations in the associated bin increases. Becker and Cleveland [3] introduced an interactive technique that named linked brushing techniques. The linked brushing technique is a collection of a dynamical method for displaying multidimensional data set. The linked brushing techniques contain for operations: highlight, shadow highlight, delete, and labeled that perform by moving a mouse control rectangle over one of the scatterplot. The effect of an operation appears simultaneously on all scatterplots that show the relationship of all selected data points in all pair of data dimensions. The brushing interactive techniques also integrated in Xmdvtool [23][3] one of the most popular tool for multidimensional data visualization. Cui et al. [5] proposed a reordering data dimensions to enhance relationship between data dimensions on scatterplot matrix and also integrated with linking and brushing techniques. Wilkinson et al. [25] proposed a scagnotics techniques for scatterplot matrix. The scagnotics of the scatterplot matrix are performed based on the geometric shape of data set on the scatterplots. The scatterplots show the convex hull, alpha hull, minimum spanning tree. Sips et al. [20] proposed some methods for measuring the quality to enhance good scaterplots in a large scatterplots. The quality of the good scatterplots is also investigated [21], [2]. Niklas Elmqvist et al. [10] introduced Rolling a dice system that support some new interactive techniques for

[1] `http://cran.r-project.org/`
[2] `http://www.ggobi.org/`
[3] `http://davis.wpi.edu/xmdv/`

scatterplot matrix. Recently, the scatterplot matrix is also integrated with parallel coordinates, the other common technique for visualizing multidimensional data [22] that is called Flowvizmenu.

Star coordinates [17] is an extension of scatterplots in which allow mapping a multidimensional data point into the two- or three-dimensional visual space. Star coordinates is also support some interactive techniques as scaling, rotation, marking, and labeled. Chen et al. [7, 8, 9] used the interactive techniques with star coordinates to labeled data that support identifying clusters. The cluster region is not automatic identified. The adjust control parameter in the star coordinates can visualize all 2D projection, that is one of advantages of the α-mapping. However, the parameters are difficult determining.

Parallel coordinates [16] is an another common information visualization technique for visualization multidimensional data set. The parallel coordinates display each multidimensional data point as a poly-line and the parallel axes. The most advantage of the parallel coordinate is the representation without loss information in the visual space. However, the poly-line can be overlap that leads a difficult to get information from the parallel coordinates. Fua et al. [12] presented a method for visualizing hierarchical clusters using hierarchical parallel coordinates. Clusters are represented at various levels of abstraction to reduce the overlap among clusters

Recently, Long and Lars [18] introduced the MultiClusterTree system that allows both clustering and visualization clusters of multidimensional data set with parallel coordinates and integrated with circular parallel coordinates. The MultiClusterTree [18] and Hierarchical Parallel Coordinates [12] visualize a hierarchical cluster tree of the multidimensional data set. The hierarchical cluster tree is created before applying visualization techniques. The VISTA [7], ClusterMap [8], iVIBRATE [9], and HD-Eye [15] used information visualization techniques to identify clusters. However, the system did not create a hierarchy cluster tree or hierarchy cluster tree did not display in an intuition.

In this paper, we present iSPLOM system in which support both create a hierarchy cluster tree through interactive with scatterplot matrix and visualization the hierarchy cluster tree in 2D radial layout.

3 Scatterplot Matrix and Radial Tree

3.1 Scatterplot Matrix

A scatterplot is a visual representation of data set that usually show the relationship between two variables. For representation multidimensional data set, the scatterplot matrix display all pair-wises of data dimensions, the matrix contain p^2 scatterplot, where p is the number of dimensions, arranged in p rows and p columns. The (i, j) entry of the matrix represent of the data set by orthogonal section projection and display the relationship between the ith and jth data dimensions.

Assume given a data set $X = [x_1, x_2, \ldots, x_n]$ consists of n multidimensional data points. Each data point $x_i = (x_{i1}, x_{i2}, \ldots, x_{ip})$ consists of p entry. The scatterplot matrix displays all possible pair of data dimensions. The diagonal scatterplot (i,i) display only one dimension $(x_{ki}, x_{ki}), 1 \leq k \leq n$ on the diagonal line. The (i,j) scatterplot display two-dimension data $(x_{ki}, x_{kj}), 1 \leq k \leq n$.

This is useful for looking at all possible two-way interactions or correlations between dimensions. The standard display quickly becomes inadequate for high dimensions and user interactions of zooming and panning are need to interpret the scatter plots effectively.

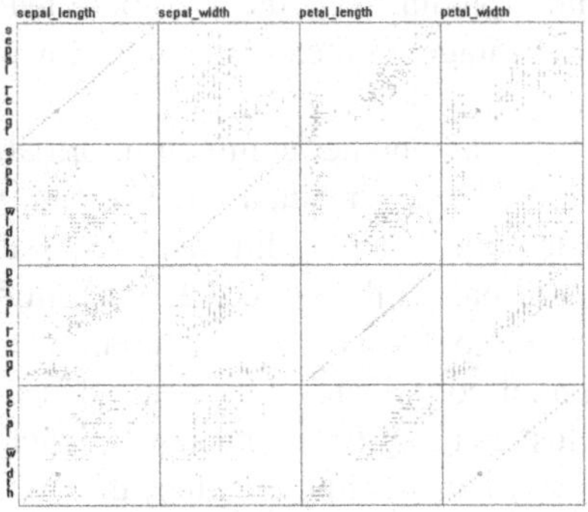

Fig. 1 A scatterplot matrix visualize the Iris data set

Figure 1 shows a scatterplot matrix for the Iris data set [23][4]. The Iris data set consists of 150 samples from three species of Iris: Iris setosa, Iris virginica and Iris versicolor. Each sample has four attributes: sepal length, sepal width, petal length, and petal width. Figure 1 shows 16 scatterplot in which contains four diagonal diaplay one dimension for each attributes and twelve scatterplot matrices symmetric. Hence, we only use the a half of scatterplot matrix that sufficient display all possible pair of data dimensions (including the diagonal scatterplot).

3.2 *Radial Tree*

The fundamental idea of our tree drawing is described as following: Considering a unit circle, the leaf nodes are placed evenly distributed on that unit circle. The root node is placed at the origin of the circle, and the internal nodes are placed on circular layers (with respect to the same origin) whose radii are proportional to the depth of the internal nodes. Hence, a mode cluster is represented by a node that placed on the unit circle. These clusters are homogeneous. All other clusters are represented by nodes placed on layers within the unit circle. These clusters are heterogeneous [18].

For the placement of internal nodes of the cluster tree, we use the notation of an annulus wedge. Given, a polar coordinate representation, an annulus wedge $W = (r, \alpha, \beta)$ denotes an unbounded region that lies outside a circle with center at the origin and radius r and is restricted by the two lines corresponding to angles α

[4] `http://archive.ics.uci.edu/ml/datasets/Iris`

and β. Figure 2 (left) shows an annulus wedge $W = (r, \alpha, \beta)$ (restricted to the unit circle).

Let tree T be the subtree of our cluster tree that is to be placed in the annulus wedge $W = (r, \alpha, \beta)$. The radius r denotes the distance of the root of T to the origin. If the root of T has depth d in the entire cluster tree, then $r = d$. Moreover, we use the notation $\ell(T)$ for the number of leaves of a tree T. Now, let $T_1, \ldots, T_k$ be those subtrees of tree T , whose root is a child node of T. For each subtree T_i, we compute the annulus wedge $W_i = (r_i, \alpha, \beta_i)$ where $r_i = d + 1$ is the radius for placing the root node T_i,

$$\alpha_i = \alpha + i\frac{\beta - \alpha}{k}, \beta_i = \alpha + (i+1)\frac{\beta - \alpha}{k}.$$

Figure 2 (middle) shows how an annulus wedge is split for a tree T with three subtrees $T_1, \ldots, T_k$. This iterative splitting of the annulus wedge is startedwith the root node of the cluster tree, which is represented by the annulus wedge $(0, 0, 2\pi)$. Finally, we can position all internal nodes of the cluster tree within the respective annual wedge. Considering sub-tree T with the corresponding annulus wedge $W = (r, \alpha, \beta)$, we place the node at position $(r\cos(\frac{\alpha+\beta}{2}), rsin(\frac{\alpha+\beta}{2}))$ with respect to the polar coordinate system. Figure 2 (right) shows the placement of nodes for the annulus wedges shown in Figure 2 (middle).

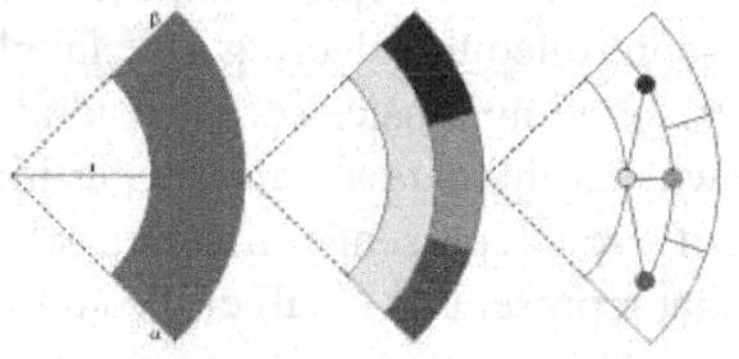

Fig. 2 Radial layout of a cluster tree. (Left) An annulus wedge domain $W = (r, \alpha, \beta)$. (Middle) Splitting the annual wedge for placing three subtrees. (Right) Placing internal nodes of cluster tree.

4 Interactive with Scatterplot Matrix

Based on the radial layout of the tree structure in Section 3, we construct a cluster tree automated based on interactive with scatterplot matrix. For drawing the nodes of the tree, we use circular disks with an automatic size and color encoding. The size of the nodes is determined with respect to the size of the respective cluster that is represented by the node. The color of the nodes is determined with respect to the position in the radial layout. Color encoding is done using the *HSV* color space. Hue H encodes the angle in our radial layout and saturation S encodes the radius (distance to the origin), while value V is constant (set to 1). Hence, the applied coloring scheme can be regarded as a slice $V = 1$ through the HSV color space.

The hierarchy cluster tree derives by a user interface for interaction with linked views. We support a linked view with the scatterplot matrix. In scatterplot matrix,

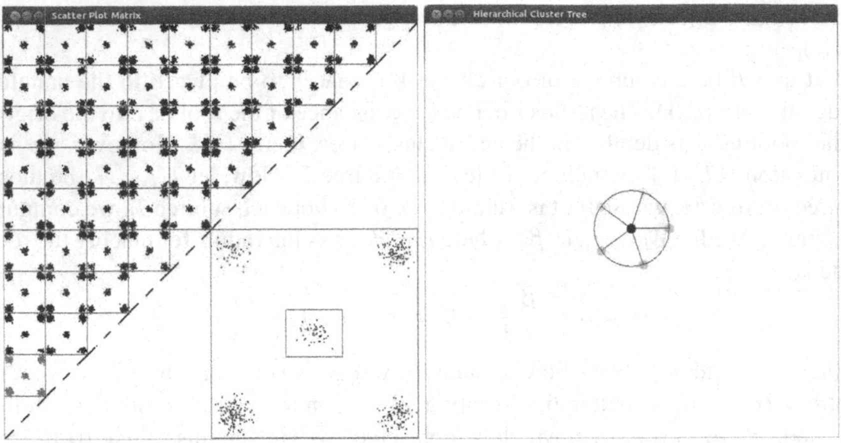

Fig. 3 Interactive with SPLOM for creating hierarchical clusters. (Left) A scatterplot is selected on the scatterplot matrix. (Right) Clusters are display as a hierarchical structure.

we have $p(p-1)/2$ scatterplots. An analyzer use keybroad to select one of the scatterplot that can separate the data set into at least two clusters. This sccaterplot is show on the a half of the diagonal scatterplot matrix as shows in Figure 3 (left). The analyzer use a mouse-controlled to selecting a rectangle region in the scatterplot matrix and create corresponding clusters on the radial tree. In Figure 3 (left), the chosen scatterplot shows that the data set can be partition into five clusters. In Figure 3 (right) the center of circle represent entire data set. Five clusters have been created on the scatterplot that represent by small circle and evenly placement on the unit circle.

We demonstrate the iSPLOM of our approach by applying it to a synthetic data set is called y14c data set consists of 480 data points in a ten-dimension. The y14c data set contains 14 clusters. Figure 3 (left) show scatterplot matrix for the y14c data set. On the scatterplot matrix we can only identify at most five clusters. Assume we select the X_1X_2 scatterplot matrix that divide data set into five clusters. Each cluster are selected and automatic represent on the Figure 3 (right).

The clusters of the data set can be detected based on interactive with partial scatterplot matrix for each subsclusters. The analyzer choose one of clusters e.g., the leaf node of the current radial tree. The scatterplot matrix display the data of this subclusters. Figure 4 (right) show that one subcluster is selected and displaying the all point in the corresponding scatterplot matrix. One of the sccaterplot is selected that divide the current subclusters into at least two small groups. Figure 4 (right) shows that this subclusters can be divide at least three small groups. Figure 4 (left) represent three this groups by a small circle and the position of this groups as in Section 3.

The linked views between the scatterplot matrix and the radial tree of hierarchy cluster tree repeated until we can identify homogeneous clusters. Figure 5 show the final results of interactive with scatterplot matrix. Each leaf node of the radial

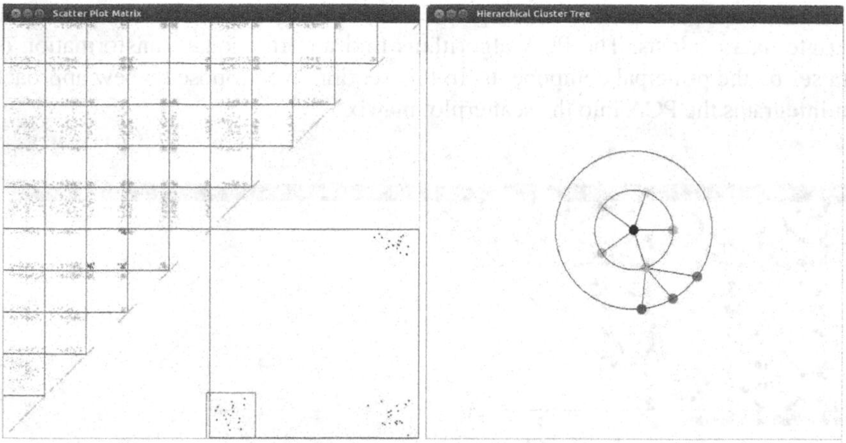

Fig. 4 Interactive with SPLOM for creating hierachy clusters. (Left) A scatterplot is selected on the scatterplot matrix. (Right) Clusters are display as a hierarchical structure.

cluster tree represents a homogeneous cluster. The radial cluster tree represents the hierarchy structure of clusters.

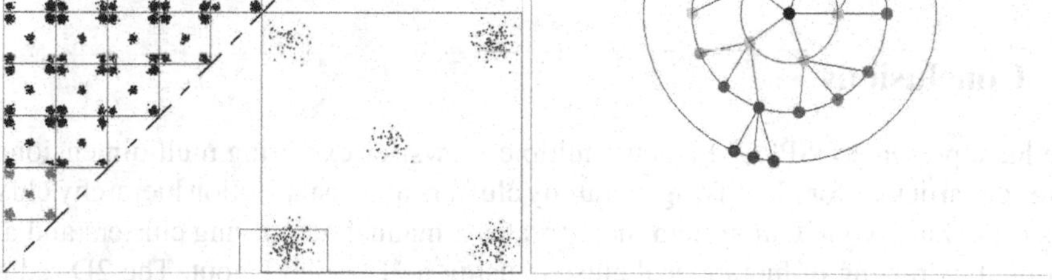

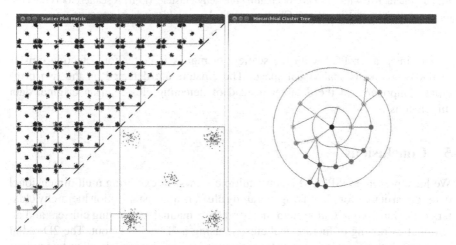

Fig. 5 Interactive with SPLOM for creating hierachy clusters. (Left) A scatterplot is selected on the scatterplot matrix. (Right) Clusters are display as a hierarchical structure.

The iSPLOM can apply for large scale and high dimensional data set. The iSPLOM system can be identify successfully 14 clusters even on the scatterplot matrix we can only at most five clusters on all scatterplots in the scatterplot matrix. For each iteration of the process, the size of data points in the subclusters is reduce and even scatterplot can be reduce e.g., the scatterplot have used to create subclusters not useful for identifying clusters any more.

The Principal Component Analysis (PCA) is the most common method for multivariate data analysis. The PCA algorithms find an orthogonal transformation of data set by the principal components. In this section, we proposed a new approach that integrates the PCA into the scatterplot matrix.

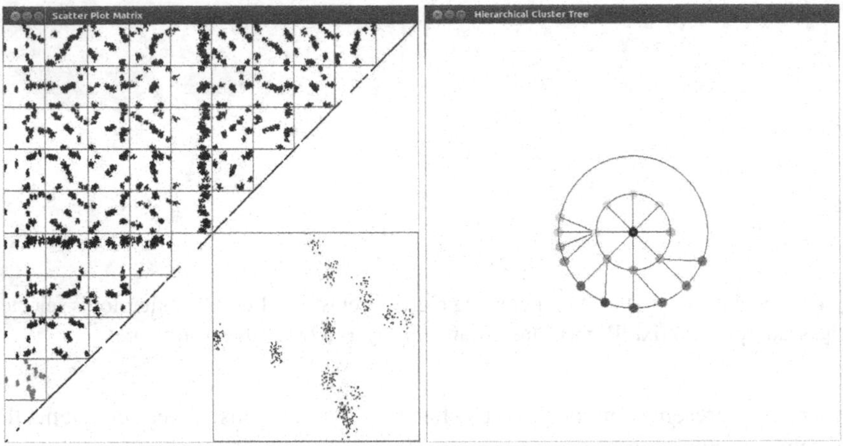

Fig. 6 Interactive with SPLOM for creating hierachy clusters. (Left) A scatterplot is selected on the scatterplot matrix. (Right) Clusters are display as a hierarchical structure.

The integration PCA with the scatterplot matrix is better for detecting cluster in the two dimensional visual space. The clusters are detected on the two principal components of the data set instead of detecting cluster in an original data dimensions.

5 Conclusions

We have presented iSPLOM a new multiple views for exploring multidimensional data set particular for identifying hierarchy clusters and visualization hierarchy clusters on radial layout. Our system incorporates a manual identifying clusters and an automatic drawing of hierarchical clusters using a 2D radial layout. The 2D radial layout of the hierarchical cluster tree supports an intuitive visualization to summarize the distribution structure of data set with subset of data dimensions.

Our approach works well for a large number of data sets and a large scale data dimensions. The 2D radial layout of the hierarchical density cluster tree provides more compact views and more flexible navigation and interaction techniques than a standard dendrogram tree layout. One aspect our approach does not detective for arbitrary shape of clusters. However, the complex shape of clusters can be identifying if we use the mouse to draw a polygonal shape. For future work, it would be investigate with other information visualization techniques as star coordinates, parallel coordinates, and Radviz.

Acknowledgments. This research is funded by Vietnam National Foundation for Science and Technology Development (NAFOSTED) under grant number 102.01-2012.04.

References

1. Agrawal, R., Gehrke, J., Gunopulos, D., Raghavan, P.: Automatic subspace clustering of high dimensional data for data mining applications. In: Proceeding SIGMOD 1998 Proceedings of the 1998 ACM SIGMOD International Conference on Management of Data, pp. 94–105 (1998)
2. Albuquerque, G., Eisemann, M., Lehmann, D.J., Theisel, H., Magnor, M.: Quality-based visualization matrices. In: Proc. Vision, Modeling and Visualization (VMV), Braunschweig, Germany, pp. 341–349 (November 2009)
3. Becker, R.A., Cleveland, W.S.: Brushing scatterplots. Technometrics 29(2), 127–142 (1987)
4. Carr, D.B., Littlefield, R.J., Nicholson, W.L., Littlefield, J.S.: Scatterplot matrix techniques for large N. Journal of the American Statistical Association 82(398), 424–436 (1987)
5. Cui, Q., Ward, M.O., Rundensteiner, E.A.: Enhancing scatterplot matrices for data with ordering or spatial attributes. In: SPIE Proceedings. Visualization and Data Analysis, vol. 6060 (2006)
6. Chambers, J.M., Cleveland, W.S., Tukey, P.A., Kleiner, B.: Graphical Methods for Data Analysis. Wadsworth Press, Belmont (1983)
7. Chen, K., Liu, L.: VISTA: Validating and refining clusters via visualization. Information Visualization 3(4), 257–270 (2004)
8. Chen, K., Liu, L.: Clustermap: Labeling clusters in large datasets via visualization. In: Proceedings of the Thirteenth ACM International Conference on Information and Knowledge Management, pp. 285–293 (2004)
9. Chen, K., Liu, L.: iVIBRATE: Interactive visualization-based framework for clustering large datasets. ACM Transactions on Information Systems (TOIS) 24(2), 245–294 (2006)
10. Elmqvist, N., Dragicevic, P., Fekete, J.-D.: Rolling the dice: Multidimensional visual exploration using scatterplot matrix navigation. IEEE Transactions on Visualization and Computer Graphics 14(6), 1539–1148 (2008)
11. Ester, M., Kriegel, H.-P., Sander, J., Xu, X.: A density-based algorithm for discovering clusters in large spatial databases with noise. Journal Data Mining and Knowledge Discovery archive 2(2), 169–194 (1998)
12. Fua, Y.-H., Ward, M.O., Rundensteiner, E.A.: Hierarchical parallel coordinates for exploration of large datasets. In: Proceedings of the Conference on Visualization 1999: Celebrating Ten Years, pp. 43–50. IEEE Computer Society Press (1999)
13. Han, J., Kamber, M., Pei, J.: Data Mining: Concepts and Techniques, 3rd edn. The Morgan Kaufmann Series in Data Management Systems. Morgan Kaufmann Publisher (2012)
14. Hinneburg, A., Keim, D.A.: An efficient approach to clustering in large multimedia databases with noise. In: Proceedings of the 4th International Conference on Knowledge Discovery and Datamining (KDD 1998), New York, NY, pp. 58–65 (September 1998)

15. Hinneburg, A., Keim, D.A., Wawryniuk, M.: HD-Eye: Visual mining of high-dimensional data. IEEE Computer Graphics and Applications 19(5), 22–31 (1999)
16. Inselberg, A.: Parallel Coordinates: Visual Multidimensional Geometry and its Applications. Springer (2009)
17. Kandogan, E.: Star coordinates: A multi-dimensional visualization technique with uniform treatment of dimensions. In: Proceedings of the IEEE Information Visualization Symposium, vol. 650 (2000)
18. Van Long, T., Linsen, L.: MultiClusterTree: interactive visual exploration of hierarchical clusters in multidimensional multivariate data. Computer Graphics Forum 28(3), 823–830 (2009)
19. Seo, J., Shneiderman, B.: Interactively exploring hierarchical clustering results. Computer 35(7), 80–86 (2002)
20. Sips, M., Neubert, B., Lewis, J.P., Hanrahan, P.: Selecting good views of high-dimensional data using class consistency. Computer Graphics Forum 28(3), 831–838 (2009)
21. Tatu, A., Albuquerque, G., Eisemann, M., Schneidewind, J., Theisel, H., Magnork, M., Keim, D.: Combining automated analysis and visualization techniques for effective exploration of high-dimensional data. In: IEEE Symposium on Visual Analytics Science and Technology, VAST 2009, pp. 59–66 (2009)
22. Viau, C., McGuffin, M.J., Chiricota, Y., Igor, J.: The flowvizmenu and parallel scatterplot matrix: Hybrid multidimensional visualizations for network exploration. IEEE Transactions on Visualization and Computer Graphics 16(6), 1100–1108 (2010)
23. Ward, M.O.: XmdvTool: Integrating Multiple Methods for Visualizing Multivariate Data. In: IEEE Conf. on Visualization 1994, pp. 326–333 (October 1994), `http://davis.wpi.edu/xmdv/`
24. Wei, W., Yang, J., Muntz, R.: STING: A statistical information grid approach to spatial data mining. In: Proceedings of the International Conference on Very Large Data Bases, pp. 186–195. Institute of Electrical and Electronics Engineers, IEEE (1997)
25. Wilkinson, L., Anand, A., Grossman, R.: Graph-theoretic scagnostics. In: IEEE Symposium on Information Visualization, INFOVIS 2005, pp. 157–164 (2005)
26. Zhang, T., Ramakrishnan, R., Livny, M.: BIRCH: an efficient data clustering method for very large databases. ACM SIGMOD Record 25(2), 103–114 (1996)

An Online Monitoring Solution for Complex Distributed Systems Based on Hierarchical Monitoring Agents

Phuc Tran Nguyen Hong and Son Le Van

Abstract. Online-monitoring solutions for activities of components (hardware, application and communication) in objects of complex distributed systems are considered very important ones. These can help administrators quickly detect potential risks as well as errors positions. Most of the online monitoring is now implemented with the tools of the operating systems and the special tools of devices, but these built-in tools can be only applied for small systems. Online monitoring for complex distributed systems with these built-in tools are not effective and administrators have to face many difficulties in the operation and handling errors. In addition, they cannot provide global information of monitored systems. This paper proposes a flexible solution and an effective monitoring model for complex distributed systems. This model based on the hierarchical monitoring agents provide administrators more important information on run-time status of hardware components and processes, and events are generated during execution of components and interaction with external objects as well.

1 Introduction

Complex distributed systems (CDS) consist of many heterogeneous devices, topologies, services and technologies; also include a large number of communication events interacting with each other on the large scale of geographical areas. These cause many potential risks in the system [1], [2], [4]. A hardware malfunction, a faulty process or an abnormal event occurs on the system may affect other events taking place at different locations in the running environment of system. These symptoms can cause a bad effect on performance and stability of the system such as congesting the network, increasing the traffic, reducing the performance of the network, so it is cause of errors of related processes and incorrect results of distributed applications. In order to ensure the effective operation of applications and improve

Phuc Tran Nguyen Hong · Son Le Van
Danang University of Education, The University of Danang, Vietnam

V.-N. Huynh et al. (eds.), *Knowledge and Systems Engineering, Volume 1*, Advances in Intelligent Systems and Computing 244,
DOI: 10.1007/978-3-319-02741-8_17,

the quality of service of distributed systems, monitoring and controlling information of the network in general and status of each object (device) in particular is the issue of primary attention. In some solutions support for system management, monitoring solution is considered as a important solution which can support administrators quickly detect errors and potential risks arise during operation of the system.

Through the survey and review some typical works such as [5], [6], [7], [13], [15], [16], we found that the advantages of software are flexibility, ease of maintenance and economic, so network monitoring software have been researched, integrated and widely deployed in most of TCP / IP network managements. Many solutions and software model have been widely deployed such as IETF's SNMP [15], Newman's Monalisa [6]... Most of these solutions and models are only interested in solving the requirements for specific monitoring problems such as performance monitoring, security monitoring, distributed computing,... but ones have not been really interested in the general activities of the objects components in distributed systems, these components are one of the important entities of distributed systems [8]. The monitoring of basic activities of the components has an important role to support for the system management before using above monitoring solutions to deeper analysis of each specific activity in CDS. Therefore, online monitoring for activities in CDS should continue to research and develop to improve the functional model for this monitoring more effective. This paper proposes a feasible monitoring solution for CDS, the monitoring work is done by hierarchical monitoring agents (local Agents, domain agents, implementation agents) and allow online monitoring general activities of objects in CDS, provide not only information about the local status and events of components within the object but also the interaction communication with external objects.

2 Technical Bases of Monitoring

2.1 Complex Distributed Systems and Monitoring Requirements

We survey the distributed systems in which consist of network architectures and distributed applications and were presented by Coulouris, Kshemkalyani in the works [1], [4]. According to this view, the distributed system consists of independent and autonomous computational objects with individual memory, application components and data distributed over network, as well as communication interactions between objects is implemented by message passing method [2]. In the context of this paper, CDS are large-scale distributed systems in TCP/IP network and have ability been configured and reconfigured during the operation of the system [9].

Due to the CDS increase rapidly in the number of inter-networks and connections, important distributed applications run on a larger scale of geographical area, more and more users and communication events interact with each other on the system. On the other hand, heterogeneous computing environment, technologies and devices are deployed in CDS. These characteristics have generated many challenges for CDS management, monitoring requirements and operation of the system are more strictly in order to ensure the quality of the system. Some challenges such as autonomous and heterogeneous; Scalability and reconfiguration; The large number of events; Large scale of geographical areas and multi levels of management. We need to consider these challenges carefully in the design of monitoring system.

Table 1 Management requirements for CDS

Aspect	Description
Application management	Presentation of operation services, distributed applications executed on the system; monitoring and controlling the status, events, behavior of applications; presentation of the historical information of the application.
Network management	Monitoring information on resources and performance; construction of network topology and network connection status; presentation of the related information between networks and distributed applications.
Scale of distributed system	The monitoring system should be designed not only suitable for the large number of applications entities and devices, but also efficient in processing large volumes of monitoring information in the system.

With basic challenges mentioned above, when we build an monitoring system for CDS, monitoring system should meet the necessary management requirements such as the application management, network management and scale of distributed system. These requirements are presented in Table 1.

2.2 Monitoring Technical Base for Distributed Systems

Through monitoring systems, administrator has full information on the activities and interactions of the components in the network, as well as global information of the system. This information actively supports for the right management decisions and appropriate actions to control system. Most of the monitoring systems focus on four basic components such as Generation, Processing, Distribution and Presentation of monitoring information on the system for the administrator. Until 1993, Mansouri Samani and Morris Sloman propose the monitoring model integrating the Implementation component [11], and this model becomes a common reference model and is cited in the most studies on monitoring system after that. According to Samani and Sloman, the five basic components should be flexibly combined depending on the specific monitoring requirements to create an effective monitoring system for distributed systems. Activities of five components are discussed in detail in the works [9], [11].

Most of the current monitoring and management system is deployed in Manager–Agent model [12], which includes three basic elements: Manager, Agents and managed objects, databases and communication protocols. The communication between the Manager and Agent can be implemented by two method such as polling (request-response) and trap.

Advantages of this model:

- Facilitating the deployment and administration.
- Easily operating and upgrading.

Disadvantages of this model:

- When Manager is crashed or Manager's network is isolated, the entire management system will not operate.
- The expanded scope of work for this model is difficult because of the processing ability of Manager is limited, so Manager's performance will rapidly decrease and it is the only weakness of the management system.

Monitoring database is widely used in the monitoring and management systems to support for analyzing, processing and updating monitoring information (monitoring reports and traces). Mazumdar and Lazar proposed database model in the work [10], this model has been seen a classical database model for organizing of the monitoring and management information. It consists of three parts: static database, dynamic and statistic database.

Through the survey and review some typical monitoring and management systems such as [5], [6], [7], [13], [15], [16], we found that:

- Most of these systems are deployed to solve the specific monitoring class such as parallel or distributed computing monitoring, configuration monitoring, performance monitoring, etc. The advantage of this class is the good deal of monitoring requirements for each problem class. However, the disadvantages of this class are that most of these products operate independently and they cannot integrate or inherit to each other. This makes it difficult to operate and manage these products for administrators and performance of the system will be greatly affected when running concurrent these products.
- The monitoring solution bases on three main groups: the hardware solution, the software solution and the hybrid solution [9]. With the advantages of flexibility, mobility, ease of maintenance and economy, therefore the software solution is used as a common way in which monitoring software was designed according to Manager-Agent model and has been widely deployed in many TCP/IP monitoring and management products.

The table 2, 3 show the results of a survey in which some typical monitoring systems are considered in 3 criteria such as monitoring function, model and implementation solution.

From the tables 2 and 3, we notice that:

- Most of the monitoring systems were designed according to Centralized and Decentralized model, distributed monitoring model system with many advantages over centralized model should be further studied and implemented in practice

Table 2 Implementation solutions and monitoring models

Monitoring system	Monitoring model		Implementation solution		
	Centralized	*Distributed*	*Hardware*	*Software*	*Hybrid*
ZM4/SIMPLE[13]	•				•
JADE[7]	•			•	
MonALISA[6]		•		•	
SNMP[15]	•			•	
MOTEL[16]	•			•	
Sniffer (NetFlow)	•		•		
Specific tools (OS, Devices)	•			•	

Table 3 Functions of the monitoring system

Monitoring system	Monitoring function			
	Computing	*Performance*	*Object*	*Activities*
ZM4/SIMPLE[13]	•			
JADE[7]	•			
MonALISA[6]		•		
SNMP[15]		•		
MOTEL[16]			•	
CorbaTrace[5]			•	
Specific tools (OS, Devices)				•

- Run-time Information about the status, events and behaviors of the components in CDS have an important role, they support administrators to know general activity information of the entire system. This information is necessary to administrators, before they go into details of other specific information. However, this general activity information is mainly based on the specific integrated tools that developed by device vendors side or operating systems side. However, these built-in tools provide discrete information on each component and independent of each device, they cannot link the components in the system and cannot solve the global problem of system information. It take a lot of time to process objects in the inter-network. Therefore, the administrators cannot effectively monitor the general activities of CDS with these tools.

In order to overcome the above exists, we propose a distributed monitoring model for CDS based on hierarchical monitoring agent system, this monitoring system consists of local agents, domain and management agents. Based on this model, we deploy a monitoring solution that effectively supports administrators in general activities of components in CDS.

3 Monitoring Model

The objective of the monitoring system is observation, collection and inspection information about the activities of the hardware components, software components and communication events. This information supports actively in system management. General monitoring architecture shows in Fig. 1.

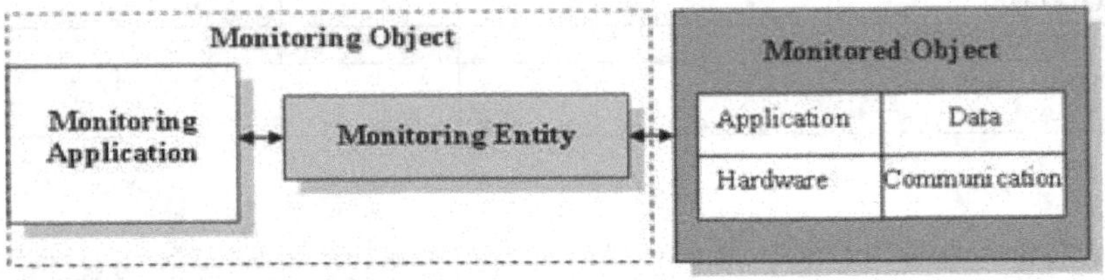

Fig. 1 General monitoring architecture

MA (Monitoring Application) is designed to support for the management Objects (administrators or other management agents). MA entity interacts with monitoring entity to support the generation of monitoring requirements and present the results of monitoring are measured from monitoring entity.

ME (Monitoring Entity) is designed to instrument the monitored objects, the instrumentation information of the system will be processed to generate the corresponding monitoring reports and send to MA.

MO (Monitored Object) consists of independent objects such as switches, routers, workstations, servers...

The monitoring system or MS in short consists of one or more MA and ME on the system. MS can be expressed as follows:

$$MS = \bigcup_{i=1}^{n} MA_i \cup \bigcup_{j=1}^{m} ME_j$$

Where n, m is number of MA, ME respectively in the system.

For the distributed monitoring system, the MAs are not fixed and independently operates on each domain, monitoring information is exchanged between the MA and each other by message passing.

Monitored system consists of many MOs, each MO consists of many components of hardware resources and software, associated with the corresponding state and the corresponding behavior. This information can be divided into two basic parts: internal part and external part [11]:

- The internal part - the local activities: these activities include processing, computing, resource requirements for process computations. The activities are locally performed within that object.

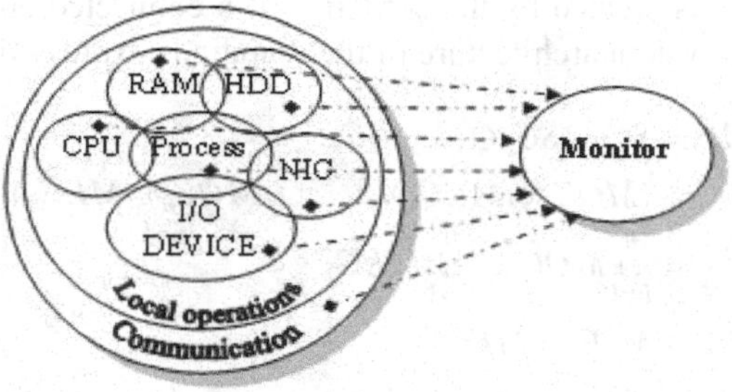

Fig. 2 General activity monitoring of the object

- The external part - the communication activities: these activities are functions that interact with other objects in the system such as controlling to interact with management system, inter-process communication. These activities are used to communicate with other objects on the system.

General activity monitoring will be implemented directly or indirectly by observing, detecting and collecting status, events appear on each monitored object or group of monitored objects. Thus, all of the local activities and the communication activities on the monitored system shall instrumented by monitoring system. According to Fig. 2, MO is a set of components contain the local activities and the communication activities with the external object. MO is expressed as follows:

$$MO = \{MO_{Proc}, MO_{Cpu}, MO_{Mem}, MO_{IOdev}, MO_{Comm}\}$$

MO's activity is denoted MOC and expressed as in (1):

$$MOC = MOC_{Proc} \cup MOC_{Cpu} \cup MOC_{Mem} \cup MOC_{IOdev} \cup MOC_{Comm} \tag{1}$$

Similar to ME:

$$ME = \{ME_{Proc}, ME_{Cpu}, ME_{Mem}, ME_{IOdev}, ME_{Comm}\}$$

ME's activity is denoted MEC, MEC monitors the activities of the components in the MO and MEC is expressed as in (2):

$$MEC = MEC_{Proc} \cup MEC_{Cpu} \cup MEC_{Mem} \cup MEC_{IOdev} \cup MEC_{Comm} \tag{2}$$

System architecture of MO is represented by SC:

$$\begin{aligned} SC &= MOC \cup MEC \\ &= (MOC_{Proc} \cup MEC_{Proc}) \cup (MOC_{Cpu} \cup MEC_{Cpu}) \cup \\ &\quad (MOC_{Mem} \cup MEC_{Mem}) \cup (MOC_{IOdev} \cup MEC_{IOdev}) \cup \\ &\quad (MOC_{Comm} \cup MEC_{Comm}) \end{aligned} \tag{3}$$

Domains activity is created by the activities of n connected objects SC_1, SC_2,... SC_n. Therefore, the system architecture of the domain is expressed as follows:

$$\begin{aligned} SD &= SC_1 \cup SC_2 \cup \ldots \cup SC_n \\ &= (MOC_1 \cup MEC_1) \cup \ldots \cup (MOC_n \cup MEC_n) \\ &= \bigcup_{i=1}^{n} MOC_i \cup \bigcup_{j=1}^{n} MEC_j \\ &= MOD \cup MED \end{aligned} \quad (4)$$

Where: MOD is the activities of monitored domain.

MED is the activities of domain monitoring entities corresponding to MOD.

Likewise, the system architecture of CDS is denoted S_{CDS} consists of m connected domains, the S_{CDS} is expressed by:

$$S_{CDS} = MO_{CDS} \cup ME_{CDS} \quad (5)$$

Where: MO_{CDS} is the activities of CDS.

ME_{CDS} is the activities of CDS monitoring entities corresponding to MO_{CDS}.

Similar to MOD, MED:

$$\begin{cases} MO_{CDS} = \bigcup_{i=1}^{m} MOD_i \\ ME_{CDS} = \bigcup_{j=1}^{m} MED_j \end{cases} \quad (6)$$

From formulas (1), (2), (3), (4), (5), (6) indicates that: To obtain full information activities in CDS, monitoring models need to be designed in accordance with the hierarchical architectural of CDS in which consists of different domains, subnets and objects in the system.

Recently, the trend of using monitoring agent has been studied and deployed in many monitoring and management systems such as [3], [14] and has good results. Therefore, our monitoring model is designed as a system of monitoring agents that have hierarchical function and been distributed throughout the monitored system. This model consists of three agent types such as TTCB is local Agents (equivalent to MEC's role), TTVM is domain agents (equivalent to MED's role) and GS is monitoring implementation agents (equivalent to MA's role).

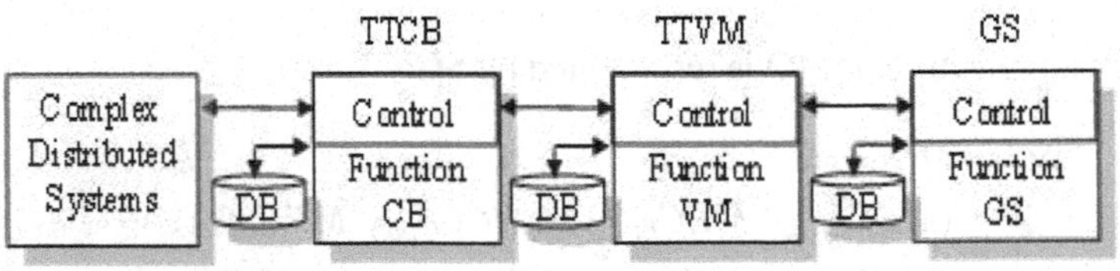

Fig. 3 System architecture of monitoring agents

TTCB is responsible for monitoring and collecting devices activities and resources and events, the executing processes in this local device. Each TTCB runs on a monitored device, one or more components are monitored by the TTCB.

TTVM is responsible for processing, synthesizing, analysis the information received from the TTCB to generate monitoring reports in which contains information about the status, events and behaviors of the components, the devices in this domain. Each group TTCB can be connected to one or more TTVM.

GS receives and analysis requests from management objects (administrators or other management agents), as well as the ability to control monitoring system, including the ability to synthesis and to presents the result of monitoring.

The function of each agent in the monitoring system is designed as Fig.4:

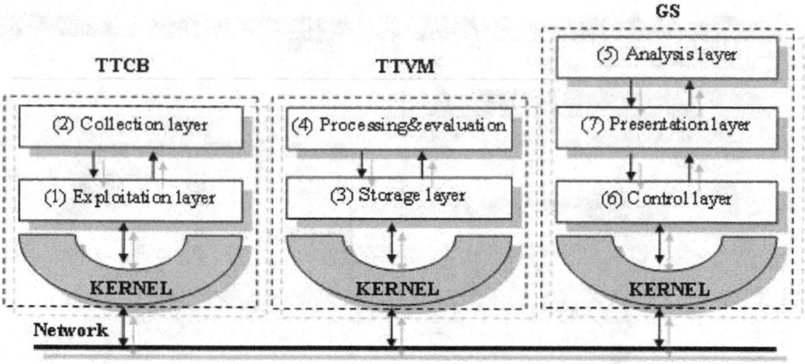

Fig. 4 The functional layer of the monitoring agents

(1) Exploitation layer: this layer performs surveys and collects information on the status, the events and the behaviors of components of monitored object by the sensors or the detectors interact with the management protocol or the kernel of the OS.

(2) Collection layer: this layer collects monitoring data, caches to data of Exploitation layer and periodically updates local monitoring data.

(3) Storage layer: this layer organizes, updates and stores monitoring data from Collection layer of the TTCB, can receive data from one or more of the TTCB in the same subnet or also from different subnets.

(4) Processing and evaluation layer: this layer processes, synthesizes, evaluates and classifies of state information, event information and behavior of the group of monitored objects on the systems.

(5) Analysis layer: this layer analyses monitoring data from Processing and evaluation layer of the TTVM to build the relationship between information, from that analysis and statistic data to determine the image of the domains and the whole system.

(6) Control layer: this layer analysis of the monitoring requirements, receive the result information and allow for starting, stopping and editing the arguments for monitoring system. In addition, this class also supports control service.

(7) Presentation layer: this layer shows monitoring results in visual form and interacts with monitoring agents (administrator or other agents).

CDS consist of the large number of heterogeneous objects in resources, operating systems and communication protocols. Therefore, TTCB function is designed corresponding to each target group-specific monitored objects and developed by using the management protocols such as SNMP or interacting with OS kernel corresponding to each specific group.

On the basis of the architectural model and functional model of the monitoring system is designed in the previous section, we deployed MCDS system to monitor activities of devices on the VMSC3 system, experimental results are shown in Fig. 5.

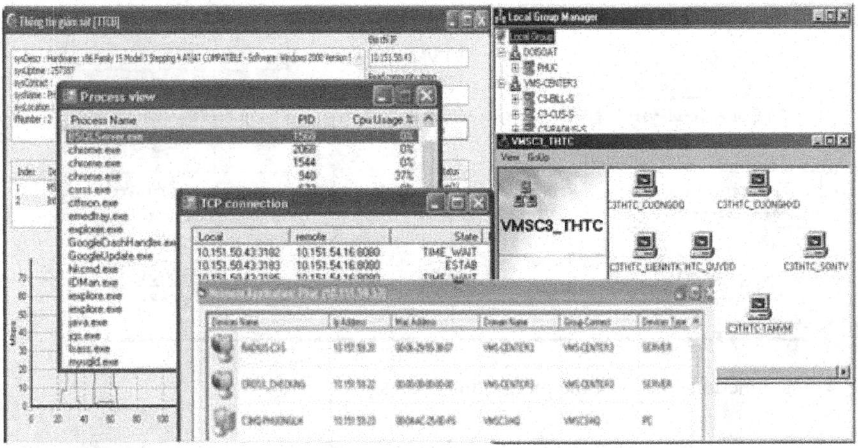

Fig. 5 Management of monitored objects

Fig. 5 shows the form of the configuration information about the local groups, in which the objects is managed by the device name, the IP address, the group name and type of devices on the monitored system. At the same time, administrators can easily track the connection status of the object by using of ICMP protocol and identify groups of devices online or offline on the system. Some forms present the information on the resources of the monitored system such as system information (descriptions, locations, OS), hardware information (Cpu, Ram, I/O,) and information on the activities of the services, the processes, the traffic, interactive communication, etc.

The monitoring system is implemented on VMSC3 network; experimental results show that our monitoring model is technically feasible solution. The local agents collect information as Fig. 5; this information will be used to send to domain agents by the group communication mechanisms. Based on the monitoring agent system, we easily develop extensions for the ability of agents to provide complete online information depending on specific monitoring requirements and effectively support the monitoring CDS.

This system overcomes the disadvantages of specific built-in tools (OS, Devices) to monitor CDS and actively support administrators on the important management information. Some actively results with the MCDS monitoring model is presented in table 4.

Table 4 Some results compare between specific built-in tools and MCDS model

Issue	Specific built-in tool	MCDS model
Implementation of monitoring requirements	Tools integrated monitored objects and OS, so Administrators must be able to use all these associated tools.	Administrators must be granted the rights in MCDS and only create monitoring requirements.
Implementation method	Tools are executed by the manual and remote connection method on each device.	Automatic Implementing with monitoring agents.
Monitoring scope	Discrete, individual, local.	Local, global.
Monitoring time	Depending on the skill of the administrator and monitored network infrastructure.	Depending on monitored network infrastructure.
Error detection	Manually	Capable of automatic warning.
Prediction and evaluation	Manually, Depending on the skill of the administrator	Automatic

4 Conclusion

Through the study of building the monitoring system for complex distributed systems, this paper proposes an architectural model and a functional model to complete the monitoring solution of online activities for complex distributed systems. Based on these models, we develop the MCDS solution to support administrators to obtain the information about local activities and communication events of objects in the system, as well as the global system information, this information actively useful for appropriate management decisions and controlling actions for the monitored system. This will help to improve the reliability and performance for important distributed applications.

References

1. Kshemkalyani, A.D., Singhal, M.: Distributed Computing Principles, Algorithms, and Systems. Cambridge University Press (2008)
2. Tanenbaum, A.S.: Modern Operating Systems, 3rd edn. Pearson Prentice Hall (2008)
3. Guo, C.G., Zhu, J., Li, X.L.: A Generic Software Monitoring Model and Features Analysis. In: Proceedings of the 2010 Second International Conference on Networks Security, Wireless Communications and Trusted Computing, pp. 61–64. IEEE express, Washington DC (2010)
4. Coulouris, G., Dollimore, J., Kindberg, T., Blair, G.: Distributed systems concepts and design, 5th edn. Addison Wesley Press (2011)

5. Corba trace, `http://corbatrace.sourceforge.net`
6. MonALISA document, `http://monalisa.caltech.edu/monalisa.htm`
7. Joyce, J., Lomow, G., Slind, K., Unger, B.: Monitoring Distributed Systems. ACM Transactions on Computer Systems 5(2), 121–150 (1987)
8. Son, L.V.: Distributed computing systems. Vietnam National University - HoChiMinh City publishing house (2002)
9. Son, L.V., Phuc, T.N.H.: Researching on an online monitoring model for large-scale distributed systems. In: Proceedings of the 13th National Conference in Information and Communication Technology, Hungyen, Vietnam (2010)
10. Mazumdar, S., Lazar, A.A.: Monitoring Integrated Networks for Performance Management. In: Proceedings of the International Conference on Communications, pp. 289–294. IEEE express, Atlanta (1990)
11. Sloman, M.: Network and Distributed Systems Management. Addison Wesley (1994)
12. Hai, N.T.: Computer Networks and Open Systems. Vietnam Education publishing house, Hanoi (1997)
13. Hofmann, R.: The Distributed Hardware Monitor ZM4 and its Interface to MEMSY. Universitat Erlangen, IMMD VII (1993)
14. Yang, S.Y., Chang, Y.Y.: An active and intelligent network management system with ontology based and multi agent techniques. Expert Systems with Applications 38(8) (2011)
15. Phuc, T.N.H., Son, L.V.: Monitoring of large scale distributed systems based on SNMP development. The Journal of Science and Technology. Danang University I(8), 79–84 (2012)
16. Logean, X.: Run-time Monitoring and On-line Testing of Middleware Based Communication Services. PhD dissertation, Swiss Federal (2000)

Incomplete Encryption Based on Multi-channel AES Algorithm to Digital Rights Management

Ta Minh Thanh and Munetoshi Iwakiri

Abstract. DRM (Digital Rights Management) systems is the promising technique to allow the copyrighted content to be commercialized in digital format without the risk of revenue loss due to piracy. However, traditional DRMs are achieved with individual function modules of cryptography and watermarking. Therefore, all digital contents are temporarily disclosed with perfect condition via decryption process in the user side and it becomes the risk of illegal redistribution. In this paper, we propose an incomplete encryption based on multi-channel AES algorithm (MAA) to control the quality of digital contents as a solution in DRM system. We employed the multi-channel AES algorithm in the incomplete cryptography to imitate multimedia fingerprint embedding. Our proposed method can trace the malicious users who redistribute the digital contents via network. We make this scenario more attractive for users by preserving their privacy.

1 Introduction

1.1 Overview

Advances in computer and network technologies have made easily to copy and distribute the commercially valuable digital content, such as video, music, picture via global digital networks. It enables an e-commerce model, that consists of selling and

---

Ta Minh Thanh
Department of Network Security, Le Quy Don Technical University, 100 Hoang Quoc Viet, Cau Giay, Hanoi, Vietnam, and Department of Computer Science, Tokyo Institute of Technology, 2-12-2, Ookayama, Meguro-ku, Tokyo, 152-8552, Japan
e-mail: `taminhjp@gmail.com`, `thanhtm@ks.cs.titech.ac.jp`

Munetoshi Iwakiri
Department of Computer Science, National Defense Academy, 1-10-20, Hashirimizu, Yokosuka-shi, Kanagawa, 239-8686, Japan
e-mail: `iwak@nda.ac.jp`

V.-N. Huynh et al. (eds.), *Knowledge and Systems Engineering, Volume 1*, Advances in Intelligent Systems and Computing 244,
DOI: 10.1007/978-3-319-02741-8_18,

delivering digital versions of content online. The main point of concern for such a business is to prevent illegal redistribution of the delivered content.

DRM systems were created to protect and preserve the owner's property right for the purpose to protect their intellectual property [1, 2, 3]. A common approach to protect a DRM system against tampering is to use a hardware based protection, often implemented in set-top-boxes. However, the biggest disadvantage of hardware based DRM systems are inflexibility and high cost. It requires a large investment cost from the service provider and time consuming for market. Additionally, hardware based DRM systems are expensive for customers. At a time where a lot of pirated contents are available on the Internet, hardware based solutions have the hard time creating value for the users. In order to reduce the developed cost, software based DRM [4, 5, 6] is proposed instead of hardware based DRM. The advantage of software based DRM is that they can be easily distributed to the users via networks and do not need to create additional installation costs. Most users would prefer a legal way to easily access content without huge initial costs or long term commitment. The problem with software based DRM system is that they are assumed to be insecure. Especially, such kind of software based DRM technologies are manipulated by encryption and watermark method separately. Therefore, original content is disclosed temporarily inside a system in the user's decryption [7]. In that case, users can save original contents without watermark information and distribute via network.

1.2 Our Contributions

In this paper, we focus on a strategy for the construction of software based DRM system that is far more secure than existing software based DRM system. We describe a design and implementation of DRM technique based on multi-channel AES algorithm (MAA). Our method will deteriorate the quality of original contents to make trial contents for distribution to widely users via network. The quality of the trial contents will be controlled with a fingerprinted key at the incomplete decoding process, and the user information will be fingerprinted into the incomplete decoded contents simultaneously.

This paper is organized as follows. The related techniques are reviewed in Section 2. The implementation of DRM based on MAA is explained in Section 3. The experimental results with JPEG (Joint Photographic Experts Group) algorithm are given in Section 4 and the conclusion is summarized in Section 5.

2 Preliminaries

2.1 Incomplete Cryptography

The incomplete cryptography consists two steps: the incomplete encoding and the incomplete decoding (Fig. 1).

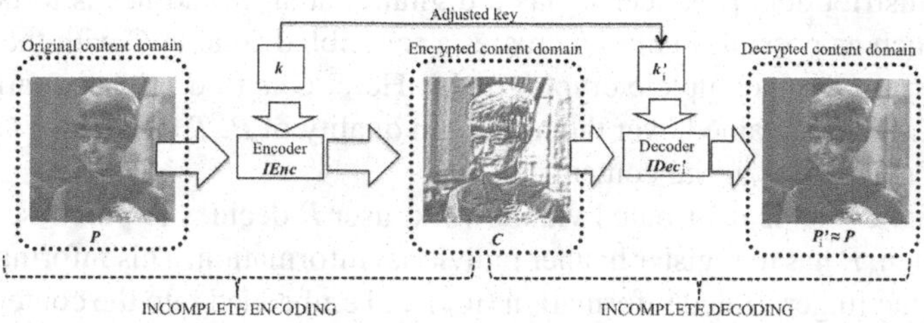

Fig. 1 Incomplete Cryptography

In the incomplete encoding process, content P is encoded based on the incomplete encoder function $IEnc$ with encoder key k to make the scrambled content C (trial content).

$$C = IEnc(k,P) \tag{1}$$

Here, C can be simply recognized as a part of P (even if C is not decoded). This feature is called *incomplete confidentiality*.

On the other hand, the incomplete decoding process is different from the complete decoding process. C is decoded by using another incomplete decryption function $IDec'_i \neq D$ and a decoded key $k'_i \neq k$. Key k'_i is adjusted based on k to allow content owner making i decoded contents P'_i.

$$P'_i = IDec'_i(k'_i, C) \tag{2}$$

Since P'_i is decoded by another decryption function $IDec'_i$ with key k'_i, it will be deferent from original content P. Therefore, the relationship of P and P'_i is $P'_i \neq P$ in incomplete cryptography system. This feature is called *incomplete decode*. Note that, D and k is the complete decryption function and complete decoded key, respectively. If C is decoded by D with k, original content P will be obtained.

The main contribution of incomplete cryptography is that the quality of P'_i can be controlled with a particular key k'_i. And when C is decoded with k'_i, P'_i is not only decoded with slight distortion, but also fingerprinted with the individual user information that is used as fingerprinting information. It is the elemental mechanism of fingerprinting based on the incomplete cryptography system.

2.2 DRM System Based on Incomplete Cryptography

The idea of the DRM system based on the incomplete cryptography is presented in this subsection. A DRM system requires to enable the distribution of original contents safely and smoothly, as well as to enable the secondary use of contents under rightful consents. When a DRM system is constructed using the incomplete cryptography to implement a content distribution system, it is not only the safety distribution method to users, but also the solution of the conventional DRM problem.

Before distribution, producer T has a digital content P and needs to be sent to users as much as possible. Thus, T creates a scrambled content C with the encoder key k based on the incomplete cryptography. Here, C is to disclose a part of P. It means that C is maintained over the minimum quality of P. T distributes C to users widely via network as a trial content.

After trial C that is distributed via network, user R decides to purchase a digital content. Then, R has to register his/her individual information. This information will be used as the fingerprinted information (w_m) and embedded into the content. When T receives the purchaser's agreement, T sends a fingerprinted key k_i' to the user R. k_i' is the incomplete decoding key and it is prepared individually to each user based on w_m.

R decodes C using k_i' and obtains the high quality content P_i'. In this decoding process, ID information (w_m) of user will be fingerprinted in P_i' as the copyright information.

Therefore, when a producer wishes to check whether the users is a legal user, he/she can extract the fingerprinting information from P_i' and compare with his user database. If the fingerprinting information matches his database, the user is a legal user. Conversely, if the fingerprinting information is a different from his database, the user is an illegal user. Furthermore, it can specify to trace the source of pirated copies. The purpose of this proposed method is to informs the producer about the existence of fingerprinting which can exactly identify users, and limit the illegal redistribution in advance.

3 The Proposed MAA Algorithm

The main idea of the proposed method is that utilizes the complete cryptography to implement the incomplete cryptography. Suppose that the complete cryptography $\mathbb{E}$ is selected and has the encryption function *CEnc* and the decryption function *CDec* with secret share key $\{k_1, k_2\}$. Here, we present the algorithm to implement the incomplete cryptography using $\mathbb{E}$. As shown in Fig. 2, the proposed MAA method consists of two processes: the incomplete encoding, the incomplete decoding.

3.1 Incomplete Encoding

The incomplete encoding is shown in Fig. 2(a). This process is implemented in the producer side by T. There are four steps in this process.

Step 1. Extract the important elements P from the digital content and split P into n blocks as shown in Fig. 2(a). Moreover, block $p_0 \sim p_{n-1}$ are grouped to make three groups $\widehat{P_A}, \widehat{P_B}, \widehat{P_C}$, respectively. The importance of three groups can be expected that is $\widehat{P_A} < \widehat{P_B} < \widehat{P_C}$.

$$P = \{p_0, \cdots, p_{i-1}, \cdots, p_{i+j-1}, \cdots, p_{n-1}\} \Rightarrow P = \{\widehat{P_A}, \widehat{P_B}, \widehat{P_C}\} \quad (3)$$

Step 2. Extract $\widehat{P_A}$ and $\widehat{P_B}$ to encode a part of P.

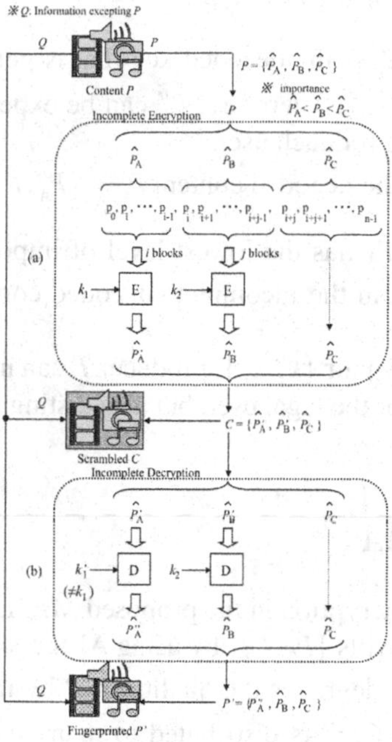

Fig. 2 Overview of MAA algorithm

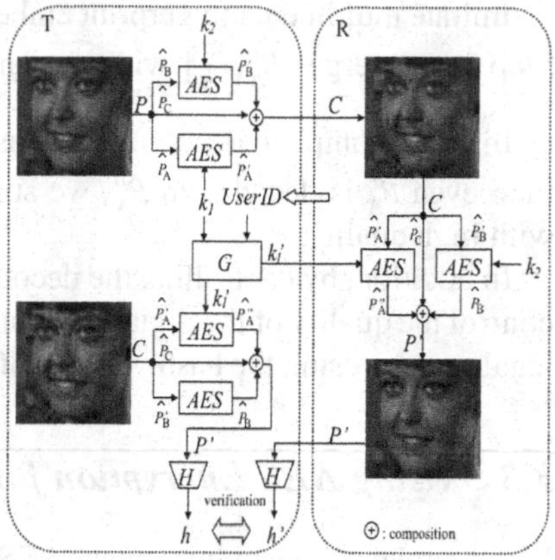

Fig. 3 Using AES for MAA algorithm

Step 3. Encode $\widehat{P_A}$ and $\widehat{P_B}$ by using the encoder function $CEnc$ with key pair $\{k_1,k_2\}$.

$$\widehat{P'_A} = CEnc(k_1,\widehat{P_A}); \quad \widehat{P'_B} = CEnc(k_2,\widehat{P_B}) \tag{4}$$

Step 4. Merge $\{\widehat{P'_A},\widehat{P'_B}\}$ with $\widehat{P_C}$ to make the scrambled content $C = \{\widehat{P'_A},\widehat{P'_B},\widehat{P_C}\}$.

In incomplete encoding, the most important group $\widehat{P_C}$ is not encoded to disclose P for users. It means that C can be simply recognized a part of P (even if C is not decoded). C is distributed widely to users as trial content via network.

3.2 Incomplete Decoding

Fig. 2(b) shows the detail of the incomplete decoding. This process is executed in the user side by R. Suppose that R had finished the purchasing process and the decoded key $\{k'_1,k_2\}$ is delivered by producer T. In order to obtain the high quality content, R uses the decoded key $\{k'_1,k_2\}$ with decoder function $CDec$ to decode the trial content C. This algorithm is shown as the following.

Step 1. Extract three groups $\{\widehat{P'_A},\widehat{P'_B},\widehat{P_C}\}$ from C.

Step 2. Decode $\{\widehat{P'_A},\widehat{P'_B}\}$ by using $IDec$ and $CDec$ with $\{k'_1,k_2\}$ to obtain $\{\widehat{P''_A},\widehat{P_B}\}$.

$$\widehat{P''_A} = IDec(k'_1, \widehat{P'_A}); \ \widehat{P_B} = CDec(k_2, \widehat{P'_B}) \quad (5)$$

Since $k'_1 \neq k_1$ then $\widehat{P''_A} \neq \widehat{P_A}$. Note that, k'_1 is the decoded key that is generated based on the individual user information w_m. Therefore, $\widehat{P''_A}$ can be expected to imitate multimedia fingerprint embedding for each user.

Step 3. Merge $\{\widehat{P''_A}, \widehat{P_B}\}$ with $\widehat{P_C}$ to make the decoded content $P' = \{\widehat{P''_A}, \widehat{P_B}, \widehat{P_C}\}$.

In the incomplete decoding, because of $\widehat{P_A}$ has the lowest level of importance, then even $\widehat{P'_A}$ is decoded to $\widehat{P''_A}$, we still obtain the incomplete decoded content P' with high quality.

In addition, by controlling the decoder key pair $\{k'_1, k_2\}$, producer T can not only control the quality of the digital content P' for the legal user, but also distinguish the legal user by using the hash value h_i of P'.

3.3 Using AES Encryption for MAA

Fig. 3 explains how to implement AES[8] encryption in the proposed MAA.

First, T extracts $\{\widehat{P_A}, \widehat{P_B}\}$ from P and encrypts $\{\widehat{P_A}, \widehat{P_B}\}$ by using AES encryption with key pair $\{k_1, k_2\}$ to obtain $\{\widehat{P'_A}, \widehat{P'_B}\}$. T degrades the quality of P by merging $\{\widehat{P'_A}, \widehat{P'_B}\}$ into $\widehat{P_C}$ to obtain the trial content C. C is distributed to many users via network.

After receiving the purchasing agreement of R, T creates the decoded key $\{k'_1, k_2\}$ based on the key generation function G with key k_1 and the UserID w_m. T sends decoded key $\{k'_1, k_2\}$ to R. In this paper, function G is bitwise XOR (eXclusive OR) and is described as $\oplus$. It means that $k'_1 = k_1 \oplus w_m$.

R uses $\{k'_1, k_2\}$ with AES decryption to decode $\{\widehat{P'_A}, \widehat{P'_B}\}$. Here, $\widehat{P'_A}$ is incomplete decoded, whereas $\widehat{P'_B}$ is complete decoded. After that, $\{\widehat{P''_A}, \widehat{P_B}\}$ will be replaced $\{\widehat{P'_A}, \widehat{P'_B}\}$ in C to obtain $P' = \{\widehat{P''_A}, \widehat{P_B}, \widehat{P_C}\}$ with high quality.

In our proposed MAA method, since k'_1 is created based on the UserID w_m and function G, it is clear that k'_1 is also changed according to UserID w_m. Therefore, the hash value h of P' can be used as the identity of the legal user.

4 Application on JPEG Algorithm

In this section, we explain the mechanism to create the scrambled content for the trial content and the incomplete decoded content. We implemented the fundamental method based on the standard JPEG (Joint Photographic Experts Group) algorithm[10].

4.1 MAA Using AES Algorithm in JPEG

Let P be the DCT table of the Y component or UV component that is extracted from JPEG image. $\{\widehat{P_A}, \widehat{P_B}\}$ are created by collecting some bits from DC coefficient of P.

$\widehat{P_C}$ is AC coefficients of P. In order to encode P, AES encryption[1] with ten rounds is employed for the implementation of MAA. The detail of MAA is explained as following.

4.1.1 Trial Content Creation

The trial content is obtained based on the incomplete encoding of MAA as following,

Step 1. Extract DC coefficient from DCT table and convert to binary array. As shown in Fig. 4, DC coefficient = "13" and its binary array is $\{0,0,0,0,1,1,0,1\}$.

Step 2. Let $\widehat{P_A}$ be i bits LSB (Least Significant Bit) of DC coefficient and $\widehat{P_B}$ be next continuos j bits. In example of Fig. 4, $i = 2$ and $j = 3$;

$$\widehat{P_A} = \{0,1\}; \widehat{P_B} = \{0,1,1\} \tag{6}$$

Step 3. Repeat Step 2 for collecting all $\{\widehat{P_A}, \widehat{P_B}\}$ to obtain 256-bit array $\{\widetilde{P_A}, \widetilde{P_B}\}$.

Step 4. Encrypt $\{\widetilde{P_A}, \widetilde{P_B}\}$ obtained in Step 3 by AES encryption with the encoded key $\{k_1, k_2\}$.

Step 5. Repeat Step 1 $\sim$ Step 4 until all DCT tables are encrypted. After encryption, the encoded bit array $\{\widetilde{P'_A}, \widetilde{P'_B}\}$ are collected.

Step 6. Split $\{\widetilde{P'_A}, \widetilde{P'_B}\}$ to make the encrypted i bits of $\widehat{P'_A}$ and j bits of $\widehat{P'_B}$, respectively. Then, restore $\widehat{P'_A}$ and $\widehat{P'_B}$ into the position of $\widehat{P_A}$ and $\widehat{P_B}$ to obtain the trial content C. In Fig. 4, since $i = 2$ and $j = 3$ then $\widehat{P'_A}$ and $\widehat{P'_B}$ became $\{0,0\}$ and $\{1,1,0\}$, respectively.

According to incomplete encoding, DC coefficient = "13" in Fig. 4 is encrypted to "24". Therefore, the trial content C is generated with low quality, then C can disclose P.

4.1.2 Fingerprinted Content Creation

In this process, user R uses the decoded key $\{k'_1, k_2\}$ to decode the trial content C. k'_1 is generated based on w_m that belongs to legal user identity. According to the AES decryption, R can decrypt $\{\widetilde{P'_A}, \widetilde{P'_B}\}$ by using $\{k'_1, k_2\}$. The detail of this process is shown in Fig. 5.

Step 1. Extract DC coefficient from DCT table of trial content C and convert to binary array. In example Fig. 5, DC coefficient = "24" and its binary array is $\{0,0,0,1,1,0,0,0\}$.

Step 2. Extract $\widehat{P'_A}$ from i bits LSB (Least Significant Bit) of DC coefficient and $\widehat{P'_B}$ from next continuos j bits. In our example, $i = 2$ and $j = 3$.

$$\widehat{P'_A} = \{0,0\}; \widehat{P'_B} = \{1,1,0\} \tag{7}$$

[1] `http://csrc.nist.gov/publications/fips/fips197/fips-197.pdf`

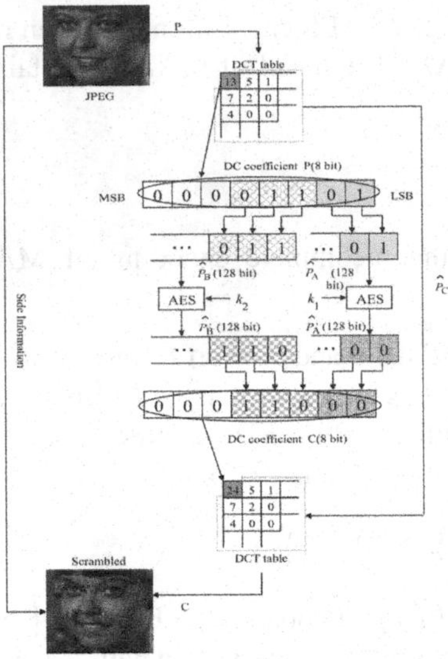

Fig. 4 The encoding a part of DC coeff

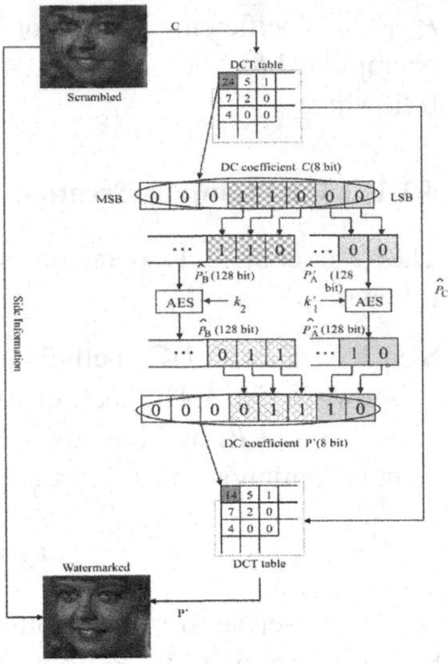

Fig. 5 The decoding a part of DC coeff

Step 3. As the incomplete encoding, repeat Step 2 for collecting all $\{\widehat{P'_A},\widehat{P'_B}\}$ to obtain 256-bit array $\{\widetilde{P'_A},\widetilde{P'_B}\}$.

Step 4. Decode the obtained $\{\widetilde{P'_A},\widetilde{P'_B}\}$ in Step 3 by AES decryption with encoded key $\{k'_1,k_2\}$. Here, since $k'_1 \neq k_1$ then $\widetilde{P''_A} \neq \widetilde{P_A}$.

Step 5. Repeat Step 1 $\sim$ Step 4 until all DCT tables are decrypted. After decryption, the decoded bit array $\{\widetilde{P''_A},\widetilde{P_B}\}$ are collected.

Step 6. Split $\{\widetilde{P''_A},\widetilde{P_B}\}$ to make the decrypted i bits of $\widehat{P''_A}$ and j bits of $\widehat{P_B}$, respectively. Then, restore $\widehat{P''_A}$ and $\widehat{P_B}$ into the position of $\widehat{P'_A}$ and $\widehat{P'_B}$ to obtain the fingerprinted content P'. In Fig. 5, since $i = 2$ and $j = 3$ then $\{\widehat{P''_A},\widehat{P_B}\}$ became $\{1,0\}$ and $\{0,1,1\}$, respectively.

According to the incomplete decoding, DC coefficient = "24" in Fig. 5 is encrypted to "14". It is clear that two LSBs ($\widehat{P_A}$) of each DC coefficient are changed according to the user identity w_m. Namely, producer T can decide the decoded key pair $\{k'_1,k_2\}$ to control the decoded content P' based on user identity w_m. Therefore, when a producer verifies the legal user of content P, he/she can extract the hash information h from P' to detect the legal user. In this paper, we use 256-bit hash value for legal user verification.

4.2 Implementation and Evaluation

4.2.1 Experimental Environment

All experiments were performed by incomplete encoding and incomplete decoding on JPEG images using the Vine Linux 3.2 system. In order to generate the encryption key k_1, k_2, k_1' of AES encryption, we used function $rand()$ of GCC version 3.3.2[2] with $seed = 1$. Additionally, the ImageMagick version 6.6.3-0[3] was used to convert and view the experimental JPEG images. The encryption keys are employed in our experiments as following,

$$k_1 = \{\texttt{53564df05d23565c31153e1b5a3b3f1a}\}$$
$$k_2 = \{\texttt{5a13584b3d62404d2d1b2a4f315d2531}\}$$
$$k_1' = \{\texttt{53564df05d23565c31153e1b5a3b3f1b}\}$$

Note that, we suppose $w_m = 1$ and $k_1' = k_1 \oplus w_m$.

4.2.2 Experimental Image

The four test images are the 8-bit RGB images of SIDBA (Standard Image Data BAse) international standard image (Girl, Airplane, Parrots, Couple) of size 256×256 pixels. Here, all images were compressed with quality 75 (the lowest $0 \leftrightarrow 100$ the highest) to make experimental JPEG images for evaluation of the proposal method. We also randomly generated the w_m to assign as the userID and the decoded key k_1' is created based on w_m.

4.2.3 Evaluation of Image Quality

We used PSNR (Peak Signal to Noise Ratio) [11] to evaluate the JPEG image quality.

The PSNR of $M \times N$ pixels images of $g(i,j)$ and $g'(i,j)$ is calculated with

$$PSNR = 20 \log \frac{255}{MSE} \quad [\text{dB}] \tag{8}$$

$$MSE = \sqrt{\frac{1}{MN} \sum_{i=0}^{M-1} \sum_{j=0}^{N-1} \{g(i,j) - g'(i,j)\}^2}$$

$(MSE : Mean\ Square\ Error)$.

In these experiments, PSNRs were calculated with RGB pixel data of original image and the JPEG image. A typical value for PSNR in a JPEG image (quality 75) is about 30dB [11].

4.2.4 Results and Analysis

Here, we present some results concerning incomplete encoding, incomplete decoding on JPEG images and discussion the power of MAA method.

[2] `http://gcc.gnu.org/`

[3] `http://www.imagemagick.org/script/`

Table 1 PSNR[dB] of experimental images

Method		P	C	P'	$h[bits]$
$AES_{1,3}$	Airplane	30.20	24.74	30.06	256
	Girl	32.71	25.99	32.51	256
	Parrots	34.25	26.29	33.97	256
	Couple	34.06	26.26	33.82	256
$AES_{2,3}$	Airplane	30.20	19.40	29.73	256
	Girl	32.71	20.91	31.83	256
	Parrots	34.25	20.60	33.13	256
	Couple	34.06	20.82	32.99	256
$AES_{3,3}$	Airplane	30.20	13.31	28.28	256
	Girl	32.71	16.87	30.12	256
	Parrots	34.25	14.09	30.55	256
	Couple	34.06	18.51	30.71	256

Let $AES_{i,j}$ indicate the algorithm of MAA in which i and j bits are extracted from DC coefficient and they are assigned to $\widehat{P_A}$ and $\widehat{P_B}$. For instance, in the example of Fig. 4 and Fig. 5, we can indicate MAA as $AES_{2,3}$. In this paper, we implemented three methods $AES_{1,3}$, $AES_{2,3}$ and $AES_{3,3}$ based on YUV component. Results and analysis

The experimental results of PSNR values are shown in Table. 1. In these experiments, we encrypted $4 \sim 6$ bits to create the trial content C, and used $1 \sim 3$ bits to fingerprint the legal user in the decoded content P'. As the results of $AES_{3,3}$ method, the PSNR values of the incomplete encoded images C are around 15dB. We recognized that the images C of $AES_{3,3}$ are not suitable for the trial image. However, the incomplete encoded images C of $AES_{1,3}$ and $AES_{2,3}$ are very suitable for the trial image. Additionally, Fig. 6 shows the relationship of i and j bits with the PSNR value of image. According these results, we could understand that in order to make the appropriate trial image, total bits $(i+j)$ can be encrypted from 2 to 5 bits (see Fig. 6(a) to confirm the PSNR: 20dB–25dB); and in order to obtain the appropriate fingerprinted image, only i bits can be incompletely decrypted from 1 to 3 bits (see Fig. 6(b) to confirm the PSNR over than 30dB). And from these results, it is clear that $AES_{1,3}$ method gave us the best results in our experiments.

Next, for confirming the effect of Y component and UV component separately, we implemented $AES_{1,3}$ method based on Y component, UV component, and YUV component, respectively. The experimental results are shown in Table. 2. According to the Table. 2, we confirmed that when we manipulated the UV component, the image deterioration was extremely more conspicuous than that when we applied to the Y component. Therefore, we can make the trial content (scrambled content) efficiently with drawing up on least UV component. However, because the image deterioration is not conspicuous when implementing Y component, there is an advantage to incompletely decode i bits into the decoded content under the maintaining its quality.

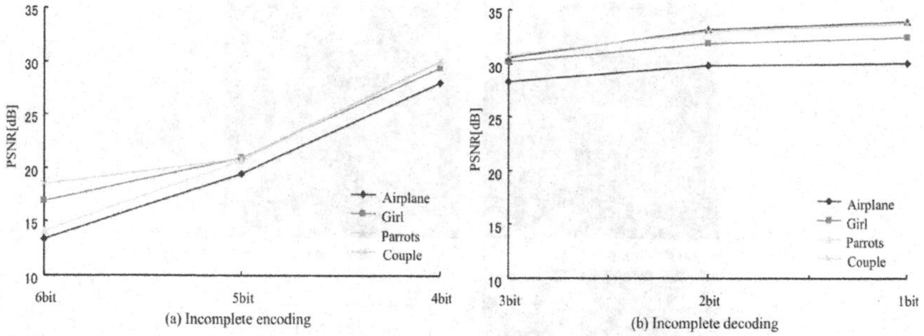

Fig. 6 The PSNR of experimental JPEG images

Table 2 PSNR[dB] of $AES_{1,3}$ method

Method		P	C	P'	$h[bits]$
$AES_{1,3}$, UV comp.	Airplane	30.20	25.79	30.10	256
	Girl	32.71	27.15	32.56	256
	Parrots	34.25	27.67	34.06	256
	Couple	34.06	27.58	33.91	256
$AES_{1,3}$ Y comp.	Airplane	30.20	27.90	30.16	256
	Girl	32.71	29.34	32.65	256
	Parrots	34.25	29.91	34.17	256
	Couple	34.06	29.98	33.97	256
$AES_{1,3}$ YUV comp.	Airplane	30.20	24.74	30.06	256
	Girl	32.71	25.99	32.51	256
	Parrots	34.25	26.29	33.97	256
	Couple	34.06	26.26	33.82	256

Fig. 7 is an experimental sample of Girl images in the UV component. According to the results in Fig. 7, it is possible to produce the trial content (see Fig. 7(b)) and the incomplete decoded content (see Fig. 7(c)) based on MAA. Furthermore, the hash value can be extracted as Fig. 7(d). It is easy to confirm that we can not distinguish the original image and the fingerprinted image. This implied that the proposed MAA method achieved image transparence.

In order to decode the trial content, producer needs to send individual k_1' to each user for incompletely decoding i bits of LSB. Based on k_1', i of LSB in the decoded image P' are randomly changed. Therefore, the hash value $h(i)$ of each P_i can be employed for identification the legal user. $h(i)$ is also saved into the producer's database for comparing with $h'(i)$ of the suspected image. In our experiment, since we used 256-bit hash value to distinguish the legal user, then our system can distinguish 2^{256} users. Here, we tried to decode the trial Girl image with 5 decoded key $k_1'(k_1'(1) \sim k_1'(5))$ and extract 5 hash values $h(1) \sim h(5)$. We obtained the following hash values

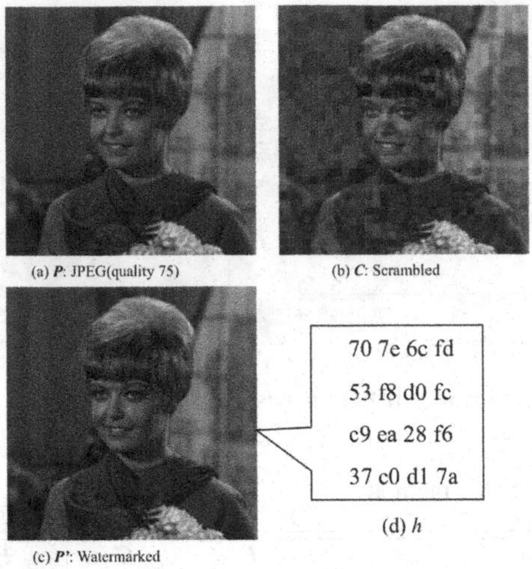

Fig. 7 An example Girl image (UV component)

$$
\begin{aligned}
h(1) &= \{\texttt{b20460b286726d1903e2ff3eba4bd677}\} \\
h(2) &= \{\texttt{315f681571049c47095f5a0322208ffd}\} \\
h(3) &= \{\texttt{dce8d9da05971e319366db5ca87e2461}\} \\
h(4) &= \{\texttt{2e6c1ca5b1439e62e4c962e3aa033770}\} \\
h(5) &= \{\texttt{f97d0109646ded9b7af798d0a24bcf53}\}
\end{aligned}
$$

Obviously, every hash values are unique. Therefore, hash value of P' can be used as the legal user's identification.

According to the above results, we have established the DRM system based on the proposed MAA method. Trial content is created to disclose the original content and distributed widely to users. In the incomplete decode process, we changed the i bits in LSB of the quantized DCT coefficient itself by a devised decryption key. Thus, the original content is not decoded temporarily inside the system. Therefore, we conclude that the above technical problem by the conventional DRM system is solved by using the incomplete cryptography system.

5 Conclusion

In this paper, we have presented a scheme of digital content distribution system based on multi-channel AES algorithm (MAA). This approach integrates the encoding process and fingerprinting progress of DRM technology. Therefore, we can eliminate the problem of the present DRM technology and manage the legal user effectively.

One of the lessons learned from this paper is that in order to make the scrambled image and the incomplete decoded image for JPEG, it is possible to process the Y component and UV component flexibly. Also, another lesson is that we can control the incomplete decoded image quality using a specialized key individually. Subsequently, the hash value is extracted from the fingerprinted image using this approach for distinguishing the legal user. The fingerprinted images are in good visual quality and have high PSNR values. The effectiveness of the proposed scheme has been demonstrated with the aid of experimental results. Therefore, we conclude that proposed MAA method is useful for the rights management technology in illegal content distribution via network.

References

1. Shapiro, W., Vingralek, R.: How to manage persistent state in DRM systems. In: Sander, T. (ed.) DRM 2001. LNCS, vol. 2320, pp. 176–191. Springer, Heidelberg (2002)
2. Kiayias, A., Yung, M.: Breaking and repairing asymmetric public-key traitor tracing. In: Feigenbaum, J. (ed.) DRM 2002. LNCS, vol. 2696, pp. 32–50. Springer, Heidelberg (2003)
3. Serrao, C., Naves, D., Barker, T., Balestri, M., Kudumakis, P.: Open SDRM? an open and secure digital rights management solution (2003)
4. Chang, H., Atallah, M.J.: Protecting Software Code by Guards. In: DRM f01: ACM CCS-8 Workshop on Security and Privacy in Digital Rights Management, pp. 160–175 (2002)
5. Emmanuel, S., Kankanhalli, M.S.: A Digital Rights Management Scheme for Broadcast Video. Multimedia System 8, 444–458 (2003)
6. Seki, A., Kameyama, W.: A Proposal on Open DRM System Coping with Both Benefits of Rights-Holders and Users. In: IEEE Conf. on Image Proceedings, vol. 7, pp. 4111–4115 (2003)
7. Lin, C., Prangjarote, P., Kang, L., Huang, W., Chen, T.-H.: Joint fingerprinting and decryption with noise-resistant for vector quantization images. Journal of Signal Processing 92(9), 2159–2171 (2012)
8. Federal Information Processing Standards Publication: Announcing the Advanced Encryption Standard (AES), FIPS (2001)
9. Tachibana, T., Fujiyoshi, M., Kiya, H.: A method for lossless watermarking using hash function and its application for broadcast monitoring. IEICE Technical Report, ITS 103(640) 103(643), 13–18 (2004) (in Japanese)
10. T.84, Digital Compression and Coding of Continuous-tone still Images - Requirements and Guidelines. International Telecommunication Union (1992)
11. Iwakiri, M., Thanh, T.M.: Fundamental Incomplete Cryptography Method to Digital Rights Management Based on JPEG Lossy Compression. In: The 26th IEEE International Conf. on AINA, pp. 755–762 (2012)

Enhance Matching Web Service Security Policies with Semantic

Tuan-Dung Cao and Nguyen-Ban Tran

Abstract. Web service security policy is a way to add some security restrictions to a web service. Matching web service security policies is hard and important to integrate web services effectively. However, the lack of semantics in WS-SecurityPolicy (WS-SP) hampers the effectiveness of matching the compatibility between security policies of different web services. To enhance matching web service security policies, we propose a semantic approach for specifying and matching the security policies. The approach uses the transformation of WS-SP into an OWL-DL ontology, the definition of a set of semantic relations that can exist between the provider and requestor security concepts, and the algorithm to determine the match level of the provider and requestor security policies. We show how these relations and the matching algorithm lead to more correct and more flexible matching of security policies.

1 Introduction

Nowadays, web service becomes a popular standard in information technology industry. To use the existing web services effectively, we need to integrate them by some modern technology and architecture like Service Oriented Architecture (SOA). Discovering a dynamic service and service selection are essential aspects of SOA. To acquire the business requirement, selecting web service must not only take the functional aspects, but also non-functional properties of the services. One of the

Tuan-Dung Cao
School of Information and Communication Technology, Hanoi University of Science and Technology, Vietnam
e-mail: dungct@soict.hut.edu.vn

Nguyen-Ban Tran
Vietnam Joint Stock Commercial Bank for Industry and Trade, Hanoi, Vietnam
e-mail: nguyenbantran@gmail.com

V.-N. Huynh et al. (eds.), *Knowledge and Systems Engineering, Volume 1*, Advances in Intelligent Systems and Computing 244,
DOI: 10.1007/978-3-319-02741-8_19,

most important non-functional properties of a web service is security. This paper focuses on web service security policy, a non-functional property of web service.

In a big company or organization, SOA is an effective way to provide and integrate its information technology services. Web service is the most adopted implementation of SOA. Message security becomes a major concern when using Web services. Therefore, message security becomes a major problem when we use SOA. Message security mainly means the confidentiality and the integrity of data transmitted through the message. Confidentiality and integrity can be assured by applying security mechanisms such as encryption and digital signature.

WS-SP [1] is widely accepted in the industry and it is currently a popular standard to be aggregated into the Web service architecture. Then matching WS-SP problem becomes more and more important while integrating Web services. However, WS-SP has a big weakness: it only allows syntactic matching of security policies. In fact, security policy matching depends on the policy intersection mechanism provided by WS-Policy [2]. The main step in this mechanism is matching the assertions specified in the service provider and requestor policies. This step only uses syntactic comparison between the security assertions, and does not use the semantics message security. Syntactic matching of security assertions restricts the effectiveness of checking the compatibility between provider and requestor policies. A simple comparison of the syntactic descriptions of security assertions maybe gets fault negative results. For example, consider syntactically different security assertions with the same meaning. Syntactic matching of security assertions only has a strict yes/no matching result. A more flexible matching is needed in order to consider subtle differences that may exist between security assertions and not bypass a potential partnership between a service requestor and a service provider. For example, in the cases when the provider and requestor security assertions have the same type but have some different proper-ties that make the provider assertion stronger, in security concepts, than the requestor assertion.

In the work [6], the authors proposed a semantic way to compare two security assertions, but they didn't compare two all security policies. The goal of this paper is to propose a semantic security approach a matching WS-SP algorithm to compare two security policies. In our approach, an ontology defined in Web Ontology Language (OWL) is used to present Web service security policies. Policy specifications offer semantic information about the requestor and provider. This information can be used to determine policy compatibility and to guarantee interoperability among requestor and provider service, with respect to security aspects

There are some ways to use semantic approach for working with security policy. You can use semantic to work with web service security configurations [7] or to man-age web service security [8]. We use semantic to get the matching level of two web service security policies .

We used a created prototype for the semantic matching of security assertions then add some relations to extend this completion to get the matching level of WS-SP. Based on the semantic interpretation of the syntactic heterogeneities that may occur between a provider assertion and a requestor assertion, our approach doesn't produce fault negative results and thus supports more correct matching. Besides, it allows introducing close match and possible match as intermediary matching degrees, which makes security assertion matching more flexible.

The remainder of this paper is organized as follows. Section 2 and 3 introduce about WS-SP and the problem : matching WS-SP. Section 4 presents the ontology and relations in this ontology with semantics. Section 5 presents our algorithm of semantics matching security policies with two cases: a simple case and complex cases. Section 6 discusses about related work. Section 7 concludes the paper with some discussion.

2 Web Service Security Policy

WS-SP is a web services specification, which is created by IBM and used to extend the fundamental security protocols specified by the *WS-Security*, *WS-Trust* and *WS-SecureConversation* by offering mechanisms to represent the capabilities and requirements of web services as policies. Security policy assertions are based on the WS-Policy framework. Policy assertions can be used to require more generic security attributes like transport layer security *¡TransportBinding¿*, message level security *¡AsymmetricBinding¿* or timestamps, and specific attributes like token types. Policy assertions can be divided in following categories: Protection assertions identify the elements of a message that are required to be signed, encrypted or existent; Token assertions specify allowed token formats (SAML, X509, Username etc.); Security binding assertions control basic security safeguards like transport and message level security, cryptographic algorithm suite and required timestamps; Supporting token assertions add functions like user sign-on using a username token.

Policies can be used to drive development tools to generate code with certain capabilities, or may be used at runtime to negotiate the security aspects of web service communication. Policies may be attached to WSDL elements such as service, port, operation and message, as defined in WS Policy Attachment. WS-SP standard

```
<sp:IssuedToken>
<sp:RequestSecurityTokenTemplate>
<wst:TokenType>...#SAMLV2.0</wst:TokenType>
</sp:RequestSecurityTokenTemplate>
</sp:IssuedToken>
```

Fig. 1 An example of WS-SP

defines a set of security policy assertions for use with the WS-Policy framework. These assertions describe how messages are secured according to the WS-Security protocol. Typically, the published policies are compliant with the WS-Policy normal form. WS-SP assertions mainly describe the token types for security tokens, the cryptographic algorithms, and the scope of protection which means the parts of the SOAP message that shall be encrypted or signed.

3 Matching Web Service Security Policies Problem

When we have many web services and want to integrate them, we have to deal with the compatibility of their security policies. It leads to a problem: how to know whether two different security policies match or not.

Syntactic matching of security assertions restricts the effectiveness of checking the compatibility between them. In order to illustrate this deficiency, suppose that a requestor is looking for a fund transfer Web service that supports the signature of the message body with a symmetric key securely transported using an X509 token. Besides, the necessary cryptographic operations must be performed using Basic256 algorithm suite. This could be formalized, based on the WS-SP standard, by adding the assertions reqAss1 and reqAss2 to the requestor security policy (SP).

Furthermore, suppose that the requestor finds a Web service that fund transfer and whose SP includes provAss1 and provAss2 assertions.

reqAss1:

```
<sp:SignedElements>
<sp:XPath>
/S:Evelope/S:Body>
</sp:SignedElements>
```

reqAss2:

```
<sp:SymetricBinding>
<sp:ProtectionToken>
<sp:X509Token/>
</sp:ProtectionToken>
<sp:AlgoritmSuite>
<sp:Basic128/>
</sp:AlgoritmSuite>
<sp:SymmetricBinding>
```

provAss1:

```
<sp:SignedParts>
<sp:Body/>
</sp:SignedParts>
```

provAss2:

```
<sp:SymetricBinding>
<sp:ProtectionToken>
<sp:X509Token/>
</sp:ProtectionToken>
<sp:AlgoritmSuite>
<sp:Basic128/>
</sp:AlgoritmSuite>
<sp:IncludeTimestamp/>
<sp:SymmetricBinding>
```

Fig. 2 Example of requestor and provider assertions

It is clear that the assertions specified in the provider SP are syntactically different than those specified in the requestor SP. Syntactic matching will then produce a no match result for these assertions. However, using semantic approach, we get a different matching result. In the above scenario reqAss1 and provAss1 assertions have the same meaning: sign the body of the message. Therefore, matching these two assertions must lead to a perfect match rather than no match. Besides, the only difference between reqAss2 and provAss2 assertions is that the *SymmetricBinding* assertion specified in the provider SP contains an extra child element which is sp:IncludeTimestamp. From security perspective, this strengthens the integrity service ensured by the message signature and makes provAss2 assertion stronger than reqAss2 assertion. Although it is not a perfect match, it is better than no match case.

From this analyzing, if the requestor can strengthen his security assertion by the inclusion of a timestamp, the perfect compatibility between both assertions will be ensured. We consider that it is more flexible to decide a possible match for this case in order to not reject a potential partnership between the service requestor and provider. Therefore, these two policies are possible match.

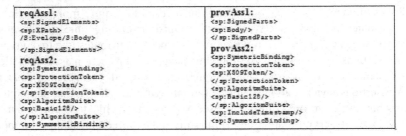

4 Semantic Matching of Web Service Security Assertions

In this section, we will show how to use semantic approach to transform a web ser-vice security policy to a form of ontology.

4.1 WS-SP Ontology

The assertions defined in WS-SP must be augmented with semantic information in order to enable semantic matching. Because this current version of WS-SP doesn't have any semantics presentation, the authors of [6] base on WS-SP and redesign an ontological representation of its assertions in order to obtain a WS-SP-based ontology that can be augmented with new semantic relations. A graphical depiction of the main parts of the ontology is shown in Fig. 3. Because some of these classes aren't designed well, we will use this main part of ontology and propose a way to extend some properties to ensure that this ontology will be used to get the matching level of two policies, not only two assertions. Web service security assertions are specified within security policies of the service provider and requestor. Typically, the structure of these policies is compliant with the WS-Policy normal form. In the normal form, a policy is a collection of alternatives, and an alternative is a collection of assertions. It is in the assertion components that a policy is specialized. Fig. 3 shows

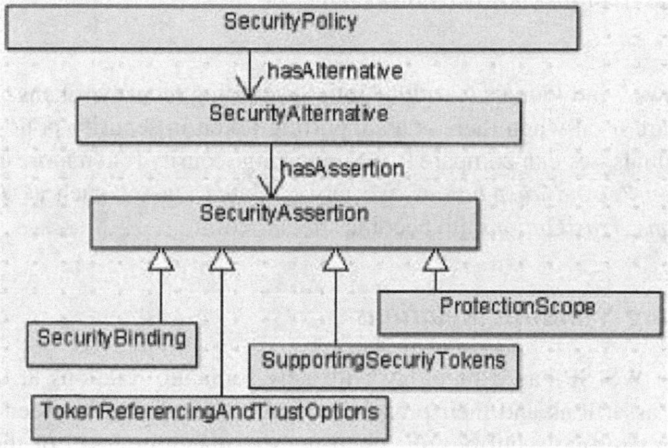

Fig. 3 Main classes of WS-SP-based ontology (from [6])

the main classes of the WS-SP-based ontology. This ontology includes the three main classes: SecurityPolicy, SecurityAlternative and SecurityAssertion in order to present security assertions within security policies. The SecurityAssertion have 4 subclasses: SecurityBinding, ProtectionScope, SupportingSecurityTokens, Token-ReferencingAndTrustOptions. The first two classes are detail in reference [6]. We will add some individuals to SupportingSecurityToken to improve its semantic.

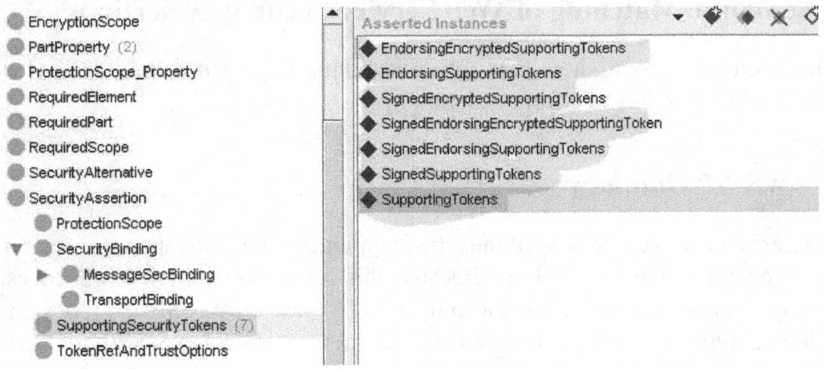

Fig. 4 Add individuals into SupportingSeuciryTokens

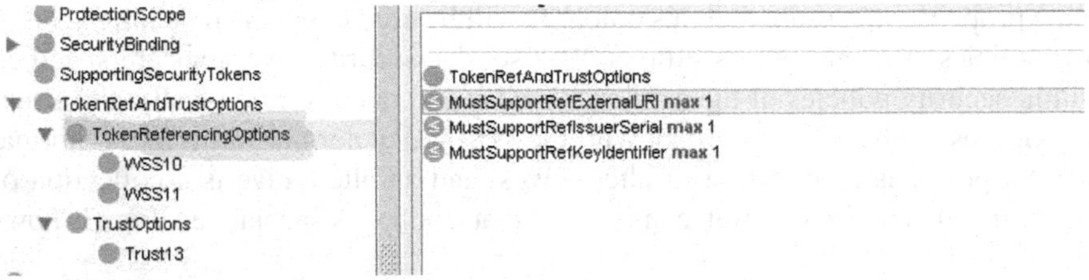

Fig. 5 Subclasses of *TokenRefAndTrustOptions* class

Fig.4 shows 7 individuals we added into SupportingSecurityTokens class. These individuals are used when there is a supporting token in security policy. By using these individuals, we can compare two SupportingSecurityToken more conrectly.

With *TokenRefAndTrustOptions*, we move related classes such as *TokenReferencingOptions, TrustOptions,* to become subclass of it.

4.2 Adding Semantic Relations

We used the WS-SP-based ontology with new semantic relations at the level of the security assertions and their properties. From the relation between assertions, we will compare two security policies to get the matching level of them. In this subsection, we epitomize all relations which are used. After we have the matching level of all assertions, we can get the matching level of a requestor policy with a provider policy.

To specify a relation between two assertions, we need to transform it into an ontology form. Subject to the security concepts, we will get the matching result, take security concepts of *ProtectionScope, SecurityBinding*, and *AlgorithmSuite*, for example. Next, we present the semantic relations.

isIdenticalTo relation: This relation allows to specify that a security concept specified by the provider is identical to a security concept specified by the requestor (no syntactic heterogeneity exists between them). There are two main cases for this relation in Table 1.

Table 1 isIdenticalTo relation

Case	isIdenticalTo
Properties of ProtectionScope, Security-Binding	- If two simple properties point to the same security concept and have equal values. - Two complex properties point to the same security concept and all their child properties is identical.
Assertions of ProtectionScope, Security-Binding	Point to the same security concept and all their properties are identical.

isEquivalentTo relation: This relation allows to specify that a security concept specified by the provider is equivalent to a security concept specified by the requestor. There are two main cases of security connects in Table 2.

Table 2 isEquivalentTo relation

Case	isEquivalentTo
Encrypted part and encrypted element or signed part and a signed element	XPath expression of message part is equivalent to the value of element. Example: '/S:Envelope/S:Body' and 'Body'.
Two algorithms suites	Algorithm suites defined in WS-SP are not disjoint and two algorithm suites can include many common elements. Example: Basic256 and Basic256Sha256.

isLargerThan relation. This relation only concerns the protection scope concept in SOAP message (in Table 3). *isSmallerThan* relation. This relation is to specify that the protection scope (in the SOAP messages) specified by the provider is smaller than the protection scope specified by the requestor. It is the opposite of *isLargerThan* relation and occurs in three cases just in an opposite manner to *isLargerThan* relation.

Table 3 isLargerThan relation

Case	isLargerThan
Part of SOAP message or XML elements	A part of message or a XML element is larger than any elements that belongs to it.
Encryption scopes, signature scopes, and required scopes	If a scope is larger than other scope.
Protection scope	If a scope has at least one property that larger than a property of other scope, and different scopes are identical or equivalent.

isStrongerThan relation. This relation is to specify that a security concept specified by the provider is stronger, in the security perspective, than a security concept specified by the requestor. There are three main cases for this relation (in Table 4).

isWeakerThan relation. This relation is to specify that a security concept specified by the provider is weaker, in the security perspective, than a security concept specified by the requestor. It is the opposite of *isStrongerThan* relation and occurs in three cases just in an opposite manner to *isStrongerThan* relation.

Table 4 isStrongerThan relation

Case	isStrongerThan
Security binding properties	Have influence on security strength, then we have *isStrongerThan* relation at the level of these properties.
Security binding assertion	If they point to the same type of security binding, and A provider assertion has at least one properties is stronger than the corresponding property of requestor assertion.
Protection scopes	If the provider scope has at least one extra scope, and the other scopes are identical, equivalent or have isLarger than relations.

hasTechDiffWith relation. In addition to concepts that allow to specify how a SOAP message is to be secured (confidentiality, integrity,), WS-SP-based ontology also includes concepts to describe technical aspects concerning how adding and referencing the security features in the message. At the level of these technical concepts, we define *hasTechDiffWith* relation to state that any mismatch between the provider concept properties and the requestor concept properties must be considered as a technical mismatch rather than a security level mismatch. *isMoreSpecificThan* relation. According to WS-SP standard, many security properties are optional to specify in a SP and WS-SP doesn't attribute default values for them. Therefore we define *isMoreSpecificThan* relation that occurs when a security concept specified by the provider is more specific (i.e., described in more detail) than a security concept specified by the requestor.

isMoreGeneralThan relation. This relation occurs when a security concept specified by the provider is more general (i.e., described in less detail) than a security concept specified by the requestor. It is the opposite of *isMoreSpecificThan* relation.

isDifferentFrom relation. This relation occurs when the security concepts specified by the requestor and the provider are semantically disparate.

5 Algorithm of Semantic Matching Web Service Security Policies

In this section, we propose an algorithm for matching provider and requestor security policies in a simple policy case and complex policies cases. This matching algorithm uses the matching level of assertions [6], which will be compared through the way mentioned below. After comparing two assertions, we will compare two security policies. There are four matching level: perfect match, close match, possible match, and no match in decreasing order of matching

5.1 Compare Two Simple Assertions

Simple assertion is a security assertion which contains only one alternative. To compare two simple assertions, we need to transform them into ontology form which bases on WS-SP ontology [6]. We will compare two assertions in ontology form, we will have many relations between them. We consider these relations to get the matching level between two assertions.

Perfect match. A perfect match occurs when *provAss* and *reqAss* are connected through *isIdenticalTo* or *isEquivalentTo* relations.

Close match. A close match occurs when *provAss* and reqAss are connected through *isMoreSpecificThan* relation. Possible match. A possible match is decided in three main cases:

Case 1. *provAss* and *reqAss* are connected through *isMoreGeneralThan* relation. This means that the information available cannot ensure that provAss can perfectly match reqAss. We assume that a potential partnership between the requestor and the provider can take place if the requestor can obtain additional information or negotiate with the provider.
Case 2. *provAss* is connected to *reqAss* through *isLargerThan, isStrongerThan*, or *hasTechDiffWith* relations. This means that the incompatibility between the two assertions doesn't negatively affect the security services and levels required in *reqAss*. We assume that a potential partnership between the requestor and the provider can take place if the requestor can strengthen his policy assertion or change some technical properties of his assertion.
Case 3. *provAss* and *reqAss* point to the same security concept and have at least one *isMoreGeneralThan, isLargerThan, isStrongerThan*, or *hasTechDiffWith* relation at the level of their properties, and their remaining properties are linked with semantic relations of type *isIdenticalTo, isEquivalentTo*, or *isMoreSpecificThan*. For example, suppose that *provAss* and *reqAss* point to the *SymmetricBinding* concept, but *provAss* has a protection token that is more general than the protection token specified in *reqAss*. In addition, *provAss* has an algorithm suite that is identical to the algorithm suite specified in *reqAss*. And finally, *provAss* has a *Timestamp* property that is stronger than the *Timestamp* property of *reqAss*. The two assertions have two heterogeneities that don't rule out the possibility of a match, so it is a possible match case.

No match. No match is decided in cases: *provAss* and *reqAss* are connected through *isDifferentFrom, isSmallerThan,* or *isWeakerThan* relations; *provAss* and *reqAss* point to the same security concept, and they have at least one *isDifferentFrom, isSmallerThan,* or *isWeakerThan* relation at the level of their properties.

In this subsection, we present the way to compare two simple assertion, so we create a function named $Compare\_Simple(A, B)$ to compare two simple assertion with the return values are: no match, possible match, close match, and perfect match. This result is sorted by increasing of value.

5.2 *Compare Two Complex Assertions*

A complex assertion has more than one alternative. In this subsection, we will proposed an algorithm to compare two complex assertions. We have a requestor A and provider B; assertions are reqA and provB. The matching level between reqA and provB is defined as the highest match of any alternative of provB with any alternative of reqA. For example, reqA includes alternatives: reqA1 and reqA2, provB includes alternatives: provB1 and provB2, with reqA1, reqA2, provB1, provB2 are simple assertions. We assume that provB1 and reqA1 have relation isIdenticalTo, and provB2 and reqA2 have relation *isDifferentFrom*. So, reqA and provB have the matching level *isIdenticalTo*.

The algorithm: Let A, B are two complex assertions. Assertion A contains alternatives: A1, A2, , An and assertion B contains alternatives: B1, B2, , Bm. Because A, B are complex assertions, so Ai or Bj maybe a complex assertion.

We present this algorithm in pseudocode:

Compare_Assertion(Assertion A, Assertion B) {
If (A is a simple assertion and B is a simple assertion) then return Compare_Simple(A, B).
/ A contains alternatives: A1, A2, , An */*
/ B contains alternatives: B1, B2, ., Bm */*
Return (Max(Compare_Assertion(Ai, Bj) with 0 <i <n+1 and 0 <j <m+1)).
}

This algorithm can compare a complex assertion with a simple assertion. In this case, a simple assertion X have one alternative is X itself.

5.3 *Compare Two Security Policies*

In above subsections, we compared two assertions in general case (simple case or complex case). Now, we will use this matching assertion algorithm to build a matching security policy algorithm. The matching process consists in checking to what extent each security assertion reqAss specified in the requestor SP is satisfied by a security assertion provAss specified in the provider SP. The matchmaker has to perform three main tasks. Firstly, it must create all possible semantic relations at the level of each pair of provider and requestor assertions and get matching level of all assertion pairs. Secondly, based on the created semantic relations, it must decide the appropriate matching degree for each provAss-reqAss pair. The final matching degree for a requestor assertion is the highest level of matching level against all of the checked provider assertions. Thirdly, after all requestor assertions have a matching level, the matching degree of requestor and provider policies is the lowest matching degree of all assertions in requestor policy.

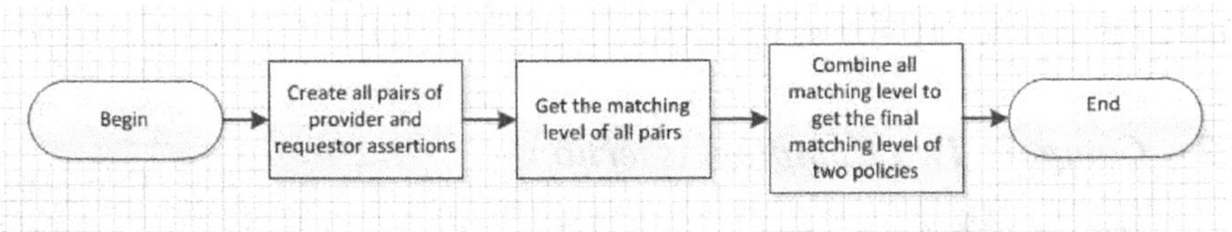

Fig. 6 Matching security policy algorithm

Matching algorithm: Let X, Y are two security policy. Policy X contains assertion X1, X2, , Xn and policy Y contains assertion Y1, Y2, , Ym. Because we only need to know a requestor policy is satisfied or not by a provider policy. Then, this algorithm compare requestor policy X with provider policy Y. With each assertion Xi, we will get the highest satisfied matching level of it with all assertion of policy Y. *Relation_-Assertion_Policy(Xi, B) = max {relation(Bj, Ai)with 0 <j <m+1)}*

With this definition, we will get the matching level of two policy is the lowest matching level of all assertion in requestor policy. Relation_policy(A, B) = min{relation_assertion_policy(Ai, B)), with 0 <i <n+1)}.

Our above proposed algorithm can help us get the more flexible and correct matching level of a requestor policy and a provider policy. This algorithm not only supports in the simple case of security policy, but also the complex case of security policy.

6 Related Work

WSPL (Web Services Policy Language) [4] is another syntactic policy language based on XML that can be used to specify a wide range of policies, including authorization, quality of service and reliable messaging etc. WSPL is of particular interest in several respects, for example, it supports merging two policies and policies can be based on comparisons other than equality allowing policies to depend on fine-grained attributes. In essence, a policy of WSPL is a sequence of rules, where each rule represents an acceptable alternative. A rule contains a number of predicates, which correspond to policy assertions in WS-Policy. Because WSPL is still based on syntactical domain models, its shortcomings are similar to WS-Policy. In a different work [5], they were interested in the work to conflict resolution semantic naming type when aligning SP in distributed multi-systems security policies. Their solution is based on ontology mediation for SP cooperation and understanding. They propose a helpful process for security experts, this process permit to mask heterogeneity and resolve conflicts between SP, using different steps.

In another work [10], the author proposed a simple ontology to compare two security policies, and then build an algorithm to compare security policies. This algorithm bases on ontology and it uses the semantic reasoner to get the final result. However, the algorithm is simple, and the comparing algorithm isn't detail.

In the work [6], the authors proposed a WS-SP-based ontology and some relations to compare two security assertions. That paper show how to get the matching level of two simple security assertions, but it lacks the comparing of all two polices and the processing in the complex security assertions case.

In addition to being compatible to WS-SP and the work [6], our approach is better than previous approaches: we extend the WS-SP-based ontology with additional semantic relations that support more correct and more flexible semantic matching of Web service security policies, not only simple assertions.

7 Conclusion and Discussion

In this paper, we used an approach to provide a semantic extension to security assertions of Web services, then from this result we compared two web service security policies. The approach is based on the transformation of WS-SP into an OWL-DL ontology, and using the relations in this ontology to get matching level of two security policies.

Our semantic approach supports more correct and more flexible security policy matching compared to syntactic matching of previous works that combined ontology and Web service security properties because our approach also compares complex security policies to get the final matching degree between them, which contain several alternatives. Besides, we plan to develop a tool that automatically transforms a WS-SP policy into our ontological representation, and show a security policy which is compatible with two requestor and provider, or print out no match result.

References

[1] OASIS: WS-SecurityPolicy 1.3, `http://www.oasis-open.org/specs/`
[2] W3C: WS-Policy 1.5, `http://www.w3.org/TR/ws-policy/`
[3] Verma, K., Akkiraju, R., Goodwin, R.: Semantic matching of Web service policies. In: Proceedings of the Second Workshop on Semantic and Dynamic Web Processes, pp. 79–90 (2005)
[4] Anderson, A.H.: An Introduction to the Web Services Policy Language. In: Fifth IEEE International Workshop on Policies for Distributed Systems and Networks, POLICY 2004 (2004)
[5] Benammar, O., Elasri, H., Sekkaki, A.: Semantic Matching of Security Policies to Support Security Experts
[6] Brahim, M.B., Chaari, T., Jemaa, M.B., Jmaiel, M.: Semantic matching of WS-Security Policy assertions (2011)
[7] Bhargavan, K., Fournet, C., Gordon, A.D., O'Shea, G.: An Advisor for Web Services Security Policies (2005)
[8] Garcia, D.Z.G., de Toledo, M.B.F.: Web Service Security Management Using Semantic Web Techniques (2004)
[9] Bhargavan, K., Fournet, C., Gordon, A.D.: Verifying Policy-Based Security for Web Services
[10] He, Z.-Q., Wu, L.-F., Zheng, H., Lai, H.-G.: Semantic Security Policy for Web Service. In: IEEE International Symposium on Parallel and Distributed Processing with Applications (2009)

An Efficient Method for Discovering Motifs in Streaming Time Series Data

Cao Duy Truong and Duong Tuan Anh

Abstract. The discovery of repeated subsequences, *time series motifs*, is a problem which has great utility for several higher-level data mining tasks, including classification, clustering, forecasting and rule discovery. In recent years there has been significant research effort spent on efficiently discovering these motifs in static time series data. However, for many applications, the streaming nature of time series demands a new kind of methods for discovery of time series motifs. In this paper, we develop a new method for motif discovery in streaming time series. In this method we use significant extreme points to determine motif candidates and then cluster motif candidates by BIRCH algorithm. The method is very effective not only for large time series data but also for streaming environment since it needs only one-pass of scan through the whole data.

1 Introduction

A time series is a sequence of real numbers measured at equal intervals. Time series data arise in so many applications of various areas ranging from science, engineering, business, finance, economic, medicine to government. There are two kinds of time series data: time series in static environment and time series in high speed data stream environment. In streaming time series database, the database changes continuously as new data points arrive continuously. Examples of the streaming time series applications are online stock analysis, computer network monitoring, network traffic management, earthquake prediction. Streaming time series have their own characteristics, compared to static time series: (1) Data are frequently updated in stream time series, thus, previous approaches applied to static time series may not work in the scenario. (2) Owing to the frequent updates, it is impossible to store all the data in memory, thus, efficient and one-pass algorithms are very important to achieve a

Cao Duy Truong · Duong Tuan Anh
Faculty of Computer Science and Engineering, Ho Chi Minh City University of Technology
e-mail: caoduytruong@hcmunre.edu.vn, dtanh@cse.hcmut.edu.vn

V.-N. Huynh et al. (eds.), *Knowledge and Systems Engineering, Volume 1*,
Advances in Intelligent Systems and Computing 244,
DOI: 10.1007/978-3-319-02741-8_20, © Springer International Publishing Switzerland 2014

real time response. Data mining on streaming environment is recently considered as one of the top ten challenging problems in the data mining research field [15].

Time series motifs are frequently occurring but previously unknown subsequences of a longer time series. One major challenge in motif discovery for time series data is the large volume of the time series data. Since the first formal definition of time series motif given by Lin et al. in 2002 [6], several algorithms have been proposed to tackle time series motif discovery. The first algorithm ([6]) defines the problem of motif discovery in time series regarding two parameters: (1) the motif length m and (2) a user-defined range parameter r. Some of these algorithms tackle the motif discovery by first applying some dimensionality reduction transformations such as PAA, PLA and some discretization techniques such as SAX, iSAX, ([1], [2], [5], [6], [14], [16]). Some of the algorithms aim to discovering motif with different lengths or discovering motif with the suitable length determined automatically ([11], [12]). However, so far surprisingly there have been very few research works on time series motif discovery in streaming environment.

In 2010, a motif discovery method for streaming time series was proposed by Mueen and Keogh, which was considered as "the first practical algorithm for finding and maintaining time series motifs on fast moving streams" [8]. However, in this work, the authors use the new *nearest-neighbor* motif definition by which time series motif is defined as a closest pair of subsequences in time series data. This definition does not take into account the frequency of the subsequences, therefore, the motif discovery algorithm proposed in [8] is not convenient to be used directly in practical applications.

In this paper, we propose an efficient method for motif discovery in streaming time series which adopts the first formal motif definition given in 2002 [6]. Our proposed method needs only one single pass over the whole data, and can update the motif results efficiently whenever there have been some new data points arriving. Our method works directly on the raw data without using any transformation for dimensionality reduction or discretization. The instances of a motif discovered by our method may be of different lengths and user does not have to predefine the length of the motif. The proposed method requires fewer user-predefined parameters and these parameters are easy to be determined through experiments. The proposed method is also not sensitive to the changes of these parameters. Our method uses significant extreme points to determine motif candidates and then clusters motif candidates to find the most significant motif by using BIRCH algorithm.

The rest of the paper is organized as follows. In Section 2 we explain briefly some basic backgrounds on time series motif and our previous work on time series motif discovery in static time series. Section 3 introduces the proposed method. Section 4 reports experimental results on the proposed method. Section 5 gives some conclusions and remarks for future work.

2 Background

In this section we introduce some useful definitions and the EP-BIRCH algorithm that can discover motif in static time series.

2.1 Time Series Motif

Definition 1. *Time Series*: A time series $T = t_1,,t_N$ is an ordered set of N real-values measured at equal intervals.

Definition 2. *Similarity distance*: $D(s_1,s_2)$ is a positive value used to measure differences between two time series s_1, and s_2, relies on measure methods. If $D(s_1,s_2) < r$, where r is a real number (called range), then $s1$ is similar to $s2$.

Definition 3. *Subsequence*: Given a time series T of length N, a subsequence C of T is a sampling of length $n < N$ of contiguous positions from T, that is, $C = t_p,...,t_{p+n-1}$ for $1 < p < N-n+1$.

Definition 4. *Time series motif*: Given a time series T, a subsequence C is called the most significant motif (or *1-motif*) of T, if it has the highest count of the subsequences that are similar to it. All the subsequences that are similar to the motif are called instances of the motif.

This definition is also the first formal definition of time series motif given by Lin et al. in 2002 [6].

Definition 5. The *motif count* (or *frequency*) of a motif M is the total number of instances that M has in time series T.

Time series motifs are typically sorted according to their motif count. The *K-motif* is the motif ranked at *K-th* position regarding number of instances.

2.2 The EP-BIRCH Algorithm

Finding Significant Extreme Points

To extract a temporally ordered sequence of motif candidates, significant extreme points of a time series have to be found. The definition of significant extreme points, given by Pratt and Fink, 2002 [10] is as follows.

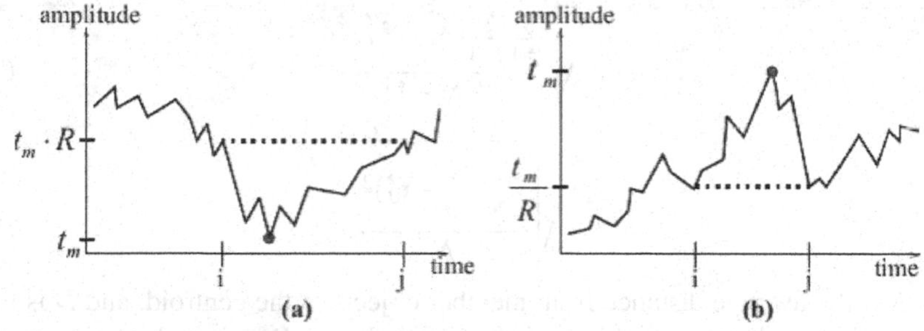

Fig. 1 Illustration of Significant Extreme Points: (a) Minimum, (b) Maximum

Definition 6. *Significant Extreme Points*: A univariate time series $T = t_1,...,t_N$ has a significant minimum at position m with $1 < m < N$, if $(t_i,...,t_j)$ with $1 \leq i < j \leq N$ in T exists, such that t_m is the minimum of all points of this subsequence and $t_i \geq R \times t_m$, $t_j \geq R \times t_m$ with user-defined $R \geq 1$.

Similarly, a significant maximum is existent at position m with $1 < m < N$, if a subsequence $(t_i, ..., t_j)$ with $1 \leq i < j \leq N$ in T exists, such that t_m is the maximum of all points of this subsequence and $t_i \leq t_m/R, t_j \leq t_m/R$ with user-defined $R \geq 1$.

Notice that in the above definition, the parameter R is called *compression rate* which is greater than one and an increase of R leads to selection of fewer significant extreme points. Fig. 1. illustrates the definition of significant minima (a) and maxima (b). Given a time series T, starting at the beginning of the time series, all significant minima and maxima of the time series are computed by using the algorithm given in [10].

The significant extreme points can be the starting point or ending point of a motif instances. Basing on the extracted significant points we can extract the motif candidates from a time series and then cluster them using BIRCH algorithm.

BIRCH Clustering

BIRCH is designed for clustering a large amount of numerical data by integration of hierarchical clustering at the initial stage and other clustering methods, such as iterative partitioning at the later stage [17]. It introduces two main concepts, *clustering feature and clustering feature tree (CF tree)*, which are used to summarize cluster representations. These structures help the clustering method achieve good speed and scalability in large databases. BIRCH is also effective for *incremental* and *dynamic clustering* of incoming objects.

Given N d-dimensional points or objects $\overrightarrow{x_i}$ in a cluster, we can define the centroid $\overrightarrow{x_0}$, the radius R, and the diameter D of the cluster as follows:

$$\overrightarrow{x_0} = \frac{\sum_{i=1}^{N} \overrightarrow{x_i}}{N} \tag{1}$$

$$D = \sqrt{\frac{\sum_{i=1}^{N} \sum_{j=1}^{N} (\overrightarrow{x_i} - \overrightarrow{x_j})^2}{N(N-1)}} \tag{2}$$

$$R = \sqrt{\frac{\sum_{i=1}^{N} (\overrightarrow{x_i} - \overrightarrow{x_0})^2}{N}} \tag{3}$$

where R is the average distance from member objects to the centroid, and D is the average pairwise distance within a cluster. Both R and D reflect the tightness of the cluster around the centroid. A clustering feature *(CF)* is a triplet summarizing information about clusters of objects. Given N d-dimensional points or objects in a subcluster, then the CF of the cluster is defined as

$$CF = (N, \overrightarrow{LS}, SS) \tag{4}$$

where N is the number of points in the subcluster, $\overrightarrow{LS}$ is the linear sum on N points and SS is the square sum of data points.

$$\overrightarrow{LS} = \sum_{i=1}^{N} \overrightarrow{x_i} \quad (5)$$

$$SS = \sum_{i=1}^{N} \overrightarrow{x_i}^2 \quad (6)$$

A clustering feature is essentially a summary of the statistics for the given subcluster: the zero-th, first, and second moments of the subcluster from a statistical point of view. Clustering features are additive. For example, suppose that we have two disjoint clusters, C_1 and C_2, having the clustering features, CF_1 and CF_2, respectively. The clustering feature for the cluster that is formed by merging C_1 and C_2 is simply $CF_1 + CF_2$. Clustering features are sufficient for calculating all of the measurements that are needed for making clustering decisions in BIRCH.

A CF tree is a height-balanced tree that stores the clustering features for a hierarchical clustering. By definition, a nonterminal node in the tree has descendents or "children". The nonleaf nodes store sums of the CFs of their children, and thus summarize clustering information about their children. Each entry in a leaf node is not a single data objects but a subcluster. A CF tree has two parameters: branching factor (B for nonleaf node and L for leaf node) and threshold T. The branching factor specifies the maximum number of children in each nonleaf or leaf node. The threshold parameter specifies the maximum diameter of the subcluster stored at the leaf nodes of the tree. The two parameters influence the size of the resulting tree.

BIRCH applies a multiphase clustering technique: a single scan of the data set yields a basic good clustering, and one or more additional scans can (optionally) be used to further improve the quality. The BIRCH algorithm consists of four phases as follows.

Phase 1: *(Building CF tree)* BIRCH scans the database to build an initial in-memory CF tree, which can be view as a multilevel compression of the data that tries to preserve the inherent clustering structure of the data.

Phase 2: *[optional]* (Condense data) Condense into desirable range by building a smaller CF tree.

Phase 3: *(Global Clustering)* BIRCH applies a selected clustering algorithm to cluster the leaf nodes of the CF tree. The selected algorithm is adapted to work with a set of subclusters, rather than to work with a set of data points.

Phase 4: *[optional]* Cluster refining

After the CF tree is built, any clustering algorithm, such as a typical partitioning algorithm, can be used in Phase 3 with the CF tree built in the previous phase. Phase 4 uses the centroids of the clusters produced by Phase 3 as seeds and redistributes the data points to its closest seed to obtain a set of new clusters.

EP-BIRCH Algorithm

The EP-BIRCH (Extreme points and BIRCH clustering) method, introduced in our previous paper [13], is an improvement of the EP-C algorithm proposed by Gruber et al. [3] for time series motif discovery. The EP-C algorithm uses hierarchical agglomerative clustering (HAC) algorithm for clustering which is not suitable to large scale time series datasets. In our EP-BIRCH method, we use BIRCH algorithm to cluster motif candidates rather than using HAC algorithm. BIRCH is especially suitable for clustering very large time series datasets. Besides, in the EP-C algorithm, each motif candidate is determined by three contiguous extreme points, but in our proposed method, motif candidate is determined by n contiguous extreme points where n is selected by user.

EP-BIRCH consists of the following steps:

Step 1: We extract all significant extreme point of the time series T. The result of this step is a sequence of extreme points $EP = (ep_1, ..., ep_l)$

Step 2: We compute all the motif candidates iteratively. A motif candidate $MC_i(T), i = 1, ..., l - n + 1$ is the subsequence of T that is bounded by the n extreme points ep_i and ep_{i+n-1}. Motif candidates are the subsequences that may have different lengths.

Step 3: Motif candidates are the subsequences that may have different lengths. To enable the computation of distances between them, we can bring them to the same length using *homothetic transformation*. The same length here is the average length of all motif candidates extracted in Step 2.

Step 4: We build the CF tree with parameters B and T. We insert to the CF tree all the motif candidates found in Step 3. We apply k-Means as Phase 3 of BIRCH to cluster the leaf nodes of the CF tree where k is equal to the number of the leaf nodes in the CF tree.

Step 5: Finally we find the subcluster in the CF tree with the largest number of objects. The *1-motif* will be represented by that cluster.

In the Step 3, to improve the effectiveness of our proposed method, we apply *homothety* for transforming the motif candidates with different lengths to those of the same length rather than spline interpolation as suggested in [3]. Spline interpolation is not only complicated in computation, but also can modify undesirably the shapes of the motif candidates. Homothety is a simpler and more effective technique which also can transform the subsequences with different lengths to those of the same length. Due to the limit of space, we can not explain the use of homothetic transformation here, interested readers can refer to [13] for more details.

3 The Proposed Method for Motif Discovery in Streaming Time Series Data

3.1 From EP-BIRCH to EP-BIRCH-STS

Our method for motif discovery in streaming time series is developed from our previous work, the EP-BIRCH algorithm for motif discovery in static time series [13].

We call the new method EP-BIRCH-STS (Extreme Points and BIRCH clustering for discovering motif in Streaming Time Series). Our method works with the features extracted from the whole time series from the beginning data point to the newest incoming data point. In our method, we can view the sequence of extracted extreme points $EP = (ep_1, ..., ep_l)$ as a simple data structure to keep all the features of the time series. In the proposed method for motif discovery in streaming time series, EP-BIRCH-STS, we require the following parameters.

- R: compression rate for computing the significant extreme points.
- *min_Length* and *max_Length*: the lower bound and upper bound for the length of the discovered motif.
- B, T: branching factor and threshold of CF tree.

EP-BIRCH-STS is mainly based on EP-BIRCH with some following modifications in order to adapt it in the streaming environment

1. In EP-BIRCH algorithm, when finding the significant extreme points, the algorithm uses two subroutines: FIND-MINIMUM(i) for finding the significant minimum starting from the $i-th$ point in the time series and FIND-MAXIMUM(i) for finding the significant minimum starting from the $i-th$ point in the time series. We can make these two subroutines incremental to accommodate the new incoming data points. So, we can make the task of extracting significant extreme points incremental.
2. We apply a *deferred update policy* that is described as follows. We delay the update of the sequence of extreme points $EP = (ep_1, ..., ep_l)$ until we identify one new significant extreme point from new coming data points and this event actually yields a new motif candidate. At that moment, we activate the motif discovery phase in order to get a new possible motif for the current time series data. When we have the new motif candidate, we apply the homothety to convert it to a suitable length and then insert it to the CF-tree. Due to the incremental nature of BIRCH clustering algorithm, the clustering step in EP-BIRCH-STS can work very efficiently to produce a new possible motif result whenever there is an incoming motif candidate.

3.2 *Post Processing*

In the paper [4], Keogh and Lin, 2005 pointed that subsequence clustering in streaming time series is meaningless when we use a sliding window to extract subsequences. The task is meaningless since there are several trivial matches (Definition 7) during clustering.

Definition 7. *Trivial Match*: Given a subsequence C beginning at position p, a matching subsequence M beginning at q, and a distance R, we say that M is a trivial match to C of order R, if either $p = q$ or there does not exist a subsequence M beginning at q such that $D(C,M) > R$, and either $q < q < p$ or $p < q < q$.

The subsequence extraction in our EP-BIRCH-STS method which based on significant extreme points does not use sliding window and therefore its clustering step does not have the problem mentioned in [4]. However to ensure there exists no trivial matches, in EP-BIRCH-STS, we perform a post processing step which excludes any trivial matches.

In our EP-BIRCH-STS method, trivial matches may arise when the compression *R* is too small. When *R* is too small, the extreme points will be not enough significant and this make the motif candidates in a subcluster to be overlapped with one another. When *R* is large enough, trivial matches can not arise since each extreme point starts a new change in the time series. Fig. 2. illustrates the possibility of trivial matches among motif candidates when R is too small.

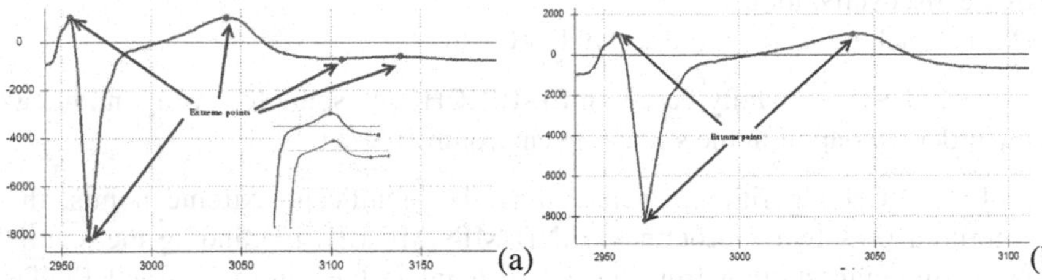

Fig. 2 Trivial matches due to the compression ratio *R*. (a) With too small *R*, motif candidates may start with the same position, but belong to the same subcluster. (b) With large *R*, there is no trivial match.

To exclude possible trivial matches in the very rare cases, after obtaining the subclusters by using BIRCH algorithm, we examine in all the subclusters to exclude any instances which have some overlap in one another. This task takes very low computation cost. Then we rank the resultant subclusters according to the number of instances in them in order to determine the top subcluster as *1-motif* and the *K-th* subcluster as *K-motif*.

4 Experimental Evaluation

In this experiment, we compare our method, EP-BIRCH-STS to the modified version of Online-MK method, the first motif discovery method for streaming time series ([8]). We implemented the two methods with Microsoft Visual C# and conducted the experiment on a Core i7, Ram 4GB PC.

First, to verify the correctness of EP-BIRCH-STS, we experiment the method on the datasets which were used in four previous works ([5], [7], [8], [9]). The test datasets consist of.

1. ECG (electrocardiogram) dataset.[1]
2. Insect behavior dataset.[2]
3. World cup dataset.[3]

[1] `http://www.cs.ucr.edu/~eamonn/iSAX/koski_ecg.dat`
[2] `http://www.cs.ucr.edu/~mueen/txt/insect15.txt`
[3] `http://www.cs.ucr.edu/~mueen/txt/wc_index.txt`

Experimental results show that when the appropriate values are selected for the parameters, the motifs found by EP-BIRCH-STS are exactly the same as those discovered in [5] and [9]. Fig. 3. reports the motifs found by our EP-BIRCH-STS method in comparison to those discovered in [5] and [9].

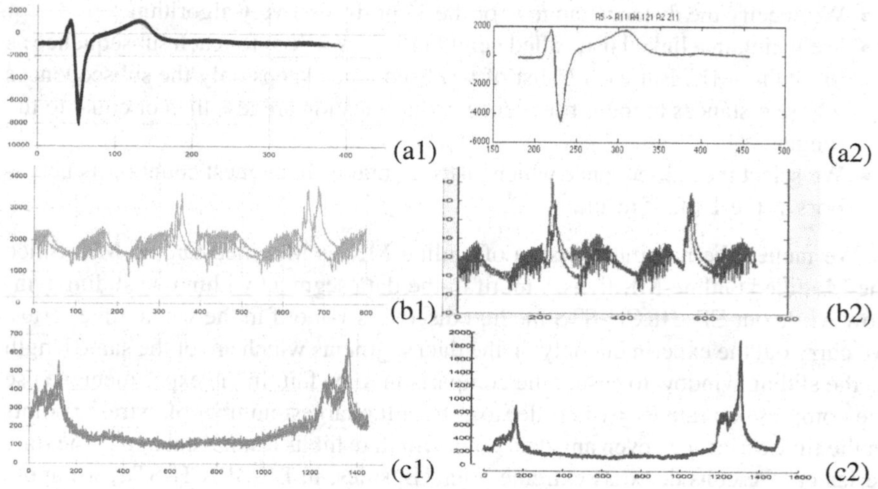

Fig. 3 Motifs discovered by the proposed method compared with those found by some previous methods. (a1), (a2) motif found in *ECG dataset* by EP-BIRCH-STS and [5]. (b1), (b2) motif found in *insect behavior dataset* by EP-BIRCH-STS and [5]. (c1), (c2) motif found in *world cup dataset* by EP-BIRCH-STS and [9].

To verify the effectiveness of our EP-BIRCH-STS in streaming time series, we conduct an experiment which accepts the above-mentioned datasets in stream manner after a fixed time period. Fig. 4. shows the 1-motifs discovered from the ECG dataset with lengths from 200 to 300 at different time points.

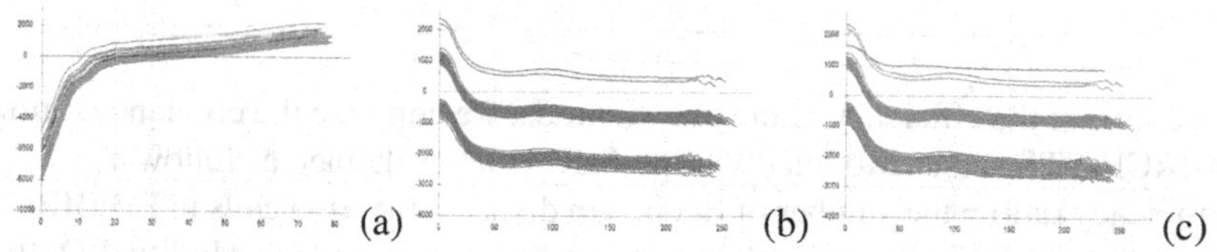

Fig. 4 Motif in *ECG dataset* at different time points. (a) At time point 71996 with 80 instances; (b) at time point 79701 with 81 instances; (c) at time point 144404 with 110 instances.

Next, we compare the performance of EP-BIRCH-STS to the modified version of Online-MK method described in [8]. We use the same datasets that were used in the experiment reported in [8]. They are EEG trace, EOG trace, insect behavior trace and a synthetic random walk (RW). Since Online-MK method aims to find the most closest pair in a streaming time series while our method aims to find the 1-motif in the sense of its first formal definition (Definition 4 in Section 2), we have to modify the Online-MK so that it can discover 1-motif (in its first formal definition). We have to add to the Online-MK the following modifications:

- We specify the distance range r for the 1-motif discovery algorithm.
- We maintain a linked list, called neighbor list ($N - list$) for each subsequence as in Online-MK. But each N-list of a subsequence keeps only the subsequences whose distances to the subsequence under question are less than or equal to the range r.
- We select the subsequence which N-list contains the highest count of its neighbors as the 1-motif result.

We named the modified version of Online-MK as Modified-Online-MK. Since the Modified Online-MK finds 1-motif in the data segment within one sliding window while our EP-BIRCH-STS method discovers 1-motif in the whole time series, we carry out the experiment only on the data segments which are of the same length as the sliding window to ensure the comparison to be fair. In the experiment we use the compression rate $R = 1$ in order to extract the largest number of extreme points in the time series, i.e., even any data point which exhibits a small change in the time series can be considered as extreme point. Besides, in EP-BIRCH-STS, we apply homothetic transformation to convert all the motif candidates to the same length of the motif we desire to discover in Modified-Online-MK.

Table 1 Efficiency ratio of EP-BIRCH-STS vs. Modified-Online-MK

Window length	1000	2000	4000	8000	10000	20000
EEG	1.64%	0.30%	0.14%	0.07%	0.06%	0.03%
EOG	3.09%	0.83%	0.37%	0.25%	0.25%	0.25%
Insect	4.14%	1.23%	0.51%	0.26%	0.23%	0.14%
RW	2.33%	0.78%	0.35%	0.19%	0.15%	0.11%

To compare the efficiency of the two methods, we compute the efficiency ratio of EP-BIRCH-STS versus Modified-Online-MK which is defined as follows:

Efficiency ratio = the number of Euclidean distance function calls in EP-BIRCH-STS * 100% / the number of Euclidean distance function calls in Modified-Online-MK.

Table 1 show the efficiency ratio of EP-BIRCH-STS vs. Modified-Online-MK on different datasets and with different data lengths. Experimental results show that in EP-BIRCH-STS the number of Euclidean distance function calls just is about 0.74% of that in Modified-Online-MK. This remarkable performance of EP-BIRCH-STS is due to the fact that EP-BIRCH-STS uses the significant extreme points to extract motif candidates and this brings out a much smaller number of motif candidates in comparison to those extracted by Modified-Online-MK. Furthermore, by using CF-tree to support in clustering, the cost of subsequence matching in EP-BIRCH-STS is reduced remarkably since each motif candidate has to be matched with a smaller number of the nodes in CF-tree.

Table 2 show the CPU time (in seconds) of motif discovery in the two methods EP-BIRCH-STS and Modified-Online-MK respectively on different datasets and with different motif lengths. We can see that in all cases, the CPU time of motif discovery in EP-BIRCH-STS is much lower than that of Modified-Online-MK.

Table 2 Run time of EP-BIRCH-STS/run time of Modified-Online-MK on different datasets and with different motif lengths

Motif length	64	128	256	396	512
EEG	0.38s/58.82s	0.47s/120.41s	0.63s/264.23s	0.75s/430.94s	0.68s/570.58s
EOG	0.75s/11.77s	0.93s/102.95s	1.35s/223.25s	1.63s/380.93s	1.58s/519.39s
Insect	0.84s/76.34s	1.16s/133.55s	1.50s/227.95s	1.63s/351.71s	1.54s/447.03s
RW	0.78s/76.92s	0.98s/133.17s	1.24s/229.64s	1.48s/350.67s	1.57s/482.21s

5 Conclusions

We haved introduced a new method for discovering motifs in streaming time series. This method, called EP-BIRCH-STS, is based on extracting significant extreme points and clustering the motif candidates by using BIRCH algorithm. This method needs only one-pass scan through the whole data and can discover 1-motif with the length in the range of *min_Length* to *max_Length*. The discovered motif some-whatdepends on the values of the compression rate R specified in the step of identifying the significant extreme points and the threshold T in the BIRCH clustering algorithm.

As for future work, we plan to find some technique to determine the two parameters R,and T automatically for any given time series dataset.

References

1. Castro, N., Azevedo, P.J.: Multiresolution Motif Discovery in Time Series. In: Proc. of the SIAM Int. Conf. on Data Mining, SDM 2010, Columbus, Ohio, USA, April 29-May 1 (2010)
2. Chiu, B., Keogh, E., Lonardi, S.: Probabilistic Discovery of Time Series Motifs. In: Proc. of 9th Int. Conf. on Knowledge Discovery and Data Mining (KDD 2003), pp. 493–498 (2003)
3. Gruber, C., Coduro, M., Sick, B.: Signature verification with dynamic RBF network and time series motifs. In: Proc. of 10th International Workshop on Frontiers in Hand Writing Recognition (2006)
4. Keogh, E., Lin, J.: Clustering of Time-Series Subsequences is Meaningless. Implications for previous and future research. Knowl. Inf. Syst. 8(2), 154–177 (2005)
5. Li, Y., Lin, J., Oates, T.: Visualizing Variable-Length Time Series Motifs. In: SDM 2012, pp. 895–906 (2012)

6. Lin, J., Keogh, E., Patel, P., Lonardi, S.: Finding Motifs in Time Series. In: Proc. of the 2nd Workshop on Temporal Data Mining, at the 8th ACM SIGKDD International Conference on Knowledge Discovery and Data Mining (2002)
7. Mueen, A., Keogh, E., Zhu, Q., Cash, S., Westover, B.: Exact Discovery of Time Series Motif. In: Proc. of 2009 SIAM International Conference on Data Mining, pp. 1–12 (2009)
8. Mueen, A., Keogh, E.: Online Discovery and Maintenance of Time Series Motif. In: Proc. of ACM SIGKDD 2010, pp. 1089–1098 (2010)
9. Mueen, A., Keogh, E., Zhu, Q., Cash, S., Bigdely-Shamlo, N.: Finding Time Series Motifs in Disk-Resident Data. In: Proc. of IEEE International Conference on Data Mining, ICDM, pp. 367–376 (2009)
10. Pratt, K.B., Fink, E.: Search for patterns in compressed time series. International Journal of Image and Graphics 2(1), 89–106 (2002)
11. Tanaka, Y., Iwamoto, K., Uehara, K.: Discovery of Time Series Motif from Multi-Dimensional Data based on MDL Principle. Machine Learning 58(2-3), 269–300 (2005)
12. Tang, H., Liao, S.: Discovering Original Motifs with Different Lengths from Time Series. Knowledge-based System 21(7), 666–671 (2008)
13. Truong, C.D., Anh, D.T.: An Efficient Method for Discovering Motifs in Large Time Series. In: Selamat, A., Nguyen, N.T., Haron, H. (eds.) ACIIDS 2013, Part I. LNCS (LNAI), vol. 7802, pp. 135–145. Springer, Heidelberg (2013)
14. Xi, X., Keogh, E., Li, W., Mafra-neto, A.: Finding Motifs in a Database of Shapes. In: SDM 2007. LNCS, vol. 4721, pp. 249–260. Springer, Heidelberg (2007)
15. Yang, Q., Wu, X.: 10 Challenging Problems in Data Mining Research. Intl. Jrnl. of Information Technology & Decision Making 5(4), 597–604 (2006)
16. Yankov, D., Keogh, E., Medina, J., Chiu, B., Zordan, V.: Detecting Time Series Motifs under Uniform Scaling. In: Proc. of the 13th ACM SIGKDD International Conference on Knowledge Discovery and Data Mining (KDD 2007), pp. 844–853 (2007)
17. Zhang, T., Ramakrishnan, R., Livny, M.: BIRCH: An efficient data clustering method for very large databases. SIGMOD Rec. 25(2), 103–114 (1996)

On Discriminant Orientation Extraction Using GridLDA of Line Orientation Maps for Palmprint Identification

Hoang Thien Van and Thai Hoang Le

Abstract. In this paper, we propose a novel Discriminant Orientation Representation, called DORIR, for palmprint identification. To extract DORIR feature, we proposed the algorithm which includes two main steps: (1) Palm line orientation map computation and (2) Discriminant feature extraction of the orientation field. In the first step, positive orientation and negative orientation maps are proposed as the input of two dimensional linear discriminant analysis (2D-LDA) for computing the class-separability features. In the second step, the grid-sampling based 2DLDA, called Grid-LDA, is used to remove redundant information of orientation maps and form a discriminant representation more suitable for palmprint identification. The experimental results on the two public databases of Hong Kong Polytechnic University (PolyU) show that proposed technique provides a very robust orientation representation for recognition and gets the best performance in comparison with other approaches in literature.

1 Introduction

Palmprint is new kind of biometric feature for personal recognition and has been widely studied due to its merits such as distinctiveness, cost-effectiveness, user friendliness, high accuracy, and so on [1]. Palmprint research employs low resolution images for civil and commercial applications. Low resolution refers to 150 dpi or less [1] (see Fig. 1a). After image acquisition, the palmprint image will be

Hoang Thien Van
Department of Computer Sciences, Ho Chi Minh City University of Technology,
Ho Chi Minh City, VietNam
e-mail: vthoang@hcmhutech.edu.vn

Thai Hoang Le
Department of Computer Sciences, Ho Chi Minh University of Science,
Ho Chi Minh City, VietNam
e-mail: lhthai@fit.hcmus.edu.vn

V.-N. Huynh et al. (eds.), *Knowledge and Systems Engineering, Volume 1*, Advances in Intelligent Systems and Computing 244,
DOI: 10.1007/978-3-319-02741-8_21,

preprocessed to align difference palmprint images and to segment the region of interest (ROI) for feature extraction and matching (Fig. 1b). Principal lines and wrinkles, called palm-lines, are very important to distinguish between different palmprints and they can be extracted from low-resolution images. There

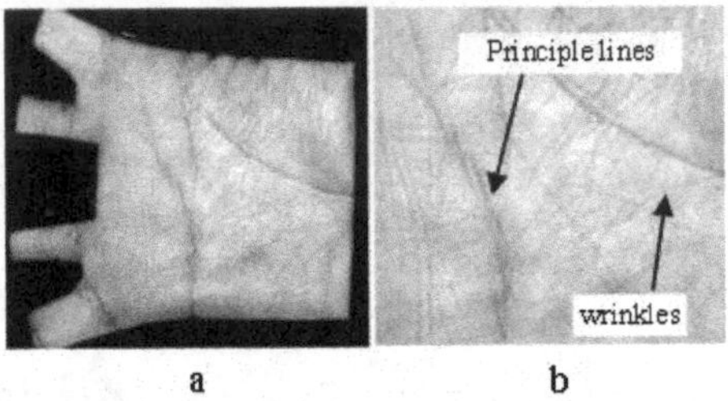

Fig. 1 a) Sample palmprint image and (b) its region of interest (ROI)

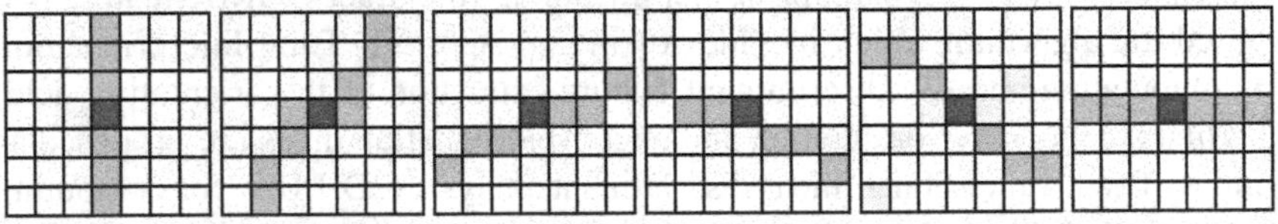

Fig. 2 Examples of the MFRAT whose size are 7 ×7, respectively, and whose lines, $L(\theta_k)$, are at directions of 0, $\pi/12$, $2\pi/12$, $3\pi/12$, $4\pi/12$, and $6\pi/12$, respectively

are many approaches exploiting palm lines for recognition such as: line-based approaches, code-based approaches, subspace-based approaches, and fusion approaches. Subspace-based approaches also called appearance-based approaches in literatures use principal component analysis (PCA), linear discriminant analysis (LDA) and independent component analysis (ICA) to project palmprint images from high dimensional space to a lower dimensional feature space [2], [3, [4]. The sub-space coefficients are regarded as features. These approaches were reported to achieve exciting results, but they may be sensitive to illumination, contrast, and position changes in real applications. Line-based approaches will extract palm lines for matching based on using or developing edge detection algorithms [5], [6], [7]. Palm lines are the basic feature of palmprint. However, the few principal lines do not contribute strongly enough to obtain a high recognition rate [3]. Therefore, principal lines can be used in palmprint classification [6]. Code-based approaches have been widely investigated in palmprint recognition area because of efficient implementation and high recognition performance. These approaches can obtain the palmprint orientation pattern by applying Gabor, Radon and transforms [8], [9], [10]. Fusion approaches utilize many techniques and integrate different features in order to provide more reliable results [11], [12], [13]. This paper proposes a novel palmprint identification algorithm based on using a robust orientation representation which is in low dimensional and discriminant feature space. The main contributions of this paper consist of the following aspects: (1) A novel method based on the Modified Finite Radon Transform (MFRAT) is proposed for computing two palm line orientation images, called Positive ORIentation Feature image (PORIF) and Negative ORIentation Feature image (NORIF), which separately describe the orientation patterns of principle lines and wrinkles. (2) GridLDA is used to project the orientation maps from the high dimensional space to lower dimensional and discriminant space. The rest of the paper is organized as follows. Section 2 presents the proposed orientation representation. Section 3 presents the GridLDA based method for compu-ting DORIR feature and identification. The experimental results are presented in sec-tion 4. Finally, the paper conclusions are drawn in section 5.

2 Our Proposed Palm Line Orientation Representation

2.1 MFRAT Background

In the MFRAT, a real function $f(x,y)$ on the finite grid Z_P^2, $Zp = \{0,1,,p-1\}$, where p is a positive integer, is defined as:

$$r[L_k] = MFRAT_f(k) = \frac{1}{C} \sum_{(i,j)\in L_k} f[i,j], \tag{1}$$

where C is a scalar to control the scale of $r[L_k]$, $f(i, j)$ represents the gray level of the pixel at (i, j) and L_k denotes the set of points that make up a line on the lattice Z_p^2 (see Figure 2):

$$L_k = \{(i,j) : j = k(i-i_0) + j_0, i \in Z_p\}, \tag{2}$$

where (i_0, j_0) denotes the center point of the lattice Z_P^2 and k means the corresponding slope of L_k. Since the gray-levels of the pixels on the palm lines are lower than those of the surrounding pixels, the line orientation $\theta$and the line energy e of the center point $f(i_0, j_0)$ of Z_P^2 can be calculated as:

$$\theta_{k(i_0,j_0)} = \arg\left(\min_k (r[L_k])\right), k = 1,2,..,N, \tag{3}$$

$$e_{(i_0,j_0)} = \left|\min_k (r[L_k])\right|, k = 1,2,..,N, \tag{4}$$

where N is the number of direction in Z_P^2. By this way, the directions and energies of all pixels are calculated if the center of lattice Z_P^2 moves over an image pixel by pixel (or pixels by pixels).

2.2 Orientation Representation of Principle Lines and Wrinkles

The orientation Code is common and robust feature for palmprint recognition [20], [7], [19]. The orientation is coded in bit representation and then uses Hamming distance to measure the difference between two orientation coding. However, the orientation code feature is still in large dimensional space and contains the redundant information. Addition to, Huang et al. [5] pointed out that the directions of most wrinkles markedly differ from that of the principal lines. For instance, if the direction of the principle lines belong to $(0°,...,\pi/2]$ approximately, the directions of most winkles will be at $[\pi/2,...,\pi)$. Therefore, we propose a robust orientation representation which separately describes the orientation maps of principle lines and wrinkles. Because the orientation of principle lines can belong to $(0°,...,\pi/2]$ or $(0°,...,\pi/2]$, the orientation representation include two planes of the orientation$\theta \in [0,180]$: positive orientation θ_{pos}, $\theta_{pos} \in [0,90]$ and negative orientaton θ_{neg}, $\theta_{neg} \in [90,180]$. These orientations of the center point (i_0, j_0) are defined based on MFRAT as follows:

$$\begin{array}{l}\theta_{pos}(i_0,j_0) = \theta_{k_p(i_0,j_0)} = \arg\left(\max_{k_p}\left(r\left[L_{k_p}\right]\right)\right), k_p = 0,1,2,3 \\ \theta_{neg}(i_0,j_0) = \theta_{k_n(i_0,j_0)} = \arg\left(\max_{k_n}\left(r\left[L_{k_n}\right]\right)\right), k_n = 3,4,5,6\end{array} \tag{5}$$

where $\theta_{pos}(\theta_{neg})$ called positive (negative) orientation because the Cosine component of θ_{pos} (θ_{neg}) is positive (negative). Then, if orientations of all pixels are computed by equations (1), (2) and (5), two new images, called Positive ORIentation Representation image (PORIR) and Negative ORIentation Representation image (NORIR) are created as:

$$PORIR = \begin{bmatrix} k_{p(1,1)} & k_{p(1,2)} & \vdots & k_{p(1,n)} \\ k_{p(2,1)} & k_{p(2,2)} & \vdots & k_{p(2,n)} \\ \cdots & \cdots & \vdots & \cdots \\ k_{p(m,1)} & k_{p(m,2)} & \vdots & k_{p(m,n)} \end{bmatrix}, k_{p(i,j)} \in \{0,1,2,3\}, i = \overline{1,m}, j = \overline{1,n} \tag{6}$$

$$NORIR = \begin{bmatrix} k_{n(1,1)} & k_{n(1,2)} & \vdots & k_{n(1,n)} \\ k_{n(2,1)} & k_{n(2,2)} & \vdots & k_{n(2,n)} \\ \cdots & \cdots & \vdots & \cdots \\ k_{n(m,1)} & k_{n(m,2)} & \vdots & k_{n(m,n)} \end{bmatrix}, k_{n(i,j)} \in \{3,4,5,6\}, i = \overline{1,m}, j = \overline{1,n} \tag{7}$$

Figures 3c and 3d show the PORIF image and the NORIF image, respectively. These two orientation maps are more class-separability than the original orientation map and can be used as the input of GridLDA to obtain projected feature matrix, called Discriminant Orienation Representation Matrix (DORIR). Finally, Euclidean

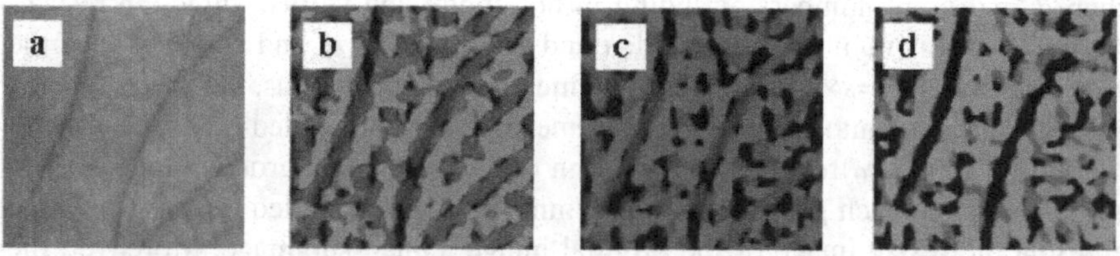

Fig. 3 (a) The original image, (b) the cosin component of the orientation map, (c) the PORIR image, and (d) the NORIR image

distance based nearest neighbor classifier is used for recognition. Next subsection presents GridLDA for extracting DORIR.

3 Grid-LDA for Computing DORIR Feature

The global discriminant feature extraction is the important stage of palmprint recognition. The aim of this stage is to find the optimal transformation by which the original data in high-dimentional space is transformed to a much lower dimensional space, and which is a representative in a low-dimensional subspace with as much class-separated information as possible. Grid sample based 2DLDA, called GridLDA, [10] is the efficient tool to compute the global discriminant feature of the palmprint image. In this section, we present the GridLDA background and our proposed me-thod for computing the DORIR feature from the NORIR and PORIR images.

3.1 GridLDA Background

Grid sampled based 2DLDA, called GridLDA, [10] is the efficient tool for extracting the discriminative and low dimensional feature for classification. GridLDA is 2DLDA with the input which is pixel-grouped images by grid-sampling strategy (see Figure 4a).

The grid-sampling is defined as: a virtual rectangular grid is overlaid on the image matrix (see Figure 4b), and the points at the intersections of gridline are sampled. The sampled pixels are packed into a subset. Then, the overlaid grid slides by one pixel in the horizontal or vertical direction. At each new position, grid-sampling is performed and new subset of random variables is obtained (see Figure 4c). Considering a $M_0 \times N_0$ image, we formulate the strategy as:

$$
\begin{aligned}
&RG(k,p) = \{rg(x_0,y_0) : x_0 = 0,..,k-1; y_0 = 0,..,p-1\},\\
&rg(x_0,y_0) = \left\{ \begin{array}{l} (x_i,y_j) : x_i = x_0 + i \times k; y_i = y_0 + j \times p;\\ i = 0,..,s-1; s = N_0/k;\\ j = 0,..,t-1; t = M_0/p \end{array} \right\} \\
&f_g(u,v) = f(x_i,y_j), u = x_0 \times k + y_0, v = i \times s + j,\\
&(x_i,y_j) \in rg(x_o,y_o), rg(x_0,y_0) \in RG(k,p)
\end{aligned} \quad (8)
$$

where k and p are numbers of sliding in horizontal and vertical direction respectively; m=$k\times p$ is number of the grid; s and t are width size and height size of the grid respectively; n=$s\times t$ is number of elements in the grid. Thus, the pixels of each image are grouped into m sets with the same size (n pixel), called $RG(k,p)$.

Each set $rg(x_0,y_0)$respects to a column of an $m\times n$ pixel-grouped matrix. Figure 4c show that each grid creates a column of the grid sampled image which can represent the resize image of the original image, called subimage. Moreover, the subimages are nearly geometric similarity. As the grid sampled image is the input of 2DLDA, 2DLDA can reduce the space dimension effectively because the columns are high correlation. Because these subimage represented for these original image has more discriminative information than that of other kind of sampling strategy (such as: Column, Row, Diagonal, and block sampling strategy), 2DLDA of the grid sampled image can extract the feature which is more discriminative than 2DLDA of all other sampling strategy. Suppose there are N training grid sampled images $A_i \in R^{m\times n}$,consisting of L known pattern classes, denoted as C_1, C_2, .., C_L,

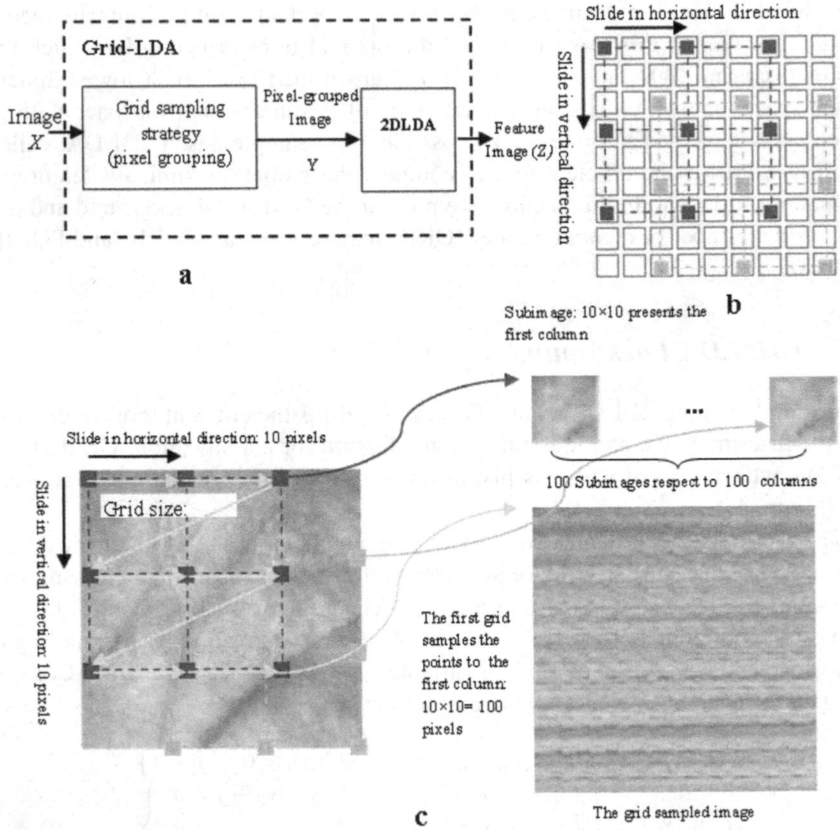

Fig. 4 (a) Block diagram of GridLDA, (b) Grid-sampling strategy, and (c) the process of grid-sampling

C_i consists of the N_i training images from the ith class and $N = \sum_{i=1}^{K} N_i$. The global centroid $\bar{A}$ of all training grid sampled image and the local centroid $\bar{A}_i$ of each class C_i is defined as $\bar{A} = (1/N) \sum_{i=1}^{N} A_i$, $\bar{A}_i = (1/N_i) \sum_{A_j \in C_i} A_j$. 2DLDA attempts to seek a set of optimal discriminating vectors to form a transform $X = \{x_1, x_2, .., x_d\}$ defined as:

$$X = \arg\max J(X) \tag{9}$$

where the 2D Fisher criterion $J(X)$ denoted as:

$$J(X) = \frac{X^T G_b X}{X^T G_W X} \tag{10}$$

where T denotes matrix transpose, G_b and G_w respectively are between-class and within-class scatter matrices:

$$S_B = \frac{1}{N} \sum_{i=1}^{L} N_i \left(\bar{A}_i - \bar{A}\right)^T \left(\bar{A}_i - \bar{A}\right) \tag{11}$$

$$S_w = \frac{1}{N} \sum_{i=1}^{L} \sum_{A_j \subset C_i} \left(A_j - \bar{A}_i\right)^T \left(A_j - \bar{A}_i\right) \tag{12}$$

The optimal projection matrices $X = \{x_1, x_2, .., x_d\}$ can be obtained by computing orthonormal eigenvectors of $G_w^{-1} G_b$ corresponding to the d largest eigenvalues thereby maximizing function J(X). The value of $d$can be controlled by setting a threshold as follow:

$$\frac{\sum_{i=1}^{d} \lambda_i}{\sum_{i=1}^{n} \lambda_i} \geq \theta \tag{13}$$

where $\lambda_1, .., \lambda_n$ is the n biggest eigenvalues of $(G_w)^{-1} G_b$ and θ is a pre-defined threshold.

Suppose we have obtained the n by d projection matrix X, projecting the m by n grid sampled image A onto X , yielding a m by d feature matrix Y:

$$Y = A.X \tag{14}$$

3.2 *DORIF Extraction for Classification*

Figure 5 shows an illustration of overall procedure of our proposed method. The processing steps of proposed method for extracting DORIR feature are summarized as follows:

1. Compute the NORIR and PORIR image of each palmprint image based on MFRAT based filter by applying equations (1), (2) and (5).
2. Based on GridLDA, compute the DORIR feature included two matrices YNORIR and YPORIR by applying equation (14) to the NORIR and PORIR image.

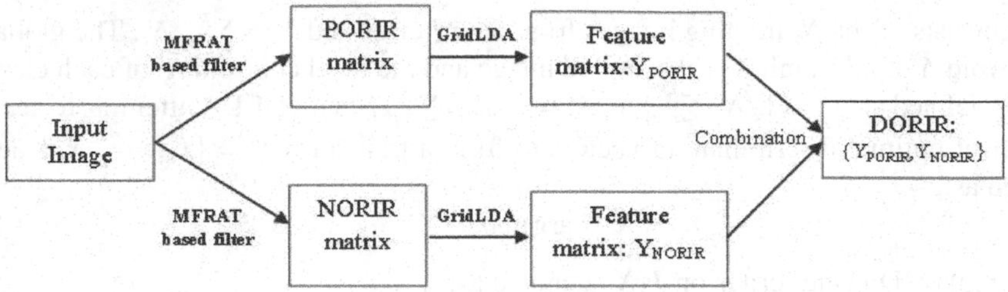

Fig. 5 An overview of our proposed method for extracting the discriminant orientation feature matrix

Figure 6 presents some results of our proposed method including: original image, NORIR image, PORIR image and some reconstructed images of these images with different dimension sizes.

Given a test palmprint image *f*, use our proposed method to obtain DORIR feature Y:$\{Y_{NORIR}, Y_{PORIR}\}$, then a nearest neighbor classifier is used for classification. Here, the distance between Y and Y_k is defined by:

$$d(Y,Y_k) = \|Y - Y_k\| = \frac{1}{6\times m\times d}\sqrt{\sum_{i=1}^{m}\sum_{j=1}^{d}(Y_{PORIR}^{(i,j)} - Y_{PORIR_k}^{(i,j)})^2} + \frac{1}{6\times m\times d}\sqrt{\sum_{i=1}^{m}\sum_{j=1}^{d}(Y_{NORIR}^{(i,j)} - Y_{NORIR_k}^{(i,j)})^2} \quad (15)$$

The distance $d(Y,Y_k)$ is between 0 and 1. The distance of perfect match is 0.

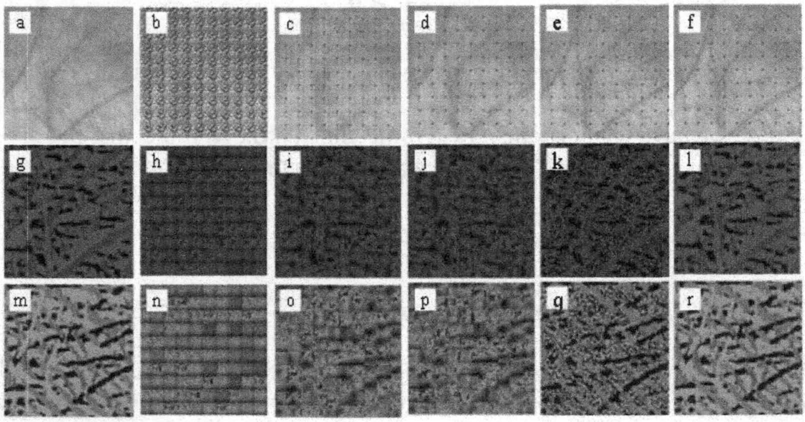

Fig. 6 Some samples which demonstrate our feature extraction method: (a) the palmprint image with size 100×100; (b)-(f) some reconstructed images of the original image by GridLDA with d={1,5, 20, 80, 99} respectively; (g) the PORIF image; (m) the NORIF image, and some reconstructed images of the PORIF image (h)-(l) and NORIF image (n)-(r) by GridLDA with d={1,5, 20, 80, 99} respectively.

4 Experimental Results

4.1 Palmprint Database

In order to evaluate the proposed method, we compare the identification performance of our method with some state of the art methods on the public palmprint databases of the Hong Kong Polytechnic University, PolyU Multispectral palmprint Databases [14]. Multispectral palmprint images were collected from 250 volunteers, including 195 males and 55 females. The age distribution is from 20 to 60 years old. We collected samples in two separate sessions. In each session, the subject was asked to provide 6 images for each palm. Therefore, 24 images of each illumination from 2 palms were collected from each subject. In total, the database contains 6,000 images from 500 different palms for one illumination. The average time interval between the first and the second sessions was about 9 days. In our experiments, we use ROI databases in which each image is downsampled with size 100×100 pixels for evaluating our feature extraction methods.

4.2 Palmprint Database

Identification is a process of comparing one image against *N* images. In the following tests, we setup two registration databases for *N=100* and *200*, which are similar to the number of employees in small to medium sized companies. The first database contains *600* templates from *100* different palms, where each palmprint has six templates in the first session. Similarly, for *N=200*, the second registration databases have *1200* templates from *200* different palms. We also setup two testing databases in the second session: Test set 1 with *1200* templates from *200* different palms and Test set 2 with *2400* templates from *400* different palms. None of palmprint images in the testing database are contained in any of the registration databases. Each of the palmprint images in the testing database is matched with all of the palmprint images in the registration databases to generate incorrect and correct identification distances. Since a registered palmprint has six templates in the registration databases, a palmprint of a registered palm in test database is matched with its six templates in the registration databases to produce six correct distances. The minimum of these six distances is taken as correct identification distance. Similarly, the palmprint image in testing database is compared all palmprint images in any registration database to produce 6*N* incorrect distances if the testing palmprint does not have any registered images, or *6N-6* incorrect distances if the palmprint has five registered images. We take the minimum of these distances as the incorrect identification distances. Table 2 presents the parameters of these datasets. If a matching distance of two images from the same palm is smaller than the threshold, the match is a genuine acceptance. Similarly, if a matching distance of two images from different palms is smaller than the threshold, the match is a false acceptance. Table 2 represents the top recognition accuracy and the corresponding feature dimensions of our method on these two datasets. Figure 7 shows a scatter plot of the correct and incorrect distance distributions obtained from palmprint features of

Table 1 Parameters of databases in identification experiments

Databases	Training set	Testing set		Number of Identification	
		Registra-tion set	Unregis-tration set	Correct distance	Incorrect distance
Set 2 (N=100)	600	600	600	600	600+600=1200
Set 3 (N=200)	1200	1200	1200	1200	1200+1200=2400

Table 2 Genuine acceptance rate of our method with false acceptance rate= 0%

Dimensions	Genuine recognition rate (%)	
	Dataset 1	Dataset 2
2×100	93.16	92.83
5×100	97.83	97.67
25×100	97.50	97.25

Compcode [8], RLOC [10] and our method. It can be observed that the distributions of our method are well separated and a linear classifier would be able to discriminate the genuine and impostor classes. The Receiving Operating Characteristic (ROC) curve of Genuine Acceptance Rate (GAR) and False Acceptance Rate (FAR) of our method and other are presented in Figure 8. From this group of figures, we can see that the recognition performance of our method is more stable and better than the state of art methods (CompCode [8] and RLOC[10]).

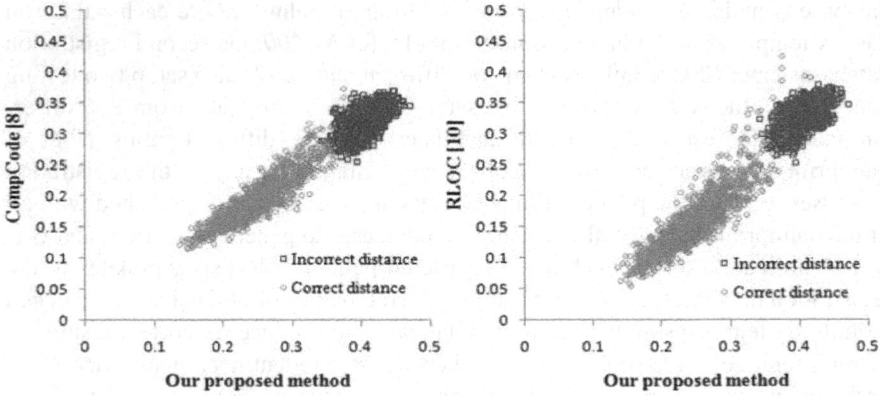

Fig. 7 Correct and incorrect distances on dataset 2

4.3 *Speed*

The proposed method and other methods are implemented using C# on a PC with CPU Intel Core2Duo 1.8 GHz and Windows 7 Professional. The average testing time of the above methods are compared in Table 3. The average execution time for the feature extraction and matching are 88 ms, and 0.12 ms, respectively. With

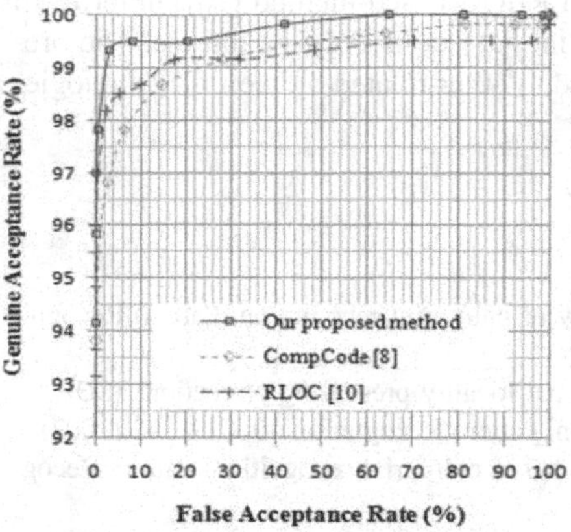

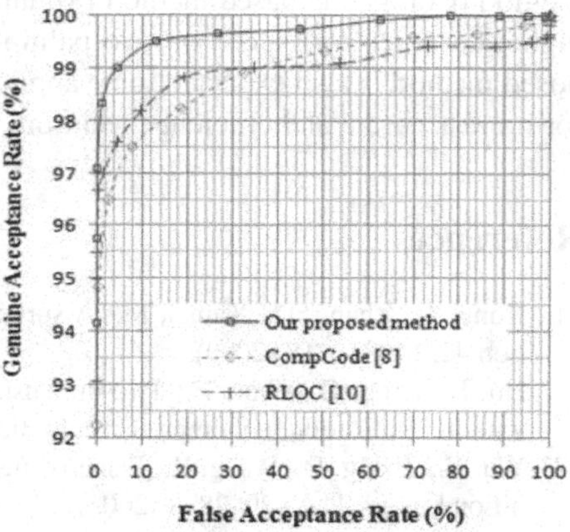

Fig. 8 The ROC curves for RLOC [10], CompCode [8] and our proposed method with d=15 on (a) dataset 1 and (b) dataset 2, respectively

Table 3 Comparison of testing time

Method	Feature extraction (ms)	Identification time for one image (ms)	
		Data set1	Data set 2
CompCode based Method [8]	305	582	823
RLOC based method [10]	81	153	211
Our proposed method	88	147	204

identification, the total identification time is about 147 ms with dataset 1 and 204 ms with dataset 2.

5 Conclusion

In this paper, we present a novel feature, called Discriminant Orientation Representation (DORIR) for palmprint identification based on using Modify Finite Radon Transform (MFRAT) and GridLDA. The MFRAT is applied to compute the negative orientation representation and positive orientation representation which describe separately the orientations of principal lines and wrinkles. Moreover, it process more quickly than Gabor filter [8], [9]. The output orientation images still contain redundant information in image searching. Therefore, the GridLDA is used to extract the discriminant feature which is more suitable for palmprint recognition. Experimental results show that our proposed method outperforms the state of art method (Compcode based method [8], and RLOC based method [10]) in terms of higher accuracy and speed on two palmprint dataset of PolyU database. The proposed method is also expected to be applied to other related biometric technologies for more accurate and efficient execution.

References

1. Kong, A., Zhang, D., Kamel, M.: A survey of palmprint recognition. Pattern Recognition 42, 1408–1418 (2009)
2. Hu, D., Feng, G., Zhou, Z.: Two-dimensional locality preserving projections (2DLPP) with its application to palmprint recognition. Pattern Recognition 40, 339–342 (2007)
3. Wu, X., Zhang, D., Wang, K.: Fisherpalms based palmprint recognition. Pattern Recognition Letters 24, 2829–2838 (2003)
4. Lu, G., Zhang, D., Wang, K.: Palmprint recognition using eigenpalms features. Pattern Recognition Letters 24, 1463–1467 (2003)
5. Huang, D.S., Jia, W., Zhang, D.: Palmprint verification based on principal lines. Pattern Recognition 41(5), 1514–1527 (2008)
6. Wu, X., Zhang, D., Wang, K., Huang, B.: Palmprint classification using principal lines. Pattern Recognition 37, 1987–1998 (2004)
7. Wu, X., Zhang, D., Wang, K.: Palm Line Extraction and Matching for Personal Authentication. IEEE Transactions on System, Man, and Cybernetics-part A: Systems and Humans 36(5), 978–987 (2006)
8. Kong, A.W.K., Zhang, D.: Competitive coding scheme for palmprint verification. In: Proceedings of International Conference on Pattern Recognition, pp. 520–523 (2004)
9. Zhang, D., Guo, Z., Lu, G., Zhang, L., Zuo, W.: An Online System of Multispectral Palmprint Verification. IEEE Transactions Instrumentation and Measurement 59, 480–490 (2010)
10. Jia, W., Huanga, D.S., Zhang, D.: Palmprint verification based on robust line orientation code. Pattern Recognition 41, 1316–1328 (2008)
11. Du, F., Yu, P., Li, H., Zhu, L.: Palmprint Recognition using Gabor Feature-based Bidirectional 2DLDA. In: Yu, Y., Yu, Z., Zhao, J. (eds.) CSEEE 2011, Part II. CCIS, vol. 159, pp. 230–235. Springer, Heidelberg (2011)
12. Van, H.T., Tat, P.Q., Le, T.H.: Palmprint verification using GridPCA for Gabor features. In: Proceedings of the Second Symposium on Information and Communication Technology SOICT 2011, pp. 217–225 (2011)
13. Van, H.T., Le, T.H.: GridLDA of Gabor Wavelet Features for Palmprint Identification. In: SoICT 2012 Proceedings of the Third Symposium on Information and Communication Technology, pp. 125–134 (2012)
14. PolyU multispectral palmprint Database, `http://www.comp.polyu.edu.hk/~biometrics/MultispectralPalmprint/MSP.htm`

Localization and Velocity Estimation on Bus with Cell-ID

Hung Nguyen, Tho My Ho, and Tien Ba Dinh

Abstract. In this study, the authors estimate the position and the velocity of a mobile terminal using only Cell-ID as input data. Unlike finger prints methods, Cell-IDs are considered as sequences and matched using a modification of Longest Common Subsequece. The study covered 693 bus stops in Ho Chi Minh City and collected 73813 transition samples as training data. 86% of the test cases in velocity estimation has error within 5 km/h. This project uses bus stops as landmarks, but can be generalized using any set of annotated landmarks.

1 Introduction

Cell-ID is the identification number of a base station. In order to send or receive calls and messages, a mobile phone must be connected a base station.

Techniques on localization and tracking with Cell-ID have been developed in the past decades [1][2]. Accuracy of localization with Cell-ID varies widely from 10m to 35km depends on beacon radius [3]. Average distance between localization with GPS sensor and localization with Cell-ID is 500m in an urban area of Italy and 800m in an urban area of United States [2]. Despite its low accuracy, Cell-ID application needs no hardware upgrade, requires minimal power usage. Cellular signal also has better penetration indoor than GPS signal.

Cell-ID data can be used alone or in hybrid mode with WiFi data. Large network operators such as AT&T and T-Mobile use Cell-ID to locate 911 callers in compliant with the E911 mandate [3].

Cell-IDs are traditionally used as fingerprints. However another approach takes Cell-ID transitions as input data and uses Hidden Markov Model (HMM) and Sequence Alignment as core algorithm [4] [5] [6]. An experiment in Sweden uses

Hung Nguyen · Tho My Ho · Tien Ba Dinh
Advanced Program in Computer Science, Faculty of Information Technology, VNU – University of Science, Ho Chi Minh City, Vietnam
e-mail: {nhung,hmtho}@apcs.vn, dbtien@fit.hcmus.edu.vn

V.-N. Huynh et al. (eds.), *Knowledge and Systems Engineering, Volume 1*, Advances in Intelligent Systems and Computing 244,
DOI: 10.1007/978-3-319-02741-8_22,

HMM and beacon map to locate the user with an error of 300m with 67% confidence level, and an error of 480m with 95% confidence level [6]. A sequence matching expirement results in the name of the route with an error rate of less than 20% [5].

Although localization and tracking with Cell-ID has been investigated thoroughly, we could not find any previous attempt to estimate velocity of a mobile phone user using only Cell-ID. This project assess the feasibility of using Cell-IDs in localization and velocity estimation for bus users. Bus is chosen because of its fixed routes and bus stops acts as identifiers for our method.

2 Method

We apply Longest Common Subsequence algorithm on cell transition data to set up a system which allows the users on bus to know where they are at, on what speed they are moving, thus the system can be applied to find the remaining time to reach a desired destination.

2.1 Data Collection

We developed Cell-Logger, an Android application, to collect our training data and testing data. The following information is collected in our dataset:

- **Main Cell information:** contains Cell-ID, Location Area Code (LAC) and Signal strength. Cell-ID is the identifier for each base station, ranging from 0x0000 to 0xffff. Since Cell-ID is not necessarily unique, both Cell-ID and LAC were taken into account to identify a station. However, collisions might still happen.
- **Neighbouring cell information:** contains a list of neighbouring cells. The mobile terminal is not currently connected with these cells but is within their range.
- **GPS information:** contains latitude, longitude, and its accuracy. GPS data is used to produce the landmark map and to test against the result.
- **Landmark information:** contains bus stop locations and bus routes. This information is retrieved from HCMC bus center website. [7]

Data for each bus route is collected. Our training data contains 19 bus routes in HCMC, estimated to be 600km of street data. The testing data contains 4 routes. The maps of coverage are shown in Figure 1a and Figure 1b.

2.2 Localization

Sequence alignment, *Longest Common Subsequence* in particular is used because it has the advantages of having low computational complexity on time and space. The accuracy we achieved on our testing set ranged from 70% to 94%.

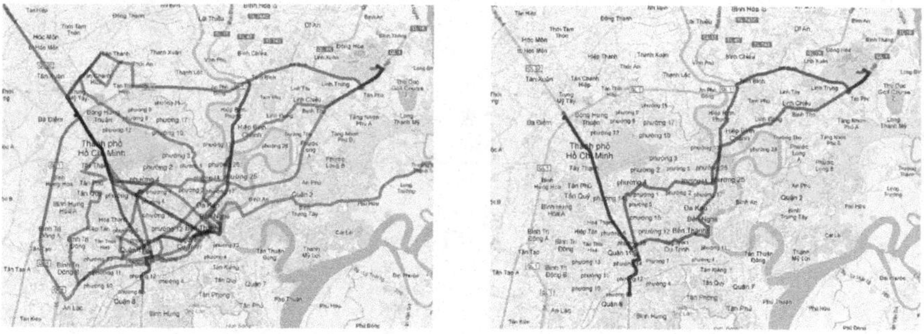

Fig. 1 Data coverage of training data (a) and testing data (b)

2.2.1 Sequence

The input data is a sequence of 2-tuples (State, Observation). At each time-frame, State is the BusStopID that the bus is moving towards, and Observation is the beacon IDs (Cell-ID) that the mobile terminal is connected to.

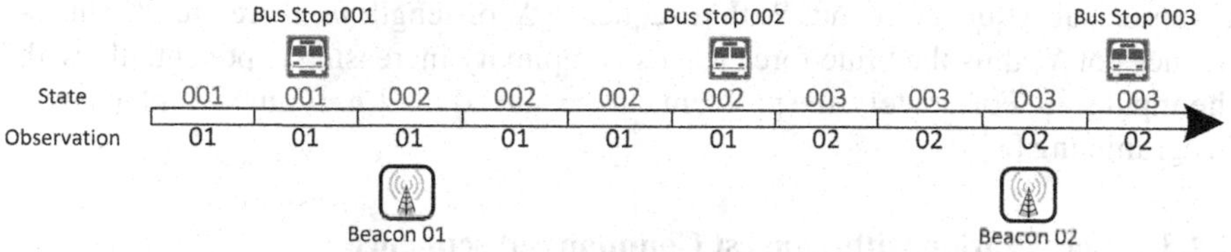

Fig. 2 The collected data is a sequence of observations (Cell-IDs) and states (BusStopIDs). The deprived sequence for this Figure is $\langle(001,01)(001,01),(002,01),(002,01),...\rangle$.

2.2.2 Longest Common Subsequence

Formal definition of *Longest common subsequence:*

A subsequence of a sequence is the given sequence with some elements (possibly none) left out. Formally given a sequence $X =< x_1,x_2...x_n >$. Z is a subsequence of X if there exist a strictly an increasing sequence $I = i_1,i_2...i_k\}$ of indices of X such that for all $j = 1,2,k$ we have $x_{ij} = z_j$

$$\text{Notation: } Z \subseteq X$$

Given two sequence X and Y. A sequence Z is a common subsequence of X and Y if and only if Z is a subsequence of X and Z is also a subsequence of Y

$$Z = CS(X,Y) \iff Z \subseteq X and Z \subseteq Y$$

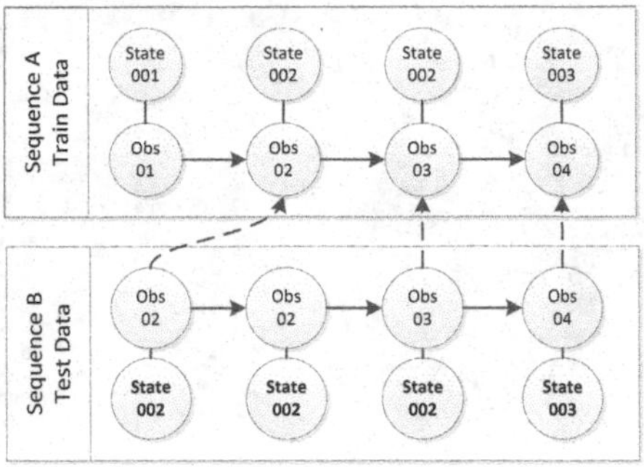

Fig. 3 An example of finding Longest Common Subsequence

Given two sequences X and Y, our goal is to find the longest possible common subsequence.

Complexity of LCS problem

A naive approach to solve *LCS* problem is to enumerate all subsequences of X and check if this sequence is also a subsequence of Y while keeping track of the longest common subsequence found. With a sequence X of length m, there are 2^m subsequences of X, thus the brute force has its complexity increasing exponentially with the input size. Fortunately, the problem can be solved in $O(mn)$ time with Dynamic Programming [8]

2.2.3 Localization with Longest Common Subsequence

As shown in Figure 3, a sequence contains a number of sequence elements. Each sequence element has two fields: state (the BusStopID we try to obtain) and observation (Cell-ID and LAC). In the training data, both fields are available. In the test data, only the observation is available. We use *LCS* algorithm to match the sequence of observation in the testing data to each of the sequence in the training set.

We use the length of the longest common subsequence as the scoring system. The last state of this sequence is our prediction, which is state 003 in Figure 3.
Here is the process of the localization algorithm with LCS:

- **Step 1:** there are many train sequences. We used LCS to output the longest common subsequences between test sequence and each train sequence.
- **Step 2:** find the longest sequence in the output of step 1. The last state of this sequence is our prediction.

The output state of the algorithm is the BusStopID of the current location. However, due to limitation of Cell-ID, a list of two or three most likely BusStopID may be returned (Table 2).

2.3 *Velocity Estimation*

Velocity of a mobile terminal is estimated based on its cell transitions. Once matched to training data, these transitions are used to calculate the distance that the mobile terminal has travelled. The velocity is calculated by distance over time.

2.3.1 Sequence Used in Velocity Estimation

The velocity estimation algorithm uses a different sequence structure compared to the sequence structure used LCS (Figure 2). In this sequence, each element is a 3-tuple of $(Observation, Repetition, State)$. *Observation* contains the beacon ID (Cell-ID) that the subject is currently connected to. *Repetition* is the number of seconds that the mobile terminal stays connected to the Cell-ID. *State* is the distance that the user has travelled while staying connected to the Cell-ID. In Figure 4, the complete sequence is $\langle(01,6,d1),(02,4,d2),...\rangle$.

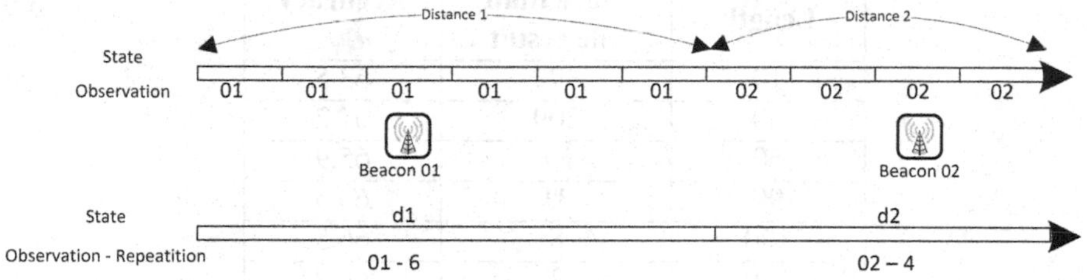

Fig. 4 New data structure for velocity estimation algorithm. The state now is the influence radius of a base beacon.

2.3.2 Velocity Estimation

Different routes may experience the same transitions but for different time intervals. Thus the sequence of each route must be processed and stored separately. To estimate the velocity of a subject, we use the following procedure:

- **Step 1:** identify the bus route ID that the bus user is travelling on. This can be done with LCS algorithm in 2.2.2.
- **Step 2:** match the Cell-ID transitions with the transitions in the identified route with LCS. The output of this step is the distance travelled.
- **Step 3:** calculate the velocity by distance over time. The time is deduced from the subject's Cell-ID transition.

3 Result

The result of localization and velocity estimation expiriments is reported in this section. An overview on collected data can be found in Apprendix section.

Table 1 Result of LCS algorithm - one bus stop tolerance. Accuracy is measured in 1000 tests.

Sequence Length	Case return more than one result	Accuracy
10	519	61.1
30	365	65.7
60	187	66.6
90	68	64.1
120	20	68.4
180	14	69.2
300	8	72.7

(a) Test result with one bus stop tolerance - Mobifone

Sequence Length	Case return more than one result	Accuracy
10	394	63.5
30	209	67.2
60	87	65.9
90	31	65.5
120	8	66.5
180	2	66.9
300	12	68.9

(b) Test result with one bus stop tolerance - Viettel

3.1 Localization Result

In all experiments, the localization accuracy increases as the sequence length increases. The number of cases that the algorithm returns more than one result also decreases as the sequence length increases (Table 1, Table 2).

Our data comes from beacons of two networks: Viettel and Mobifone. The maximum accuracy we achieve is 72.7% in Mobifone data set (Table 1). But generally, Viettel with denser beacon network results in better results. Table 2 shows the result when LCS returns two bus stops (the current bus stop, and the next bus stop) or three bus stops (the previous bus stop, the current bus stop, and the next bus stop).

With two bus stop tolerance, we achieve a maximum accuracy of 87.2% in the Mobifone data set. With three bus stop tolerance, we achieve a maximum accuracy of 93.9% in the Viettel data set(Table 2).

From one to two bus stop tolerance, the accuracy increases about 15%. From two to three bus stop tolerance, the accuracy increases about 5%.

Table 2 Result of LCS algorithm with two bus stop and three bus stop tolerance. Accuracy is measured in 1000 tests.

Seq length	Accu -racy
10	68.9
30	75.3
60	77.7
90	79.6
120	80.8
180	84.2
300	87.2

(a) Test result with two bus stop tolerance Mobifone

Seq length	Accu -racy
10	69.2
30	76.8
60	81.5
90	85.2
120	85.1
180	92.1
300	91.5

(b) Test result with three bus stop tolerance - Mobifone

Seq length	Accu -racy
10	73.2
30	76.3
60	78.2
90	79.8
120	80.9
180	80.7
300	84.2

(c) Test result with two bus stop tolerance - Viettel

Seq length	Accu -racy
10	73.3
30	78.2
60	83.2
90	88.5
120	88.2
180	91.4
300	93.9

(d) Test result with three bus stop tolerance - Viettel

3.2 *Velocity Estimation Result*

Table 3 shows the accuracy of the velocity estimation algorithm in part 2.3.2. The estimated velocity is compared against the real velocity acquired by GPS data. We observe that the accuracy of the algorithm increases as the sequence length increases. Although the accuracy can be 92.5% (error radius less than 5km/h) with sequence length 300, this setting is not recommended because in 300 seconds the user may have passed 3-4 bus stops (Figure 5).

4 Discussion

In this project, we provided new insight into Cell-ID techniques. We developed a complete solution to locate buses' position with respect to its bus stops and to estimate the buses' velocity using a combination of Cell-ID and sequence alignment techniques.

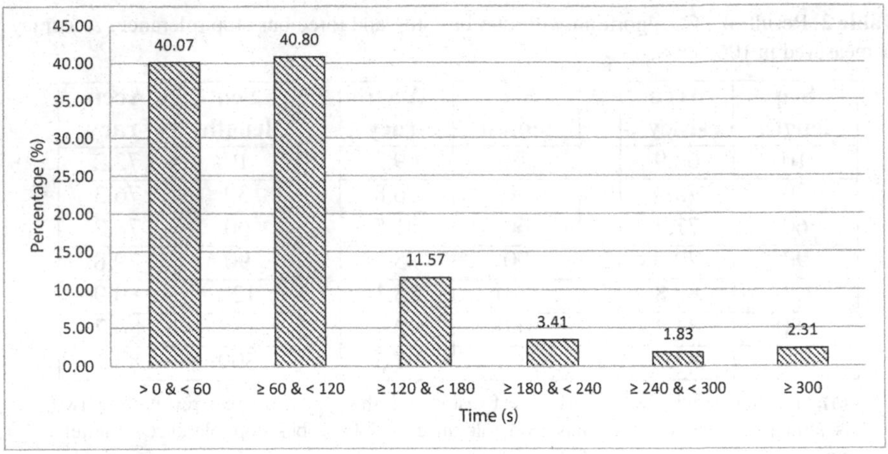

Fig. 5 Time interval between adjacent bus stops in training set

Table 3 Velocity estimation accuracy. Each case are tested 1000 times.

Sequence Length	Accuracy with 100% of correct matching	Real Accuracy
10	57	31.7
30	72.1	36
60	84.7	47.3
90	88.4	56.4
120	92.5	61.6
180	92.3	74.7
300	93.2	86

- **Sequence length** is the length of test sequence. A sequence with length 60 contains cell data in 60 seconds.
- **Accuracy:** Estimated velocity is compared against true velocity. We counted the percentage that estimated velocity fallen within 5km/h range to the true velocity.
- **Accuracy with 100% correct matching:** No matching were done in these experiments. We input random sequence from the training set but knew exactly from which route the sequence was, where it started and ended.
- **Real accuracy:** We input random sequence from the training set but used the matched information from LCS to estimate the velocity.

Localization result

We implemented LCS to localize bus stop on its route. LCS works on a sequence as a whole. LCS would perform well even if some Cell-IDs are unprecedented in the training set. With most of the settings, localization accuracy are greater than 68% confidence level, which makes the system is applicable and provides meaningful information to the users.

Velocity estimation result

Two different tests are conducted in the velocity estimation part.

Firstly, we used the original sequence from the training set as input, in other words BusRouteID, the start and the end of the sequence are provided. This experiment shows the upper limit of velocity estimation accuracy and is used as a reference to the later test.

Secondly, we used a testing set that is different from the training set. This experiment shows the accuracy of velocity estimation in real life. With sequence length 300, 86% of the estimated velocity is fallen within 5km/h radius to the true velocity.

In all experiments, the accuracy increases as the sequence length increases because we have more information with longer sequence. Additionally, outside inferences and noises from vehicle changing its velocity are reduced over longer time span.

Application of the project

With the help of this paper, location and velocity of buses is ready to the end users. Base on this information, a traffic congestion control system can be employed. Future research and insight about traffic and user movement trend can be studied in this system.

Future work

There are still some limitations needed to be improved in future research.

First, the localization accuracy is low due to the limitation of Cell-ID. This can be reduced by using main Cell-ID along with neighbouring cells or in hybrid mode with WiFi signal. We choose Cell-ID along with neighbouring cells because WiFi cards take considerable time and power to scan surrounding environments. Retrieving neighbouring cells may be limited by mobile phone manufacturers but this approach is promising. Dual SIM card phones also falls into this category. We also considered using signal strength of Cell-ID but the data was too noisy to add useful pattern of the user's movement.

Second, the beacon map of a network may be changed without any notice by the network operator. To apply this project into real life, an automated update server that uses anonymous data of users to build a consistent beacon map. The consistency would allow us to build up a system to track the movement of each bus and then alarm the user that is currently waiting for that bus. Besides cellular has better penetration than GPS signal, and thus it may be possible to expand this system on train and subways system.

Third, the user must wait a considerable amount of time (2-5 minutes) to achieve the desired accuracy in velocity estimation. The wait time can be reduced to be

Table 4 Distance and time interval of transitions of Viettel and Mobifone

	Viettel	Mobifone
Average distance (m)	131.5146	161.6895
Average time interval (s)	35.00046	42.80436
Max distance (m)	1317.346	1682.351
Max time interval (s)	559	374
Min distance (m)	0.541486	0.267081
Min time interval (s)	1	1
Number of samples	2188	1789

reasonable in trade off with accuracy. To find the balance between wait time and accuracy is also an important problem to be solved before applying this project into real life.

Overall this paper provides a cheaper alternative to GPS sensor since the method can be applied without additional modification to infrastructures. Cell-ID techniques should be applied in a small area with dense network activity.

Appendix

Data Analysis Result

In total, the collected data covered 19 bus routes consists of two main network providers in Vietnam: Viettel and Mobifone. Table 4, Figure 6 shows the overview of main cell in collected data. The phones main cell fluctuates dramatically. Although the maximum distance between transitions can be extreme (about 1.5km), these cases occurs infrequently and mostly on remote area. About 50% of the transitions occur within 100m and about 75% occurs within 200m. 68% of transitions occur within 50s for Mobifone network and 78% of transitions occur within 50s for Viettel network. The average distance and time for Viettel network to change main cell is shorter than Mobifone. These could be the result of Viettel network having denser coverage than Mobifone network.

The fluctuation of main cell can act as both our advantage and disadvantage. The faster the main cell changes, the more information we have about the distribution of cells within that area. If the main cell remains unchanged while driving through a street, we cannot distinguish where the street begin or end since in both case all information we get is the same cell-ID. However, if main cell fluctuates back and forth between the same set of cells, these fluctuations do not indicate the users movement since the users position remains within the same region. Thus, these fluctuations are noises and hard to predict.

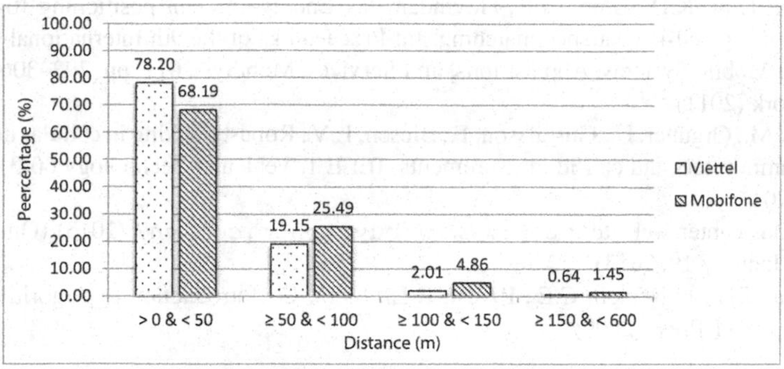

Fig. 6 Distance between Cell-ID transitions

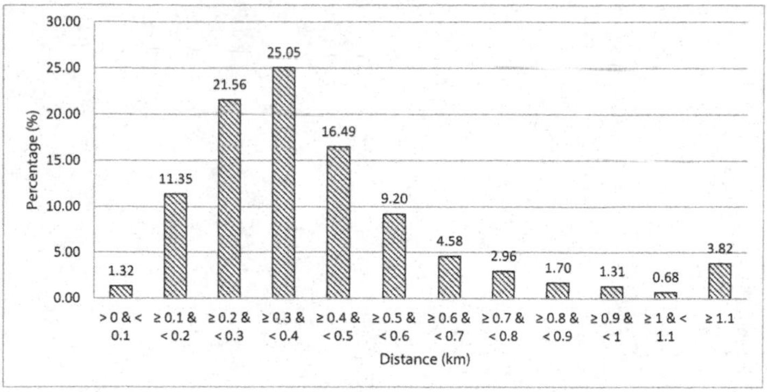

Fig. 7 Distance between bus stops

Acknowlegments. This research is supported by research funding from Advanced Program in Computer Science, University of Science, Vietnam National University - Ho Chi Minh City.

References

1. Ahson, S., Ilyas, M.: Location-Based Services Handbook: Applications, Technologies, and Security. Taylor & Francis (2011)
2. Trevisani, E., Vitaletti, A.: Cell-id location technique, limits and benefits: An experimental study. In: Proceedings of the Sixth IEEE Workshop on Mobile Computing Systems and Applications, WMCSA 2004, pp. 51–60. IEEE Computer Society, Washington, DC (2004)
3. Willaredt, J.: Wifi and cell-id based positioning - protocols, standards and solutions (2011)
4. Laasonen, K., Yliopisto, Tietojenkäsittelyopin Laitos, H.: Mining cell transition data. Department of Computer Science, series of publications A, report A-2009-3. University of Helsinki (2009)

5. Paek, J., Kim, K.H., Singh, J.P., Govindan, R.: Energy-efficient positioning for smartphones using cell-id sequence matching. In: Proceedings of the 9th International Conference on Mobile Systems, Applications, and Services, MobiSys 2011, pp. 293–306. ACM, New York (2011)
6. Bshara, M., Orguner, U., Gustafsson, F., Biesen, L.V.: Robust tracking in cellular networks using hmm filters and cell-id measurements. IEEE T. Vehicular Technology 60(3), 1016–1024 (2011)
7. Hcmc bus center website, `http://www.buyttphcm.com.vn/` (2013) (Online; accessed January 19, 2013)
8. Cormen, T.H., Leiserson, C.E., Rivest, R.L., Stein, C.: Introduction to Algorithms, 3rd edn. The MIT Press (2009)

A New Improved Term Weighting Scheme for Text Categorization

Nguyen Pham Xuan and Hieu Le Quang

Abstract. In text categorization, term weighting is the task to assign weights to terms during the document presentation phase. Thus, it affects the classification performance. In this paper, we propose a new term weighting scheme $logtf.rf_{max}$. It is an improvement to *tf.rf* – one of the most effective term weighting schemes to date. We conducted experiments to compare the new term weighting scheme to *tf.rf* and others on common text categorization benchmark data sets. The experimental results show that $logtf.rf_{max}$ consistently outperforms *tf.rf* as well as other schemes. Furthermore, our new scheme is simpler than *tf.rf*.

1 Introduction

The task of text categorization is to classify documents into predefined categories. Text categorization has been studied extensively in recent years [12]. In the vector model, each document is presented as a vector of terms. Each vector component contains a value presenting how much the term contributes to the discriminative semantics of the document. The goal of a term weighting method is to assign appropriate weights to terms in order to achieve high classification performance.

Term weighting methods can be divided into two categories, namely, supervised term weighting method and unsupervised term weighting method [8]. The traditional term weighting methods such as *binary*, *tf*, *tf.idf* [11], belong to the unsupervised term weighting methods. The other term weighting methods (for example, $tf.\chi^2$ [3]), that make use of the prior information about the membership of training documents in predefined categories, belong to the supervised term weighting methods.

Nguyen Pham Xuan · Hieu Le Quang
Faculty of Information Technology, VNU University of Engineering and Technology, Vietnam
e-mail: {nguyenpx.mi10,hieulq}@vnu.edu.vn

V.-N. Huynh et al. (eds.), *Knowledge and Systems Engineering, Volume 1*, Advances in Intelligent Systems and Computing 244,
DOI: 10.1007/978-3-319-02741-8_23, © Springer International Publishing Switzerland 2014

The supervised term weighting method *tf.rf* [8] showed consistently better performance than many other term weighting methods in experiments using SVM and kNN, two of the most commonly-used algorithms for text categorization [14]. The OneVsAll approach used in [8] transforms the multi-label classification problem of N categories into N binary classification problems, each of which associates with a different category. For each term, *tf.rf* requires N *rf* values, each for a binary classification problem.

In this paper we present a new term weighting scheme $logtf.rf_{max}$, an improvement to *tf.rf*. Our scheme requires a single *rf* value for each term for a multi-label classification problem. Moreover, it uses $logtf = log_2(1.0 + tf)$ instead of *tf*. Our experimental results show that our scheme is consistently better than *tf.rf* and others.

The rest of paper is organized as follows. Section 2 reviews related works. Our new improved term weighting method is described in Section 3. Section 4 and Section 5 report our experimental settings and results, as well as our disscusion. Finally, we draw conclusions in Section 6.

2 Related Works

In this section, we give a brief overview of the existing term weighting methods. We also describe the feature selection method CHI_{max} which relates to an improvement in our new term weighting scheme.

2.1 Traditional Term Weighting Methods

Generally, the traditional term weighting methods are rooted from the information retrieval field and they belong to the unsupervised term weighting methods. The simplest *binary* term weighting method assigns 1 to all terms in a document in the vector representation phase. The most widely-used term weighting approaches in this group is *tf.idf*. *tf* is the frequency of a term in a document and *idf* is the inverse document frequency of a term. *tf* has various variants which use the logarithm operation such as *log(tf)*, *log(1 + tf)*, *1 + log(tf)* [9]. The goal of logarithm operation is to scale down the effect of noisy terms.

2.2 Supervised Term Weighting Methods

The supervised term weighting methods are the ones that make use of the prior information about the membership of training documents in predefined categories to assign weights to terms. One way to use this known information is to combine *tf* and a feature selection metric such as χ^2, Information Gain, Gain Ratio [3], [2].

tf.rf is the supervised term weighting method that combines *tf* and *rf* (relevance frequency) factor which proposed by Lan et al. [8]. As said in Introduction, OneVsAll transforms a multi-label classification problem into N binary classification problems, each relates to a category which is tagged as the positive category and all other categories in the training set are grouped into the negative category. Each term t requires one *rf* value in each category C_i, and this value is computed as follows:

$$rf(C_i) = \log_2(2 + \frac{a}{c}). \tag{1}$$

where, a is the number of documents in category C_i which contains t and c is the number of documents not in category C_i which contains t. The purpose of *rf* is to give more weight to terms which help classify documents into the positive category.

2.3 Feature Selection Method CHI_{max}

CHI is a feature selection method using the χ^2 statistic, which is used to compute score of a term t in a category C_i as bellow:

$$\chi^2(t, C_i) = N * \frac{(a*d - b*c)^2}{(a+c)*(b+d)*(a+b)*(c+d)}. \tag{2}$$

a is the number of documents in category C_i which contain t, b is the number of documents in category C_i which do not contain t, c is the number of documents not in category C_i which contains t, d is the number of documents not in category C_i which do not contain t, N is the total number of documents in the training set.

For the multi-label classification problem which is transformed by the OneVsAll method, each feature is assigned a score in each category as described in 2. Then all these scores are combined into a single final score base on a function such as *max* or *average*. Finally, all features are sorted by final scores in descending order and top p highest score features are selected. The levels of p are optional. CHI_{max} and CHI_{avg} are two CHI feature selections that use *max* and *average* to combine all scores, respectively. Following [10], CHI_{max} performs better than CHI_{avg}.

3 Our New Improved Term Weighting Method

As said before, for each term in a multi-label classification problem, *tf.rf* uses N *rf* values, each of them for a different binary classifier. Meanwhile, our scheme, which also uses OneVsAll method, assigns a single rf_{max}(maximum of all *rf*) for each term for all binary classiffiers. By this way, the weight of a term is now corresponded to the category that this term repesents the most. This approach is similar to the one used in the CHI_{max} feature selection method described in Section 2. Our experimental results (see Section 5) show that this improvement helps to increase classification performance.

One consequence of making use of the highest value is that our scheme is simpler than *tf.rf*. For a *N-class* problems, the *tf.rf* requires N vector representations for each document, one for each binary classificatier. Meanwhile, our scheme need only one presentation for all N binary classifiers.

Our other observation is that *tf.rf* has the lower result than *rf* in some cases as described in [8]. We believe that the reason of the problem is the impact of noisy terms that repeated many times in a document. Table 1 illustrates this point.

Table 1 Examples of two terms which have different *tf* and $log_2(1 + tf)$

term	*tf*	$log_2(1+tf)$
song	2	1.58
the	10	3.45

Clearly, *benefit* (a common word occuring in many categories) is a noisy term. According to *tf* scheme, the ratio of weight of *the* and *song* is 5:1. Meanwhile, this ratio becomes 3.45:1.58 by $log_2(1+tf)$ scheme. In other words, the effect of *the* is scaled down by $log_2(1+tf)$ scheme. We tried to combine *rf* or rf_{max} with *logtf* instead of *tf* as to improve the classification results. In our experiments, schemes using *logtf* show better performance than others using *tf* on two data sets.

To sum up, our new improved term weighting method computed weight for a term *t* as follows:

$$logtf.rf_{max} = log_2(1.0 + tf) * \max_{i=1}^{N}\{rf(C_i)\} \qquad (3)$$

where *tf* is the frequency of *t*, *N* is the total number of categories, $rf(C_i)$ is defined in equation 1 Section 2.

4 Experiments

This section describes the experimental settings, including data corpora, inductive learning algorithm and performance measures.

4.1 Data Corpora

We used the Reuters News corpus and the 20 Newsgroups corpus - the two commonly used benchmark data sets. We choose these data sets so as to our results can be compared with others, especially those reported in [8].

4.1.1 Reuters News Corpus

[1]. The Reuters-21578 corpus contains the 10794 news stories, including 7775 documents in the training set and 3019 documents in the test set. There are 115 categories that has at least 1 training documents. We have conducted experiments on Reuter

[1] Reuters-21578 corpus can be downloaded from `http://www.daviddlewis.com/resources/testcollections/reuters21578/`

top ten (10 largest categories in this corpus) and each document may be categorized in more than one category. In text preprocessing phase, 513 stop words, numbers, words containing single char and words occurring less than 3 times in the training were removed. The resulting vocabulary has 9744 unique words (features). By using CHI_{max} for feature selection, the top $p \in \{500, 2000, 4000, 6000, 8000, 10000, 12000\}$ features are tried. Besides, we also used all words in the vocabulary.

The categories in the Reuters News corpus have the skewed distribution. in the training set, the most common category (*earn*) accounts for 29% of the total number of samples, but 98% of the other categories have less than 5% samples.

4.1.2 20 Newsgroups Corpus

[2]. The 20 Newsgroups corpus (20NG) is a collection of roughly 20,000 newsgroup documents, divided into 20 newsgroups. Each newsgroup corresponds to a different topic. After removing duplicates and headers, the remaining documents are sorted by date. The training set contains 11314 documents (60%) and 7532 documents (40%) belong to the test set. In text preprocessing phase, 513 stop words, words occurring less than 3 times in the training or words containing single char were removed. There are 37172 unique words in vocabulary. We used CHI_{max} for feature selection, the top $p \in \{500, 2000, 4000, 6000, 8000, 10000, 12000, 14000, 16000\}$ were selected.

The 20 categories in the 20 Newsgroups corpus have the rough uniform distribution. This distribution is different from the category distribution in the Reuters News corpus.

4.2 *Inductive Learning Algorithm*

We used linear SVM algorithm since it has shown better performance than other algorithms in the prior studies [4], [7]. In addition, for SVM methods, linear kernel is simpler but as good as other kernels like RBF [6]. The linear SVM library we used is LIBLINEAR 1.93 [5].

4.3 *Performance Evaluation*

Measures for a category are generally *precision* (p), *recall* (r) and F_1 [16]. F_1 is a combination of *precision* and *recall* and it is defined as bellow:

$$F_1 = \frac{2.p.r}{p+r}. \tag{4}$$

F_1 is used to balance out *precision* and *recall* since we can normally not have both high *precision* and *recall* at the same time. To evaluate the performance of all

[2] The 20 Newsgroups corpus can be downloaded from
`http://people.csail.mit.edu/jrennie/20Newsgroups/`

categories in a multi-label classification problem, we have two averaging methods for F_1, namely $micro-F_1$ and $macro-F_1$. $Micro-F_1$ is dependent on the large categories while $macro-F_1$ is influenced by the small categories as described in [12]. By using these measures, our results are comparable with other results, including those in [8].

5 Results and Discussion

We will describe the experimental results and discussion in this section. In addition to *binary, tf, tf.rf, rf, logtf.rf*, we used *tf.rf_max, rf_max, logtf.rf_max* that schemes use rf_{max} instead of *rf*. The experimental results of these eight term weighting methods with respect to $micro-F_1$ and $macro-F_1$ measure on the Reuter News corpus and the 20NG corpus reported from Figure 1 to Figure 4. Each line in the figures shows the performance of each term weighting method at different of features selection levels.

5.1 Results on the 20NG Corpus

Figure 1 shows the results in term of $micro-F_1$ on the 20NG corpus. Generally, the $micro-F_1$ values of all methods increase when the number of selected features increases. $logtf.rf_{max}$ and rf_{max} are consistently better than others at all feature selection levels. Almost all term weighting methods achieve their peak at a feature size around 16000 and the best three $micro-F_1$ values 81.27%, 81.23% and

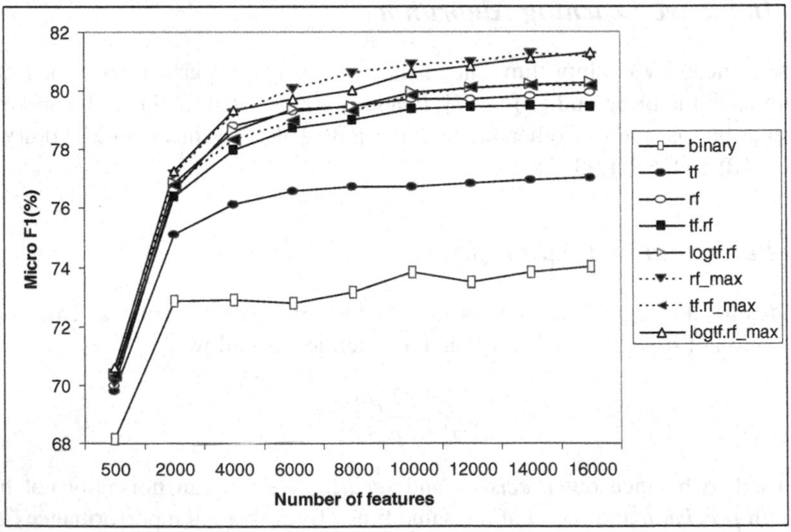

Fig. 1 $micro-F_1$ measure of eight term weighting schemes on 20NG with different numbers of features

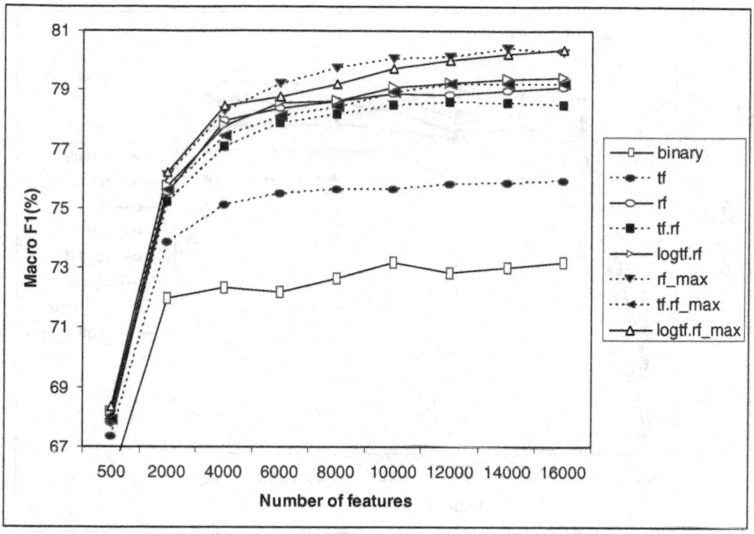

Fig. 2 $macro-F_1$ measure of eight term weighting schemes on 20NG with different numbers of features

80.27% are reached by using $logtf.rf_{max}$, rf_{max} and *logtf.rf*, respectively. *tf.rf* and *rf* reach their peak of 79.46% and 79.94%.

Figure 2 depicts the results in term of $macro-F_1$ on the 20NG corpus. The trends of the lines are similar to those in Figure 1. $logtf.rf_{max}$ and rf_{max} are better than other schemes at all different numbers of selected features.

5.2 Results on the Reuter News Corpus

Figure 3 shows the results with respect to $micro-F_1$ on the Reuters News corpus. From 6000 features onwards, the $micro-F_1$ values generally increase. $logtf.rf_{max}$ and $tf.rf_{max}$ are consistently better than others as the level of feature selection is bigger than 8000. Almost all term weighting methods achieve their peak at the full vocabulary. The best three $micro-F_1$ values 94.23%, 94.20% and 94.03% achieved by using $tf.rf_{max}$, $logtf.rf_{max}$ and *tf.rf*. Scheme rf_{max} and *rf* account for 93.50% and 93.10% at the full vocabulary.

Figure 4 depicts the results in term of $macro-F_1$ on the Reuters News corpus. The performances of eight schemes fluctuate as the number of selected features is smaller than 8000. From this point onwards, $logtf.rf_{max}$ and *logtf.rf* are schemes that are consistently better than others.

Our experimental results confirm the classification results of *tf.rf* and *rf* (the peaks and trends) as reported in [8]. Firstly, *tf.rf* shows consistently better than *rf*, *tf* and *binay* on the Reuter News corpus (Figure 3). Moreover, the performance of *rf* is better than *tf.rf*, *tf* and *binary* on the 20NG corpus.

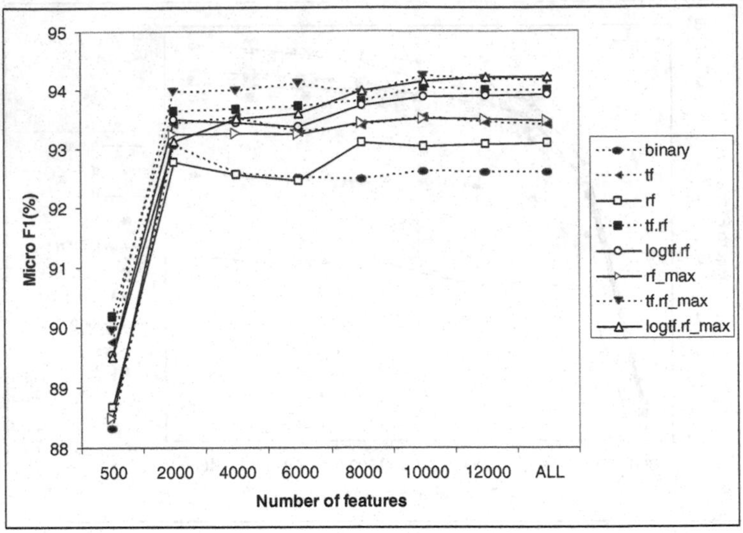

Fig. 3 $micro-F_1$ measure of eight term weighting schemes on Reuter with different numbers of features

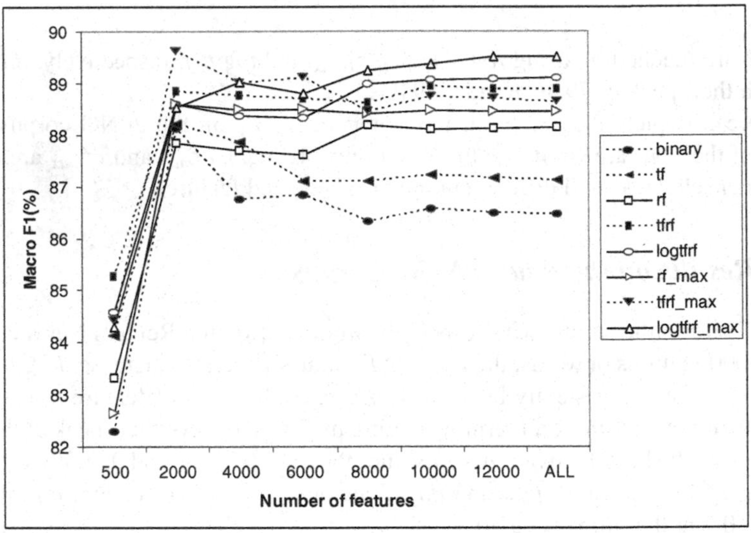

Fig. 4 $macro-F_1$ measure of eight term weighting schemes on Reuter with different numbers of features

5.3 Observations

There are some our observations of schemes with our proposed improvements:

- The schemes used rf_{max} factors are better than other schemes with *rf* factor. Specifically, $tf.rf_{max}$, $logtf.rf_{max}$ and rf_{max} are better than *tf.rf*, *logtf.rf* and *rf*, respectively in all Figures.
- The schemes used *logtf* factor yield better performance than others used *tf* factor on the 20NG corpus (Figure 1 and Figure 2). On the Reuter News corpus, the schemes used *logtf* have a comparably good performance as the schemes used *tf* (Figure 3 and Figure 4).
- $logtf.rf_{max}$, a combination of two improvements, has a comparably good performance as $tf.rf_{max}$ and rf_{max}, two best schemes on the Reuter News corpus and the 20NG corpus, respectively.
- $logtf.rf_{max}$ shows significantly better than *tf.rf* on the 20NG corpus and consistently better than *tf.rf* on the Reuter News corpus as the level of feature selection exceed 6000.

In brief, $logtf.rf_{max}$ steadily has higher performance than other shemes in our experiments.

6 Conclusions

In this study, we have introduced a newly term weighting scheme $logtf.rf_{max}$ that apply two improvements to *tf.rf* - one of the best term weighting schemes to date. Firstly, our scheme requires a single *rf* value for each term while *tf.rf* requires many *rf* values in a the multi-label classification problem. Second, our scheme used *logtf* instead of *tf*. The experimental results show that our newly term weighting scheme is consistently better than *tf.rf* and others on two data sets with the different category distribution.

For future works, we will use other classification methods (for example kNN) as well as text corpuses to further validate $logtf.rf_{max}$.

References

1. Buckley, C., Salton, G., Allan, J., Singhal, A.: Automatic Query Expansion Using SMART: TREC 3. In: NIST SPECIAL PUBLICATION SP, pp. 69–69 (1995)
2. Debole, F., Sebastiani, F.: Supervised term weighting for automated text categorization. In: Sirmakessis, S. (ed.) Text Mining and its Applications. STUDFUZZ, vol. 138, pp. 81–97. Springer, Heidelberg (2004)
3. Deng, Z.-H., Tang, S.-W., Yang, D.-Q., Li, M.Z.L.-Y., Xie, K.-Q.: A comparative study on feature weight in text categorization. In: Yu, J.X., Lin, X., Lu, H., Zhang, Y. (eds.) APWeb 2004. LNCS, vol. 3007, pp. 588–597. Springer, Heidelberg (2004)
4. Dumais, S., Platt, J., Heckerman, D., Sahami, M.: Inductive learning algorithms and representations for text categorization. In: The Seventh International Conference on Information and Knowledge Management (1998)
5. Fan, R.E., Chang, K.W., Hsieh, C.J., Wang, X.R., Lin, C.J.: LIBLINEAR: A Library for Large Linear classification. The Journal of Machine Learning Research 9, 1871–1874 (2008), Software available at `http://www.csie.ntu.edu.tw/~cjlin/liblinear`

6. Hsu, C.W., Chang, C.C., Lin, C.J.: A practical guide to support vector classification (2003)
7. Joachims, T.: Text categorization with support vector machines: Learning with many relevant features. In: Nédellec, C., Rouveirol, C. (eds.) ECML 1998. LNCS, vol. 1398, pp. 137–142. Springer, Heidelberg (1998)
8. Lan, M., Tan, C.L., Su, J., Lu, Y.: Supervised and traditional term weighting methods for automatic text categorization. IEEE Transactions on Pattern Analysis and Machine Intelligence 31(4), 721–735 (2009)
9. Leopold, E., Kindermann, J.: Text categorization with support vector machines. How to represent texts in input space? Machine Learning 46(1-3), 423–444 (2002)
10. Rogati, M., Yang, Y.: High-performing feature selection for text classification. In: The Eleventh International Conference on Information and Knowledge Management, pp. 659–661. ACM (2002)
11. Salton, G., Buckley, C.: Term-weighting approaches in automatic text retrieval. Information Processing and Management 24(5), 513–523 (1988)
12. Sebastiani, F.: Machine learning in automated text categorization. ACM Computing Surveys (CSUR) 34(1), 1–47 (2002)
13. Wu, H., Salton, G.: A comparison of search term weighting: term relevance vs. inverse document frequency. ACM SIGIR Forum 16(1), 30–39 (1981)
14. Yang, Y., Liu, X.: A re-examination of text categorization methods. In: The 22nd Annual International ACM SIGIR Conference on Research and Development in Information Retrieval, pp. 42–49. ACM (1999)
15. Yang, Y., Pedersen, J.O.: A comparative study on feature selection in text categorization. In: Machine Learning-International Workshop Then Conference, pp. 412–420. Morgan Kaufmann Publishers, Inc. (1997)
16. Yang, Y.: An evaluation of statistical approaches to text categorization. Information Retrieval 1(1-2), 69–90 (1999)

Gender Prediction Using Browsing History

Do Viet Phuong and Tu Minh Phuong

Abstract. Demographic attributes such as gender and age of Internet users provide important information for marketing, personalization, and user behavior research. This paper addresses the problem of predicting users' gender based on browsing history. We employ a classification-based approach to the problem and investigate a number of features derived from browsing log data. We show that high-level content features such as topics or categories are very predictive of gender and combining such features with features derived from access times and browsing patterns leads to significant improvements in prediction accuracy. We empirically verified the effectiveness of the method on real datasets from Vietnamese online media. The method substantially outperformed a baseline, and achieved a macro-averaged F1 score of 0.805. Experimental results also demonstrate the effectiveness of combining different feature types: a combination of features achieved 12% improvement of F1 score over the best performing individual feature type.

Keywords: Gender prediction, browsing history, classification.

1 Introduction

The effectiveness of many web applications such as online marketing, recommendation, search engines largely depends on their ability to provide personalized services. An example of such personalization is in behavioral targeting. Behavioral targeting is a technique for online advertising that helps advertisers to match advertisements

Do Viet Phuong
R&D Lab, Vietnam Communication Corporation, Hanoi, Vietnam
e-mail: phuongdoviet@hotmail.com

Tu Minh Phuong
Department of Computer Science, Posts and Telecommunications Institute of Technology, Hanoi, Vietnam
e-mail: phuongtm@ptit.edu.vn

V.-N. Huynh et al. (eds.), *Knowledge and Systems Engineering, Volume 1*,
Advances in Intelligent Systems and Computing 244,
DOI: 10.1007/978-3-319-02741-8_24, © Springer International Publishing Switzerland 2014

to proper users based on the user behaviors when using Internet. According to a recent study [21], behavior-based ads have gained a significant business traffic improvement over simple web ads. These systems rely heavily on prior search queries, locations, and demographic attributes to provide personalized targeting.

For personalized services, many user features have been explored, among them demographic attributes such as gender and age have been shown to be of great value. In many cases, however, such information is not explicitly available or is incomplete because users are not willing to provide. Prediction of these features is an alternative way to get the missing information, which has generated significant research interest and practical values. Previous studies on demographic prediction have focused on analyzing blogs or reviews to predict the gender and age of their authors [8, 15]. However, a major part of Internet users do not write blogs and reviews, thus limiting the applicability of such methods. Another approach is to predict user gender and age from web page click-through log data. The main advantage of this approach is that such data is available for most Internet users, who browse pages for news, music, videos, products, etc. [10, 11]. To predict gender and age from browsing data, a common strategy is to treat webpages and their elements (such as words, categories, and hyperlinks) as hidden attributes, and use these attributes to propagate demographic information between users.

In this paper, we present a method for predicting user gender from browsing history data. We cast the problem as an obvious binary classification problem. We extract different features from browsing log data and use these features to represent users. In [10], Hu and colleges noted that training a classifier in the user side using low-level textual features from visited webpages may not provide accurate predictions. In this study we show that by using high-level content features, namely topics and categories of webpages, and combining these features with access times and order of viewing pages, the user centered classification approach is quite competitive and can achieve state-of-the-art performance. We conducted experiments on a dataset collected from Vietnamese web-sites to verify the effectiveness of the proposed method. The results show that combining different feature types leads to significant improvement over baseline algorithms. The method achieved a macro-average F1 score of 0.805.

The rest of the paper is organized as follows. Section 2 reviews related work. Section 3 describes the proposed method in detail. Section 4 present experimental study and results. Section 5 concludes the paper.

2 Related Work

There are two main groups of methods for predicting demographic attributes. The first group relies on textual data users have written such as reviews, blogs, comments, tweets, emails. Methods of this group focus on the analysis of writing style or speaking style of the user and its association with demographic attributes. Herring and Paolillo studied the prediction of gender from traditional, well written documents [8]. Herring et al. [9], and Yan et al. [23] investigated the identification of gender from blogs. Nowsom *et al.* used n-gram features of the user's post to the forums with 1,400 posts of 47 men and 450 women [14]. Burger *el al*, predicted gender of Twitter users by analyzing content of tweets and additional features [2]. They use several types of features including n-grams, name and descriptions. Another work that used other features, in addition to the text content of blogs, was proposed by Rosenthal et al. [19]. Otterbacher combined writing style, words, and metadata for predicting gender of movie review users [15]. Recently, researchers started to investigate the identification of gender for Twitter users [1, 16]. In addition to writing style, speaking style has also been exploited to infer user gender from conversational data [6, 7].

Methods of the second group are based on browsing behaviors to predict demographic attributes of the user. Pages or multimedia content the user has visited are used to propagate demographic information from users for which such information is known to other users viewing the same content. Filippova studied the detection of user gender from both user comments on videos and the relationship of the people watching the same video [5]. Hu *et al.* used data from browsing log data of users whose gender is known to infer the demographic characteristics associated with elements of webpages the users visited [10]. They then used these demographic characteristics to predict the gender and age of users visiting the same webpages. Kabbur *et al.* followed a similar approach: they used machine learning techniques to predict demographic attributes associated with different elements of web-sites such as words, HTML elements, hyperlinks, but not requiring information from web-sites' users [11].

Our method presented here belongs to the second group of methods in that it relies on browsing behaviors of users. However, in contrast to methods presented in [10] and [11], we do not learn demographic attributes of web-sites but instead train a classifier in the user side directly with appropriately chosen features.

3 Methods

3.1 General Framework

We assume that we are given a set of users for which we know the gender. Also, for each of those users as well as users whose gender is unknown we are given the user's browsing history in forms of a sequence of visited webpages $(p_1, \ldots, p_n)$ along with times when the pages were visited $(t_1, \ldots, t_n)$ (number n of visited pages can be different from user to user). Note that such a sequence can be derived from multiple browsing sessions of the user by concatenating the browsing sequences of all sessions. The problem is to predict the gender for new users, whose gender is unknown.

We cast the problem of predicting user gender as a binary classification problem with two class labels: male and female. From the sequence of accessed pages we derive a set of features to represent the user. We choose *Support Vector Machines* (SVM) as the classification algorithm due to its superior performance in a number of applications. In the next sections we will describe features for use with SVM in detail.

3.2 Features

An important step in classification-based methods is to select features to represent users. A straightforward way is to represent each user by the word content of all the pages the user has visited. However, as noted by Hu *et al.* [10] and confirmed by our experiments, using only words as features gives poor prediction accuracy. Thus, for accurate predictions, we need to consider other sets of features.

In this study, we use the content of webpages visited to represent users. However, instead of using word features directly, we generate several types of higher-level content based features, namely the category to which a page belongs to and the main topic of the page. We also augment the content-based features by considering the access times and the order of pages (or more precisely their categories) visited. In this section, we describe each type of features in detail.

3.2.1 Category-Based Features

Most news websites organize their pages in categories, for examples sports, political, entertainments. Many other types of web pages such as forums, e-commerce pages are also classified into categories for convenient browsing. It is often observed that men and women are biased toward different categories [3]. For example, sport category has more male readers whereas fashion category has more female readers. To utilize this observation in predicting user gender, we first determine the categories of pages a user has visited and create category based features.

Mapping webpages to standard categories.

In practice, each web-site uses its own collection of categories which may be different from those of other websites. Thus, we need to define a standard unified collection of categories and map pages from different publishers into this standard collection. In this work, we use the category scheme provided by Wada (`http://wada.vn`) for this purpose. Wada is a search engine and web directory specially developed for Vietnamese webpages and currently is the best maintained web directory in Vietnam. Wada's directory has two levels: the first level consists of 12 categories which are further divided into 126 subcategories in the second level. Table 1 shows the categories provided by Wada web directory. Due to space limit, the table shows only subcategories from "Sports" as an example of second level subcategories.

We use the following procedure to map a webpage to one of standard categories/ subcategories:

- If the page has already been assigned a category/subcategory by the web-site it comes from, and the original category/subcategory is equivalent to one of standard categories/subcategories then the page is assigned the equivalent category/subcategory. Note that if a page is mapped to a standard subcategory then

Table 1 The standard collection of 12 categories. Subcategories are shown only for "Sports"

Category
Th thao (Sports); X hi (Social); Kinh t Kinh doanh (Economy - Business); Sc khe (Health) ; Khoa hc Gio dc (Science - Education); Vn ha Ngh thut (Culture and Arts); Thi trang Lm p (Fashion Beauty); Hi-Tech; Gia nh (Family); t Xe my (Cars Motorcycles); Khng gian sng (Living space); Gii tr Du lch (Entertainment)
Subcategories of *Th thao* (Sports)
Bng (Football); Bng Chuyn(Volleyball); V thut (Martial arts); Golf; Yoga; Billiards; Bng bn (Table tennis); Qun vt (Tennis); Bi li (Swimming); Cu lng (Badminton); C vua (Chess); ua xe (Racing); Mo him (Adventure); Others

its category is the parent category of the mapped subcategory. Table 2 shows examples of categories from Vietnamese sites which can be mapped to their equivalent categories from the standard collection.

- If the page does not have its own category/subcategory or the original page category/subcategory does not have its equivalence from the standard category collection then we use an automated classifier to classify the page into one of standard (sub) categories. Specifically, we retrieved 694628 articles from (`http://www.soha.vn`). These articles have already been classified by Wada, thus we used them as training data. We used the TF-IDF scheme to represent articles and used a Nave Bayes classifier implementation from library Minorthird (`http://minorthird.sourgeforge.net`) for training and prediction.

Table 2 Examples of original (sub)categories and their equivalent standard (sub)categories

Standard category	Original category	Standard sub-category	Original subcategory
Sports	thethao.vnepxress.net/*	Football	*/bongdaplus.vn/*
	thethao24.tv/*	Football	*/tin-tuc/bong-da-quoc-te/*
	dantri.com.vn/the-thao/*	Tennis	*/tin-tuc/tennis/*
	kenh14.vn/sport/*	Tennis	*/tennis-dua-xe/*
Hi-Tech	dantri.com.vn/suc-manh-so/*	Computer	*/vi-tinh/*
	duylinh.vn/*	Computer	*/thiet-bi-vi-tinh/*
	www.ictnews.vn/*	Devices	*/dien-thoai/*
	http://vozforums.com/*	Devices	*/may-tinh-bang/*

Creating category-based features.

Using standard categories assigned to the pages a user has visited, we create category-based features for the user as follows. Assume the user has visited n pages denoted by $(p_1, \ldots, p_n)$ and their categories are $(c_1, \ldots, c_n)$, where c_i can be one of the 12 standard categories. We count the number of times each of the 12 standard categories occurs in $(c_1, \ldots, c_n)$ and put the counts together to form a vector of 12 elements. We normalize the vector so the elements sum to one and use the normalized vector as category-based features.

We apply a similar procedure to standard subcategories assigned to the pages to create additional 126 subcategory-based features for the user.

3.2.2 Topic-Based Features

The categories and subcategories used in the previous section are created manually and provide only a coarse-grained categorization of webpage contents. It is useful to consider other methods, preferably automated ones, to organize pages into categories at finer levels of granularity. Here, we adopt *latent Dirichlet allocation* (LDA) [1], a widely used topic modeling technique, to infer topics from textual content of web pages and use the topics to create additional features. We use features created this way to augment the feature set generated from categories as described in the previous section.

In LDA, a topic is a distribution over words from a fixed vocabulary, where highly probable words tend to co-occur frequently in documents about this topic. The main idea of topic modeling is to see documents as mixtures of topics. A topic model defines a topic structure for a collection of documents (or a corpus), and a stochastic procedure to generate documents according to the topic structure. Documents in a collection are generated using the following procedure. First, the model generates a set of topics, which will be shared by all the documents in the collection. Second, the model generates each document in the following two-step process. In the first step, the model randomly selects a distribution θ over topics. Then, for each word of the document, the model randomly picks a topic according to θ and randomly draws a word from the word distribution of the selected topic. Since only the textual content of documents are observed while the other information such as topics, the proportion of topics in a given document, and topic assignment for each word are hidden, the central computational problem of LDA is to infer those structures from observed document content. This process is often known as learning topic model and can be performed by using approximate algorithms such as variational inference [1] or Gibbs sampling [22].

We use the following procedure to assign topics for web pages accessed by a given user:

- First, we retrieve a set of web pages to form a corpus. We use LDA as the model and use the Gibbs sampling algorithm described in [22] to learn topic structures from the corpus. Note that this step is performed only once and the learned model will be used for all users.
- Next, for each web page accessed by the user, we use the topic model learned in the previous step to infer the topic proportion and topic assignment for words within this page.
- Finally, for each page, we use the topic with the highest probability from the topic distribution inferred in the previous step as the topic of the page.

This procedure maps each accessed page into one of K topics, where K is the pre-chosen topic number. For the given user, we count the number of times the user has accessed each of K topics. The vector of these counts is then normalized to form a distribution, i.e. the vector elements sum to one. We use the normalized count vector elements as topic-based features for the given user.

3.2.3 Time Features

The next type of features to consider is time features. The motivation for using time features is based on the assumption that men and women have different time patterns when surfing web. For example, because women usually spend more time for preparing dinner than men, it can be expected that women access Internet less frequently then men at 6 pm – 8 pm time slot.

For each page a user has accessed, we record the access time. We use one-hour intervals to represent time, thus access time can get an integer value from [0,...,23] corresponding to 24 hours of a day. For the given user, we count the number of times the user has clicked a page in each of 24 time intervals, resulting in a vector with 24 elements. We normalize this vector so that the elements sum to one and used the normalized vector as time features.

3.2.4 Sequential Features

Besides the actual pages (or more precisely, their categories) a user has viewed, we hypothesize that the order of viewing is also influenced by the user's gender. For example, men tend to change between categories more frequently while browsing than women do. To verify this hypothesis, we create features to represent the order of page categories a user has viewed and experimentally study the impact of such feature on the prediction accuracy.

Given a sequence of pages $(p_1,\ldots,p_n)$ a user has viewed, we first determine their categories $(c_1,\ldots,c_n)$ as described in one of previous sections. From this sequence, we extract all k-grams, where each k-gram is a subsequence of length k with k being a pre-specified parameter. Formally, a k-gram starting at position i is the subsequence $(c_i, c_{i+1}, \ldots, c_{i+k-1})$. Note that if the user's access history has been recorded from multiple sessions then we do not consider k-grams that cross session borders, i.e. only k-grams belonging to single sessions are counted.

From 12 standard categories, it is possible to generate 12^k k-grams. To reduce the complexity and exclude irrelevant k-grams, we use a feature selection technique based on mutual information to select m k-grams that have maximum correlation with gender classes. The mutual information between k-gram S and gender class G is computed as follows:

$$I(G,S) = \sum_{g \in G} \sum_{s \in S,\ s \neq 0} p(g,s) \log \left(\frac{p(g,s)}{p(g)p(s)} \right) \tag{1}$$

where G ={male, female}, and we only sum over users with nonzero values of S, i.e. users with k-gram S occurring in their access sequences.

We select m (m is a parameter of the method) k-grams with the highest values of mutual information to create so called *sequential features* as follows. For a given user, the number of times each of the m k-grams occurs in her browsing history is counted. This results in a vector of size m, which is then normalized to form a distribution. The elements of this vector are then used as sequential features for SVM.

3.3 Combining Features

We experimented with different combinations of features. There are a number of ways to combine features of different types for SVM classifiers. For example, one can compute a separate kernel for each type of features and then take a (weighted) linear combination of kernels. Another method is to train a separate classifier for each type of features and use a Bayesian framework to combine the classifiers' output. In this study, we simply concatenate vectors of different feature types to form a unified feature vector.

We recall that feature vector of each type has been normalized so that the elements sum to one. This guarantees equal contributions of all feature types and allows avoiding the excessive influence of features with large values. We also normalize the final feature vector to have unit norm.

4 Experiments and Results

4.1 Experimental Setup

4.1.1 Data

We collected data for our experiments from users of different services offered by VCcorp (`http://vccorp.vn`). VCcorp is one of the biggest online media companies in Vietnam which operates several popular websites including news (`http://kenh14.vn`, `http://dantri.com.vn`), e-commerce websites (`http://enbac.com`, `http://rongbay.com`), social media, digital content (`http://soha.vn`). VCcorp is also an advertising network with online advertising system Admicro.

We used a set of users with known gender to create training and test sets for the experimented methods. The information about those users was collected via MingId (`http://id.ming.vn`). MingId is a *Single Sign On* system that serves as a unified account management service for entire VCcorp eco-system. A user with a MingId account can log in to any account-required service offered by VCcorp. MingId has more than two million registered users, among them about 200 thousands users log in regularly. MingId uses several rules to check if information provided by registered users is reliable. We removed users with unreliable account information, as detected by the system, and users who rarely use Internet (who visit less than 20 pages in a month), thus retained 150 thousands users for experiments. Among those users, 97 thousands are male and the remaining 53 thousands are female.

We collected browsing history of these users for a period of one month and used the collected data to create features as described in the previous sections. The same set of 694628 news articles from `http://www.soha.vn`, as used for training the Nave Bayes classifier in section 3.2, was used for training LDA.

4.1.2 Evaluation Metrics

We judged the performance of the proposed and a baseline method in terms of precision p, recall r, and F1 score. For each class (male, female), precision is defined as the number of correctly predicted cases divided by the number of all predictions of this class. Recall is defined as the number of correctly predicted cases divided by the number of all cases of this class. F1 is the harmonic mean of precision and recall:

$$F1 = \frac{2pr}{p+r}$$

We also report macro-averaged F1 score, which is the average of F1 scores of two classes: male and female.

4.1.3 Evaluation and Settings

We used 10-fold cross-validation to evaluate the merits of each method under test. The set of all users was divided into 10 subsets with equal numbers of male and female users. One subset was held out as the test set and the remaining subsets were used as the training set. We repeated this procedure ten times for each subset held out as the test set and reported averaged results.

We used the implementation of SVM provided in LIBSVM library (`http://www.csie.ntu.edu.tw/~cjlin/libsvm/`) with RBF kernel. Two parameters C and γ were selected by running the tool grid.py provided with LIBSVM on a held-out dataset and then used for all experiments. For LDA, we used the Matlab topic modeling toolbox by Steyvers and Griffiths (`http://psiexp.ss.uci.edu/research/programs_data/toolbox.htm`) for training and inference. Following [22], parameters α and β of LDA were set to 50/K and 0.01 respectively, where K is the number of topics. The selection and effect of K will be discussed in the next section. We used bi-grams for sequential features and selected 31 bi-grams with the highest mutual information values. The number of bi-grams was chosen using the same small held-out set that was used to select LDA parameters.

4.1.4 Comparison and Baselines

We tested our method with different combinations of feature types and compared the method with the one proposed by Hu *et al.* [10] using the same cross-validation setting. Hu and colleges described a method consisting of two parts: First, they represented webpages by different content-based features and used support vector regression to predict demographic tendencies for each page. The predicted tendencies of webpages are then used to predict gender for users by aggregating over visited webpages. Second, they leveraged similarity among users and webpages to deal with data sparseness. When applying only the first part, they achieved a macro-averaged F1 score of 0.703. Adding the second part improved the F1 score up to 0.797.

Due to lack of data, we could not re-implement the second part. In our experiments, we re-implemented the first part of their method and used it for comparison.

4.2 *Results*

4.2.1 Effect of Topic Number for Topic-Based Features

We first studied the effect of number of topics K on the performance of topic-based features. We experimented with different values of K ranging from 50 to 200. For each value of K we learned the topic structure and used only topic features to represent users. Fig. 1 shows the macro-averaged F1 scores of the method when using only topic-based features with different number of topics.

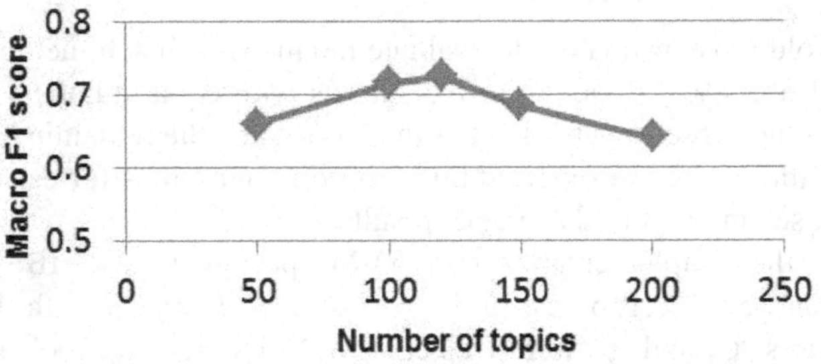

Fig. 1 F1 scores for different topic numbers when using only topic-based features

The results show that $K = 120$ gives the highest F1 score. Interestingly, this number of topic is very close to the number of subcategories in our standard category collection, although an inspection of important words in each topic shows that not all topics have subcategories with the same meaning. In all experiments reported in the following section, we used $K = 120$, which corresponds to 120 topic-based features.

4.2.2 Performance of Individual Feature Types

In the next experiment, we compared the performance of each feature type when used individually. We used features of each type as input for SVM and reported the recalls, precisions, and F1 scores averaged over 10 folds in table 3.

Table 3 Performance of individual feature types. The numbers of feature are in parentheses

Type of feature	Gender	p	r	F1	Macro F1
Sequential features (31)	Male	0.6	0.561	0.58	1
	Female	0.62	0.61	0.61	
Time features (24)	Male	0.684	0.667	0.68	1
	Female	0.661	0.64	0.65	
Category-based features (12)	Male	0.66	0.67	0.66	1
	Female	0.67	0.653	0.66	
Category-based and sub category-based features (12+126)	Male	0.72	0.71	0.71	1
	Female	0.7	0.68	0.69	
Topic-based features (120)	Male	0.73	0.708	0.72	1
	Female	0.74	0.726	0.73	

As shown in the table, sequential features gave the lowest F1 scores (macro F1 is only 0.595) while the highest macro-averaged F1 score of 0.725 was achieved with 120 topic-based features. The relatively low F1 score of sequential features show that the order of page views is not very relevant to user gender. Simple time features performed surprisingly well, achieving macro F1 of 0.665, higher than F1 score of 0.66 achieved when using 12 features based on categories. This result supports the intuition that male and female users browse the Web at different times. It is interesting to note that topic-based features, which have been generated automatically by LDA, resulted in more accurate predictions than features from manually created categories (0.725 vs 0.7 macro F1 scores).

4.2.3 Combining Features and Comparison with Method by Hu et al.

In the next experiment, we measured the performance of different combinations of feature types and compared with that of method by Hu *et al.* (without the second part), which serves as the baseline in this experiment. Since topic-based features have shown the best performance in previous experiments, we used them as the base features and gradually added features of other types. The results are summarized in table 4. The results show that our method outperformed the baseline in term of F1 score when using any combination of topic–based and other features. It is also observed that adding more features resulted in accuracy improvement. The best macro F1 score of 0.805 was achieved when combining all the four types of features or when using just topic–based, sequential, and time features. This is 12% improvement over the best individual feature type and 15% improvement over the baseline. Note that, Hu *et al.* achieved F1 score of 0.703 in their dataset when using only the first part of their method which is the same as implemented in our baseline. When adding the second part (smoothing by leveraging similarity among users and webpages), they achieved F1 score of 0.797. In our dataset, the baseline achieved F1 score of 0.695 (very close to their reported 0.703), while our method achieved the F1 score of 0.805. We note that it is not possible to compare the proposed method and full implementation of method by Hu *et al.* in this way and the reported results just show how our method improves over the baseline.

Table 4 Performance of different combinations of feature types

Type of feature	Gender	p	r	F1	Macro F1
Baseline (first part of method by Hu *et al.* [10])	Male	0.68	0.72	0.7	1
	Female	0.67	0.71	0.69	
Topic-based + Time	Male	0.762	0.758	0.76	1
	Female	0.74	0.736	0.74	
Topic-based + Sequential	Male	0.747	0.72	0.73	1
	Female	0.737	0.74	0.74	
Topic-based + Time + Sequential	Male	0.821	0.8	0.81	1
	Female	0.81	0.79	0.8	
All features	Male	0.82	0.807	0.81	1
	Female	0.81	0.8	0.8	

5 Conclusion

We have proposed a method for predicting the gender of Internet users based on browsing behavior. From browsing log data of users the method extracts several types of features including high-level content features and time features. The features are used as input for SVM based classification. Experimental studies with real data show that our method outperformed a baseline method by a large margin. The experimental results also show the usefulness of combining different types of features for improved prediction accuracy. Although we experimented only with gender prediction, the method can be extended to predict other demographic information such as age.

Acknowledgments. Financial support for this work was provided by FPT Software. We thank VCcorp for providing data.

References

1. Blei, D.M., Ng, A.Y., Jordan, M.I.: Latent Dirichlet Allocation. Journal of Machine Learning Research 3, 993–1022 (2003)
2. Burger, J.D., Henderson, J., Kim, G., Zarrella, G.: Discriminating gender on Twitter. In: Proc. of EMNLP 2011, pp. 1301–1309 (2011)
3. Computerworld Report: Men Want Facts, Women Seek Personal Connections on Web, `http://www.computerworld.com/s/article/107391/Study_Men_want_facts_women_seek_personal_connections_on_Web`
4. Ellist, D.: Social (distributed) language modeling, clustering and dialectometry. In: Proc. of TextGraphs at ACL-IJCNLP 2009, pp. 1–4 (2009)
5. Filippova, K.: User demographics and language in an implicit social network. In: Proceedings of EMNLP-CoNLL 2012 Proceedings of the 2012 Joint Conference on Empirical Methods in Natural Language Processing and Computational Natural Language Learning, pp. 1478–1488 (2012)
6. Garera, N., Yarowsky, D.: Modeling latent biographic attributes in conversational genres. In: Proc. of ACL-IJCNLP 2009, pp. 710–718 (2009)
7. Gillick, D.: Can conversational word usage be used to predict speaker demographics? In: Proceedings of Interspeech, Makuhari, Japan (2010)

8. Herring, S.C., Paolillo, J.C.: Gender and genre variation in weblogs. Journal of Sociolinguistics 10(4), 710–718 (2010)
9. Herring, S.C., Scheidt, L.A., Bonus, S., Wright, E.: Bridging the gap: A genre analysis of weblogs. In: HICSS 2004 (2004)
10. Hu, J., Zeng, H.J., Li, H., Niu, C., Chen, Z.: Demographic prediction based on user's browsing behavior. In: Proceedings of the 16th International Conference on World Wide Web, pp. 151–160 (2007)
11. Kabbur, S., Han, E.H., Karypis, G.: Content-based methods for predicting web-site demographic attributes. In: Proceedings of ICDM 2010 (2010)
12. MacKinnon, I., Warren, R.: Age and geographic inferences of the LiveJournal social network. In: Statistical Network Analysis: Models, Issues, and New Directions Workshop at ICML 2006, Pittsburgh, PA (June 29, 2006)
13. Mulac, A., Seibold, D.R., Farris, J.R.: Female and male managers' and professionals' criticism giving: Differences in language use and effects. Journal of Language and Social Psychology 19(4), 389–415 (2000)
14. Nowson, S., Oberlander, J.: The identity of bloggers: Openness and gender in personal weblogs. In: Proceedings of the AAAI Spring Symposium on Computational Approaches for Analyzing Weblogs, Stanford, CA, March 27-29, pp. 163–167 (2006)
15. Otterbacher, J.: Inferring Gender of Movie Reviewers: Exploiting Writing Style, Content and Metadata. In: Proceedings of CIKM 2010 (2010)
16. Pennachiotti, M., Popescu, A.M.: A machine learning approach to Twitter user classification. In: Proceedings of AAAI 2011 (2011)
17. Phuong, D.V., Phuong, T.M.: A keyword-topic model for contextual advertising. In: Proceedings of SoICT 2010 (2012)
18. Popescu, A., Grefenstette, G.: Mining user home location and gender from Flickr tags. In: Proc. of ICWSM 2010, pp. 1873–1876 (2010)
19. Rosenthal, S., McKeown, K.: Age prediction in blogs: A study of style, content, and online behavior in pre- and post-social media generations. In: Proc. of ACL 2011, pp. 763–772 (2011)
20. Schler, J., Koppel, M., Argamon, S., Pennebaker, J.: Effects of age and gender on blogging. In: Proceedings of the AAAI Spring Symposium on Computational Approaches for Analyzing Weblogs, Stanford, CA, March 27-29, pp. 199–205 (2006)
21. Search Engine Watch Journal, Behavioral Targeting and Contextual Advertising, `http://www.searchenginejournal.com/?p=836`
22. Steyvers, M., Smyth, P., Rosen-Zvi, M., Griffiths, T.: Probabilistic author-topic models for information discovery. In: Processing KDD 2004. ACM, New York (2004)
23. Yan, X., Yan, L.: Gender classification of weblogs authors. In: Proceedings of the AAAI Spring Symposium on Computational Approaches for Analyzing Weblogs, Stanford, CA, March 27-29, pp. 228–230 (2006)

News Aggregating System Supporting Semantic Processing Based on Ontology

Nhon Do Van, Vu Lam Han, Trung Le Bao, and Van Ho Long

Abstract. The significant increase in number of the online newspapers has been the cause of information overload for readers and organizations who occasionally deal with the news content management. There have been several systems designed to manually or automatically take information from multiple sources, reorganize and display them in a single place, which relatively makes it much more convenient for readers. However, the methods used in these systems for the aggregating and processing are still limited and insufficient to meet some public demands, especially those relate to the semantics of articles. This paper presents a technical solution for developing a news aggregating system where the aggregation is automatic and supports some semantic processing functions, such as categorizing, search for articles, etc. The proposed solution includes modeling the information structure of each online newspaper for aggregation and utilizing Ontology, along with keyphrase graphs, for building functions related to the semantics of articles. The solution is applied to build an experimental system dealing with Vietnamese online newspapers, with the semantic processing functions for articles in the field of Labor & Employment. This system has been tested and achieved impressive results.

1 Introduction

At its most basic, a news aggregator, also termed a feed aggregator, feed reader or simply aggregator, is a website which aggregates information from multiple sources into one location for easier viewing. The information should be news headlines, blogs, video blogs or, mostly, articles published in online newspapers. While the concept is simple in theory, some news aggregators might take many forms in practice as indicated in [9].

Nhon Do Van · Vu Lam Han · Trung Le Bao · Van Ho Long
Computer Science Faculty, University of Information Technology,
Vietnam National University – Ho Chi Minh City, Vietnam
e-mail: nhondv@uit.edu.vn,
{vulh91,trunglb.cs,vanholong}@gmail.com

V.-N. Huynh et al. (eds.), *Knowledge and Systems Engineering, Volume 1*, Advances in Intelligent Systems and Computing 244,
DOI: 10.1007/978-3-319-02741-8_25,

Different methods are currently used for the aggregation in popular aggregating systems: those sites where content were still entered by humans are Huffing Post, Drudge Report, NewsNow, etc. or sites such Google News, Yahoo! News, Viet Bao, Bao Moi on the other hand, are based on RSS or web crawling techniques, as described in [2], filling the content from a range of either automatically or manually added sources. A common limitation of the current systems is the search and categorization techniques mainly based on humans or data processing, therefore their available functions are still quite insufficient to meet the increasing demands especially those relate to the semantics, such as the demand of semantic search for articles, the demand of periodically gathering articles whose semantics are related to some specific topics of concern, etc.

Semantic or conceptual approaches attempt to implement to some degree of syntactic and semantic analysis; in the other words, they try to reproduce to some degree of understanding of the natural language text that users would provide corresponding to what they think. While many other proposals and efforts aimed at improving handling the semantics of documents have yet to be widely implemented, the methods of using Ontology seem to be the most suitable in practice. Ontology is a terminology, which is used in Artificial Intelligence with different meanings, and some of its definition can be found in [3], [7] and [8]. Ontology is also widely used in developing several intelligent applications, such as: intelligent problem solvers in educations [11], document retrieval system [10], etc. In particular, the approach based on Ontology is considered modern and most appropriate for the representation and handling of the content and meaning of documents as also be described in [4], [5], [6] and [10].

The main goal of this paper is to introduce models, algorithms, and techniques for building a news aggregating system where the aggregation is entirely automatic and supports some semantic processing functions, such as search, categorization articles in a particular field, e.g. the field of Labor & Employment. The system aims to deal with newspapers within a country or in a language. The information structures of the newspapers, which are first modeled and stored in the database, will be used for the scheduled aggregation. The advantages of this approach are flexible in adding new sources, easy for maintenance and highly accurate in news aggregation. The semantic processing functions in this system are based on an Ontology and the keyphrase graphs representing the semantic content of the associated articles; this solution are well described in [10] and has achieved impressive results when it is applied in a document retrieval system in the field of Information Technology, with the semantic search function serving students, teachers, and manager as well.

2 Newspaper Model

2.1 Basic Concepts

The definitions about a website and a webpage in this paper are indifferent to the particulars. A webpage belonging to a particular website is designed for some specific purposes with a specific layout. This layout is considered as the content structure of the webpage, which characterizes the way it presents information and is specified by a set of the HTML elements. In the context of a single website, many webpages might have the same content structures as others.

Definition: An HTML element is a special object representing a specific node in the DOM tree (Document Object Model tree) associated with the HTML document of the webpage, and has following attributes:

- Type: The type of the HTML element.
- Address: The HTML element's address in the HTML document.

In the context of a particular webpage, the 'Type' attribute of HTML element characterizes its function in presenting content of the webpage and is assessed by its inner content (element content), e.g. the HTML element whose content is the title of the article, the HTML element whose content is the abstract of the article, etc. The 'Address' attribute of an HTML element is used to specify its position when it's applied in a specific HTML document. The value of HTML element's Address can be easily represented by Xpath, which is detailed in [1]. The number of HTML element Types in a specific webpage or website relies on their kinds and the design purposes of our system, i.e. in a news aggregating system, all Types of HTML element used for modeling an online newspaper are listed as follows:

- HE-List: The HTML element contains the list of URL address of articles. This HTML element type has an additional attribute which is 'Priority' ($1 \leq Priority \leq 5$) representing importance weight of the articles it covers.
- HE-Pagination: The HTML element contains the paging function of the webpage.
- Other HTML Element types representing parts of a particular article including: HE-Title, HE-ReleasedDate, HE-Abstract, HE-Author, HE-Tags, HE-RelArticles, HE-Content.

2.2 *Modeling a General Newspaper*

Modeling the information structure of an online newspaper is the prerequisite for the automatic news aggregation. In the context of a news aggregating system, the model a general online newspaper is composed of the seven components:

$$(\text{Attrs}, \text{H}, \text{F}, \text{A}, \text{F}_{Str}, \text{A}_{Str}, \text{Rels})$$

The components are described as follows:

- **The set of attributes: Attrs**
  The newspaper has following attributes used for management purpose:
  - Name: the name of the newspaper, e.g., BBC Vietnam, Dan Tri, VietNam-Net, etc..
  - URL: the URL address of the newspaper.
  - Collecting (true/false): Specifies whether or not the newspaper is currently chosen for aggregating process.

- **Home Page: H**
  H is the default webpage returned when accessing the newspaper by its URL address. It contains the latest articles of all categories of the newspaper at the access time. H is used only for evaluating the importance weight of each article. The content structure of H includes the set of HE-List HTML elements, each of which contains the list of articles associated with a particular category webpage *f* in F.

- **Set of category webpages: F**
  Each $f \in$ F represents a category webpage of the newspaper, which posts only articles related to its domain of knowledge. Each newspaper has its own perspective in categorizing articles, which is considered subjective. Each $f \in$ F has following attributes: Name (e.g. Sport, Education, Business, etc.), URL, and Collecting (true/false, specifies whether or not it is currently chosen for the news aggregating process).

- **Set of collected articles: A**
  Each article $a \in$ A is an instance of the article posted on the newspaper, and has following attributes: URL (the URL address of the associated webpage publishing the article), Title, Abstract, ReleasedDate, CollectedDate, Author, Tags, Content (the detail content of the article) and SemanticRepresentation (the keyphrase graph representing semantic content of the article).

- **Set of content structures associated with each category webpage: $\mathbf{F}_{Str}$**
  Each $\mathrm{f}_{Str} \in \mathrm{F}_{Str}$ is a content structure associated with one or more category webpages $f \in$ F, and is specified by the set of HTML elements:
  - Some HE-List HTML elements which contain lists of articles.
  - The HE-Pagination HTML element that contains paging function of the category webpage.

- **Set of content structures of the article pages: $\mathbf{A}_{Str}$**
  Each $\mathrm{a}_{Str} \in \mathrm{A}_{Str}$ is a content structure of some article webpages, and has following HTML element: the HE-Title HTML element, the HE-Abstract HTML element, the HE-ReleasedDate HTML element, the HE-Author HTML element, the HE-Tags HTML element, the HE-Content HTML element, the HE-Related Articles HTML element, the HE-Pagination HTML element.

- **Set of internal relations within the newspaper: Rels**
  Rels includes some relations between components of the newspaper:

  - The hierarchical relation "is sub-category webpage of" r_{HYP} on set F, i.e. $r_{HYP} \subseteq F \times F$: each category webpage $f \in F$ represents a domain of knowledge with different level of specifications. For this reason, a category webpage might include other sub-category pages. r_{HYP} constitutes a hierarchy tree on F.
  - The relation "belongs to" r_{BL} between two sets A and F.

$r_{BL} \subseteq A \times F$: depending on the newspaper's own perspective, an article might belong to one or more category webpages.

- The fuzzy related relation "relate to" $r_{REL} : A \times A \to [0,1]$.
  i.e. $r_{REL} = \{((a_1,a_2),w) \mid (a_1,a_2) \in A \times A, w \in [0,1]\}$
  Assuming $((a_1,a_2),w) \in r_{REL}$, we say that a_1 "relate to" a_2 by the importance weight w. If w = 1, we can say that a_1 is "identical to" a_2.
- The relation "has structure" r_{Str} between two sets F and F_{Str}.
  i.e. $r_{Str} \subseteq F \times F_{Str}$: each category webpage $f \in F$ has its own content structure, which is specified by a member of the set F_{Str}.
  i.e. $\forall f \in F, \exists! f_{Str} \in F_{Str} : (f, f_{Str}) \in r_{Str}$.

3 CK-ONTO Model and Keyphrases Graph

The CK-ONTO (Classed Keyphrase based Ontology) and the document representation model, which are introduced in [10], are used to build some functions related to semantic processing in the field of Labor & Employment.

3.1 Ontology Model

The CK-ONTO model is a system composed of six components:

$$(\text{K}, \text{C}, \text{R}_{KC}, \text{R}_{CC}, \text{R}_{KK}, \text{label})$$

In which the components are described as follows:

- **K** is a set of keyphrases
  Keyphrases are the main elements to form the concept of Ontology. There are two kinds of keyphrases: single keyphrase and combined keyphrase. For example, single keyphrases are *labor, labor force, unemployment*; combined keyphrases are *unemployed labor, labor and employment, characteristic of labor force.*

- **C** is a set of classes of keyphrases
  Each class c ∈ C is a set of keyphrases related to each other by a certain semantics. A keyphrase may belong to different classes. The classification of K depends on the specialization of concepts. For example, class *ECONOMY SECTOR* = {*agriculture, industry, forestry, fishery, service, etc.*}.

- **R**$_{KC}$ is a set of relations between keyphrase and class
  A binary relation between K and C is a subset of $K \times C$ and $R_{KC} = \{r \mid r \subseteq K \times C\}$. In this paper, R_{KC} only includes a relation called "belongs to" between keyphrase and class, which is defined as a set of pairs (k,c) with $k \in K, c \in C$.

- **R**$_{CC}$ is a set of relations between classes
  A binary relation on C is a subset of $C \times C$ and $R_{CC} = \{r \mid r \subseteq C \times C\}$. There are two types of relations between classes are considered: hierarchical relation

and related relation. In hierarchical relation, a class can include multiple sub classes or be included in other classes. With Related relation, a keyphrase may belong to many different classes. They are called "related to each other" but not in meaning of inclusion or containment.

- **R_{KK}** is a set of relations between keyphrases
  A binary relation on K is a subset of $K \times K$, i.e. a set of ordered pairs of keyphrases of K, $R_{KK} = \{r \mid r \subseteq K \times K\}$. The number of relations may vary depending on considering the knowledge domain. These relations can be divided into three groups: equivalence relations, hierarchical relations, non-hierarchical relations.

- **label** is labeling function for classifying keyphrase

3.2 *Document Representation Model*

In the Document Representation Model, each individual article is considered as a document. Thus, each article can be represented by a graph of keyphrases in which keyphrases are connected to each other by semantic relations.

Definition: A keyphrase graph (KG) defined over CK-ONTO, is a triple (G_K, E, l) where:

- $G_K \subset K$ is the non-empty, finite set of keyphrases, called set of vertices of the graph.
- E is a finite set with elements in $G_K \times G_K$, called set of directed arcs of the graph.
- $l : E \to R_{KK}$ is a labeling function for arcs. Every arc $e \in E$ is labeled by relation name or relation symbol.

In case these graphs are used for representing a document, keyphrase vertices represent keyphrases of CK-ONTO ontology treated in the document (reflect the main content or subject of the document), and the labeled arcs represent semantic links between these keyphrases.

Definition: An extensive keyphrase graph, denoted as G_e, derived from keyphrase graph $G = (G_K, E, l)$, is a triple (G_K, G_R, E') satisfying the following conditions:

- (G_K, G_R, E') is a bipartite, finite and directed graph.
- $G_K \subset K$ is a non-empty keyphrase vertex set.
- $G_R \subset R_{KK}$ is a relation vertex set which represents the semantic relations between keyphrases. The vertex set of the graph is $N = G_K \cup G_R$ and $G_K \cap G_R = \emptyset$. Each arc $e \in E$ is correspond to a vertex $\tilde{r} \in G_R$ with $\tilde{r} = (e, lab(e))$.
- E' is a non-empty set with elements in $G_K \times G_R \cup G_R \times G_K$, called set of arcs of the graph. Vertices of the bipartite graph are divided into two nonempty, disjoint sets G_K and G_R, with two different kinds of vertices. All arcs then connect exactly one vertex from G_K and one vertex from G_R.

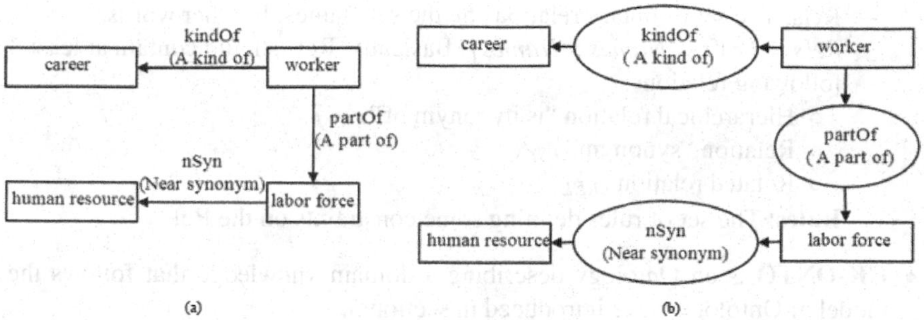

Fig. 1 An example of keyphrase graph (a) and extensive keyphrase graph (b)

Definition: Let $G = (K,R,E)$ be a keyphrase graph (in extensive form). A sub keyphrase graph (*subKG*) of G is a keyphrase graph $G' = (K',R',E')$ such that:

$$K' \subseteq K, R' \subseteq R, E' \subseteq E, \text{ and } (i,j) \in E' \Rightarrow i,j \in K' \cup R'$$

A *subKG* of G can be obtained from G only by repeatedly deleting a relation vertex (and arcs incident to this vertex) or an isolated keyphrase vertex.

4 System Model

The news aggregating system serves the purpose of managing all the defined newspapers and supporting some semantic processing functions, such as categorizing articles and search for articles with high customization capabilities. The system includes following components:

(E, GF, CK-ONTO, GRels)

Each component is described as follows:

- **E** is the set of newspapers of interest which we need to manage and collect articles from. Each $e \in E$ represents a single newspaper which has been manually defined the information structure by following the model of a general newspaper that we have proposed in section 2.
- **GF** is the Ontology for all categories that any articles might belong to. Categorizing articles of different newspapers is considered different and subjective. Therefore, all collected articles need to be reorganized and recategorized. GF consists of three components:

  (Names, Rels, Rules)

  The components are described as follows:

  - **Names**. The set of names of any fields, subjects, categories, for example, Economics, Finance and Banking, Politics, etc.

- **Rels**: The set of binary relations on the set Names. In other words, $Rels = \{r \mid r \subseteq Names \times Names\}$. Basically, Rels should contain at least 3 following relations:
  - o Hierarchical relation "is hyponym of" r_{HYP}.
  - o Relation "synonym" r_{SYN}.
  - o Related relation r_{REL}.
- **Rules**: The set of rules defining some constraints on the Rels.

- **CK-ONTO** is an Ontology describing a domain knowledge that follows the model of Ontology weve introduced in section 3.

- **GRels** is the set of general binary relations between the components. At its most basic, we suggest it should have at least following relations:

  - The hierarchical relation "is a specialized newspaper of" r_{HYP} on the set E, i.e. $r_{HYP} \subseteq E \times E$. For some reasons, some newspapers might consist of several specialized sub-newspapers which are put under one domain name. In the context of a news aggregating system, they are considered as individual newspapers but with hierarchical relationships, which is represented by relation r_{HYP}. For example, *BBC (http://www.bbc.com)* and *BBC Travel (http://www.bbc.com/travel)* are considered as two distinct newspapers, and *(BBC Travel, BBC)* $\in r_{HYP}$.
  - The binary relation "belong to" r_{BL} between two sets $GF.Names$ and $E(F)$, in which $E(F) = \cup_{e \in E} e.F$, i.e. $R_{BL} \subseteq E(F) \times GF.Names$: this relation is used for articles categorization purpose. Each category webpage of any newspapers publishes articles that are considered belonging to a specific category or topic which is the name of the category webpage.

5 Technical Problems and Algorithms

The structure of the newspapers, along with their respective category webpages, which have been defined and put into the management system, will be used for the scheduled aggregation. The aggregating process must collect all fresh articles published in these newspapers as quickly and accurately as possible. Based on the Ontology, the keyphrase graph, constituted by keyphrases and relations, associated with the collected articles will be generated and stored in the database automatically. The semantic processing problems, including categorizing and search for articles, are related to the measure of the semantic similarity between the keyphrase graphs.

5.1 Aggregating Algorithm

Let f be a category webpage of a particular online newspaper $e \in E, f \in e.F$. The basic unit of aggregating process within e is the aggregating within each category webpage f of e. The algorithm for aggregating all fresh articles published in f is described as follow:

```
Step 1: htmlDocument  ← GetHtmlSource(f.URL);
Step 2: articleLinks ← GetArticleLinks(f, htmlDocument);
Step 3: Count ← 0;
    for URL in articleLinks{
      if(CollectArticle(URL) = true)
        Count++; }
Step 4: if(StoppingCondition(f) ≠ true){
      htmlDocument ← NextPage(f,htmlDocument);
      Go to Step 2; }
Step 5: Record the information and stop the aggregating
        process within f;
```

All related functions are described as follows:

- **GetHtmlSource**(URL): send an http request, wait for the response and return the Html Document associated with the given URL.
- **GetArticleLinks**(f, htmlDocument): in the context of htmlDocument, get all distinct URLs appearing in the inner content of each HE-List which are members of the set f.FStructure – the content structure associates with f.
- **CollectArticle**(f, URL): collect a fresh article belonging to f by the given URL and generate the keyphrase graph associated with this article.
- **StoppingCondition**(f): specify the stopping condition for the aggregating process in f by using the number of articles successfully collected in current paging of f.
- **Nextpage**(f, htmlDocument): return a new HTML document associated with subsequent page compared to the current page of f.

The news aggregating is the process of repeating the above function for each category webpage of the newspapers in the system. The news aggregating for each of the newspaper is independent; therefore, we can implement some multi-threading techniques to optimize the aggregating speed.

5.2 *Semantic Processing*

All semantic processing functions implemented in the systems are mainly based on measuring the semantic similarity between 2 keyphrase graphs, which is also detailed in [10]. Given 2 keyphrase extensional graphs $H = (KH, RH, EH)$ and $G = (KG, RG, EG)$ defined over CK-ONTO, the $Rel(H, G)$ reflecting the semantic similarity between two keyphrase graphs is evaluated by the following formula:

$$Rel(H,G) = Max\{v(\Pi) \mid \Pi \text{ is partial projection from H to G}\}$$

v(Π) is a valuation pattern of a projection Π(f,g) from H to G, defined as follows:

$$v(\Pi) = \frac{\sum_{k \in KH} tf(g(k),G) * \alpha(k,g(k)) * ip(g(k),G) + \sum_{r \in RH} \beta(r,f(r))}{|KH| + |RH|}$$

In which, $\alpha : K \times K \to [0,1]$ and $\beta : R_{KK} \times R_{KK} \to [0,1]$ is two mappings to measure the semantic similarity between two keyphrase graphs and two relations defined in the CK-ONTO. The value β are selected manually based on the opinions of experts in the field. The "Term frequency", denoted as $tf(k,d)$, is the frequency of the given keyprhase in the given document, and $ip(k,d)$ is the importance weight of the keyphrase k in the document d based on the location of occurrence of k in the document. The process of the semantic search for articles includes following steps:

- **Step 1**: Read the search query (q) input.
- **Step 2**: Process and generate the keyphrase graph KG(q) representing (q).
- **Step 3**: Search for collected articles in the database which are appropriate for the search query (q) and return a list of sorted articles.
  The set of all collected articles $E(A) = \cup_{e \in E} e.A$ are represented as a set of indexed keyphrase graph $KG(E(A)) = \{G_1, G_2, \ldots, G_k\}$.
  (3.1) Find in KG(E(A)) those articles matching KG(q) by measuring the semantic similarity between keyphrase graphs. A keyphrase graph g is consider matching KG(q) if $Rel(g, KG(q)) \geq \delta$:

```
Result ← ∅;
for g in KG(E(A))
    if Rel(g,KG(q)) ≥ δ then
        Result ← Result ∪ {(g, Rel(g, KG(q))};
```

  (3.2) Sort the articles in the Result by Rel values.
- **Step 4**: Show the search result and suggest modifying the search query (q). The result includes a list of sorted suitable articles and a list of keyphrases relating to the search query to help user having more satisfying search result.
- **Step 5**: Modify the search query and go back to step 2 until user expectation is satisfied.

6 Implementation and Testing

6.1 Implementation

The proposed solution is applied to build a news aggregating system for Vietnamese online newspapers with requirements to meet real demands, especially the efficiency in news aggregating and other advanced functions related to semantic processing for articles in the field of Labor & Employment. Besides, we also design some tools supporting defining the structure of new newspapers. The system is geared towards the needs of users who wish to periodically aggregate or semantic search for articles related to some specific topics or subjects of concern. The system includes 3 main modules:

- Management module: manage newspapers structures, manage the categorization of articles, manage and schedule for news aggregation. All the management works should be done through a friendly web interface.

- News aggregating module: scheduled aggregation for article using some multi-threading techniques for optimization, categorizing articles and generating the associated keyphrase graph for each of them based on the CK-ONTO.
- Semantic processing and search module: implementing some basic search techniques and the semantic search in the field of Labor & Employment.

The data storage of system is created based on the relational database model including: a database for information structure of online newspaper and articles collected, database for the Ontology for the field of Labor & Employment, and some XML files for configuration. Currently, the Ontology for Labor & Employment includes over 1000 keyphrases, 32 classes and 11 relations on keyphrases.

6.2 Testing and Results

Currently, there have been 42 Vietnamese online newspapers defined and put into the system for scheduled aggregation. All the management works including defining and managing each newspaper, managing and scheduling for news aggregating process, managing the news categorization, managing the CK-ONTO, etc. are easily done through a friendly web interface. Besides, the tools supporting the defining newspaper structures are considered helpful as it makes the defining works become much more easily and accurately.

The accuracy of the defining structures of newspapers mostly decides the precision of the articles aggregation. Besides, the speed of the aggregation relatively depends on the hardware infrastructure. On the average, over 2700 fresh articles a day are aggregated and processed by this system, which fairly meets the real requirements. The precision and the recall for aggregation are used to evaluate the effectiveness of this module and calculated by following formulae:

$$Precision_{Aggregation} = \frac{|A_{correct}|}{|E(A)|} \qquad Recall_{Aggregation} = \frac{|A_{correct}|}{|T(A)|}$$

In which, $A_{correct}$ is the set of correctly collected articles. A correctly collected article must be exactly the same with the original one posted in the newspaper, including all parts: title, abstract, content, author, released date, etc. $E(A)$ is the set of all collected articles, i.e. $E(A) = \cup_{e \in E} e.A$ and $T(A)$ is the set of all articles of those newspapers which the system should have collected correctly. According to our statistics on 10 selected newspaper, the precision and the recall for the aggregation are approximately 97.5% and 95.2% respectively.

Semantic search for articles in the field of Labor & Employment is mainly based on CK-ONTO model and keyphrase graphs. Measures of the precision and the recall are also used to evaluate the effectiveness of this function. According to our test results of 100 selected search queries in the field of Labor and Employment on a fixed set of 50000 collected articles of different categories, the threshold of semantic similarity $\delta = 0.7$ gives the best results: the average precision of the semantic search function is approximately 85.20% and the average recall is 87.10%. Besides, the semantic search function has been highly appreciated by invited users as it is much more intelligent than other traditional search methods mainly based on data processing.

Conclusions

We have described a solution for developing a news aggregating system where the aggregation is entirely automatic and supports some tasks related to the semantic content of articles, such as article categorization and semantic search for articles based on Ontology and keyphrase graphs. The proposed solution is a model that shows the integration of the components, such as model of a particular newspaper, Ontology model for a particular field, semantic representation model for article and the model of the whole aggregating system with semantic processing and searching techniques. The strategies for news aggregation and search technique based Ontology proposed in this solution are considered flexible, easy for maintenance and relatively appropriate for demands in reality.

The solution is applied to build an experimental system for Vietnamese online newspapers, which has achieved good results in both aggregating efficiency and semantic processing. In fact, implementing a news aggregator supporting semantic processing in a particular field for Vietnamese or any other languages is not different as the information structures of suitable newspapers must be defined and the CK-ONTO must be made up by the keyphrases in such language. In the foreseeable future, we're going to add new sources for aggregation, generalize and improve current Ontology model for more general and effective in semantic processing. Besides, we're investing in improving the storage method for much better meet the increasing demands in reality. We hope this research result will be the basis and a tool for building many news aggregators supporting other languages and countries with the semantic processing in various different fields.

References

1. Bergeron, R.: XPath - Retrieving Nodes from an XML Document. SQL Server Magazine (2000), `http://sqlmag.com/xml/xpath151retrieving-nodes-xml-document`
2. Brawer Sascha, B., Maximilian, I., Michael, K.R., Narayanan, S.: Web crawler scheduler that utilizes sitemaps from websites. United States Patent 8417686 (2011)
3. Gruber, T.R.: Toward Principles for the Design of Ontologies Used for Knowledge Sharing. International Journal Human-Computer Studies 43(5-6), 907–928 (1995)
4. Styltsvig, H.B.: Ontology-based Information Retrieval. A Dissertation Presented to the Faculties of Roskilde University in Partial Fulfillment of the Requirement for the Degree of Doctor of Philosophy (2006)
5. Eriksso, H.: The semantic-document approach to combining documents and ontologies. International Journal of Human-Computer Studies 65(7), 624–639 (2007)
6. Zhong, J., Zhu, H., Li, J., Yu, Y.: Conceptual Graph Matching for Sematic Search, pp. 92–106. Springer, Heideberg (2002)

7. Sowa, J.F.: Knowledge Representation: Logical, Philosophical and Computational Foundations. Brooks/Cole (2000)
8. Stojanovic, L., Schneider, J., Maedche, A., Libischer, S., Suder, R., Lumpp, T., Abecker, A., Breiter, G., Dinger, J.: The Role of Ontologies in Autonomic Computing Systems. IBM Systems Journal 43(3) (2004)
9. Chowdhury, S., Landoni, M.: News aggregator services: user expectations and experience. Online Information Review 30(2), 100–115 (2006)
10. Do, V., Huynh, T.T., PhamNguyen, T.: Sematic Representation and Search Techniques for Document Retrieval Systems. In: Selamat, A., Nguyen, N.T., Haron, H. (eds.) ACIIDS 2013, Part I. LNCS, vol. 7802, pp. 476–486. Springer, Heidelberg (2013)
11. Do, N.V.: Intelligent Problem Solvers in Education - Design Method and Applications. In: Koleshko, V.M. (ed.) Intelligent Systems, InTech (2012) ISBN: 978-953-51-0054-6

Inference of Autism-Related Genes by Integrating Protein-Protein Interactions and miRNA-Target Interactions

Dang Hung Tran, Thanh-Phuong Nguyen, Laura Caberlotto, and Corrado Priami

Abstract. Autism spectrum disorders (ASD) are a group of conditions characterized by impairments in social interaction and presence of repetitive behavior. These complex neurological diseases are among the fastest growing developmental disorders and cause varying degrees of lifelong disabilities. There have been a lot of ongoing research to unravel the pathogenic mechanism of autism. Computational methods have come to the scene as a promising approach to aid the physicians in studying autism. In this paper, we present an efficient method to predict autism-related candidate genes (autism genes in short) by integrating protein interaction network and miRNA-target interaction network. We combine the two networks by a new technique relying on shortest path calculation. To demonstrate the high performance of our method, we run several experiments on three different PPI networks extracted from the BioGRID database, the HINT database, and the HPRD database. Three supervised learning algorithms were employed, i.e., the Bayesian network and the random tree and the random forest. Among them, the random forest method performs better in terms of precision, recall, and F-measure. It shows that the random forest algorithm is potential to infer autism genes. Carrying out the experiments with five different lengths of the shortest paths in the PPI networks, the results show the advantage of the method in studying autism genes based on the large scale network. In conclusion, the proposed method is beneficial in deciphering the pathogenic mechanism of autism.

Dang Hung Tran
Hanoi National University of Education, Hanoi, Vietnam
e-mail: `hungtd@hnue.edu.vn`

Thanh-Phuong Nguyen · Laura Caberlotto · Corrado Priami
The Microsoft Research - University of Trento Centre
for Computational Systems Biology, Italy
e-mail: `{nguyen,caberlotto,priami}@cosbi.eu`

Corrado Priami
Department of Mathematics, University of Trento, Trento, Italy

V.-N. Huynh et al. (eds.), *Knowledge and Systems Engineering, Volume 1*,
Advances in Intelligent Systems and Computing 244,
DOI: 10.1007/978-3-319-02741-8_26, © Springer International Publishing Switzerland 2014

1 Introduction

Autism is a neurodevelopmental disorder which belongs to a group of conditions known as autism spectrum disorders (ASD), autism in short, which share common characteristics including impairments in reciprocal social interaction and communication. In recent years, the strong role of genetics in autism raised hopes that by understanding the genetic cause of autism, it might be possible to treat the disease. However finding specific genes has proven to be very difficult. Autism is defined by a set of distinct behavioral symptoms that can be extremely variable, that can be triggered by multiple genetic changes and environmental factors, and that may even represent more than one disease. Numerous reports have been published supporting a significant role for miRNAs in the pathophysiology of neuropsychiatric disorders [7]. The functions attributed to miRNAs overlap with the dysfunctions found in the autism brain including growth abnormalities and delays and disorganization in neuronal maturation. Moreover, alterations in the control over the multiple mRNAs targeted by each miRNA could lead to the phenotypes observed in autism spectrum disorders that depend on genetic background and different factors including environment. We are currently still far from unraveling the molecular mechanisms of the autism, and thus developing efficient methods to uncover autism genes remains a great challenge.

The recent blooming of public proteomic and genomic databases allows researchers to advance computational methods to study disease genes. Early work on this problem typically used sequences [1] or annotations [27], and investigated disease genes as separate and independent entities. However, it is well known that biological processes are not those of single molecules but the products of complex molecular networks, especially protein-protein interaction network (PPI network). Several research groups have been exploiting the human PPI network to predict human disease genes via their corresponding product proteins, which are intuitively called disease proteins [13]. The PPI-based methods differ on the interpretation of the commonly accepted assumption that "the network-neighbor of a disease gene is likely to cause the same or a similar disease" [15, 23, 4, 11].

In addition to PPI data, other disease-related data spread out in a large amount of public databases [15]. It is a challenge to integrate multiple data into PPI-based methods to better predict disease genes. In [30], the authors built a training set by selecting the positive samples from the known disease genes in the Online Mendelian Inheritance in Man database (the OMIM database, [12]) and the negative samples from known non-disease genes in the Ubiquitously Expressed Human Genes database (the UEHG database, [26]). Lage *et al.* presented the phenomic ranking of protein complexes linked to human diseases for predicting new candidates for disorders using a Bayesian model [20]. Borgwardt and Kriegel combined graph kernels for gene expression and human PPI to do the prediction [5]. Radivojac *et al.* combined various data sources of human PPI network, known gene-disease associations, protein sequence, and protein functional information with a support vector machines framework [24]. Other works were based on integrating PPI network data with gene expression data [16], or with disease phenotype similarities [29]. Nguyen and Ho proposed a semi-supervised learning method to predict disease genes by exploiting data regarding both disease genes and disease gene neighbors via PPI network [22].

Instead of predicting genes that cause some diseases, several works have concentrated on using PPI to discover genes associated with specific diseases, e.g., Alzheimer's disease, using heuristic score functions [18].

Because autism-related data has been limited, the combination of a large scale PPI network and a miRNA-target networks provide us a better prediction for autism candidate genes.

In this paper, we present a novel method to predict autism-related candidate genes (autism genes in short) by integrating the PPI network and the miRNA-target interaction network. There are two main contributions: (1) a new technique to represent the topological features replying on shortest path calculation to combine two networks of PPIs and miRNA-target interactions; (2) a supervised learning framework to predict autism genes. This is the first work that uses supervised learning for integrating the miRNA-target and the PPI networks to study autism. The idea of combining miRNA data is based on the assumption that miRNA are responsible for regulating the expression not only of their target genes in certain cell types or during specific developmental periods, but also to "finetune" the levels of co-expressed genes [3, 19].

To demonstrate the high performance of our method, we run several experiments on three different PPI networks extracted from the BioGRID database [8], the HINT database [9], and the HPRD database [17]. We then employed three supervised learning methods, i.e., the Bayesian network, the random tree, and the random forest, for predicting autism candidate genes. Among them, random forest method performs better in terms of precision, recall, and F-measure. It shows that the random forest algorithm is potential to infer autism genes. Carrying out the experiments with five different lengths of the shortest paths in the PPI networks, the results show the advantage of the method in studying autism genes based on the large scale network. In conclusion, the proposed method is beneficial in deciphering the pathogenic mechanism of autism.

The remainder of the paper is organized as follows. In Section 2, we present the materials and methods for predicting autism genes. Then the results are given in Section 3. Finally Section 4 summarizes some concluding remarks.

2 Materials and Method

Our proposed method consists of four main steps: (1) Extracting multiple data sources for autism-related gene prediction, (2) Constructing the integrated network of protein interactions and miRNA-target interactions, (3) Representing topological features of proteins in the integrated network, and (4) Predicting autism-related genes using supervised learning. We detail these four steps in the next four sub sections.

2.1 *Extracting Multiple Data Sources for Autism-Related Gene Prediction*

To build the human PPI network, we selected the HINT database, the BIOGRID database, and the HPRD database because of their comprehensiveness and reliability. The HINT (High-quality INTeractomes) database is a database of high-quality PPIs in different organisms [9]. HINT has been compiled from different sources and then filtered both systematically and manually to remove erroneous and low-quality interactions leading to 27,493 binary and 7,629 co-complex interactions for human. The BioGRID database is an online interaction repository with data compiled through comprehensive curation efforts with 131,624 non-redundant human protein interactions [8]. The HPRD database is known as one of the most complete human databases, in which all of data were manually curated and integrated a wealth of information relevant to the function of human proteins in health and disease [17]. The latest version of HPRD contains 41,327 human protein interactions. Based on extracted data from the HINT database, the BioGRID database, and the HPRD database, three PPI networks are separately reconstructed to obtain a comprehensive view of the human interactome.

The miRNA-target interactions were extracted from the TarBase database, which is a manually curated collection of experimentally tested miRNA targets, in human/mouse, fruit fly, worm, and zebrafish [25]. Note that the set of miRNA-target interactions in this study consisted of only experimentally confirmed ones to ensure the high quality of data.

Because there does not exist a complete data source of autism, we combined a number of public data sources including literature for obtaining the autism-related data. The autism-related genes were extracted from the Online Mendelian Inheritance in Man (OMIM) database, and the Comparative Toxicogenomics Database (CTD) database [10]. In the OMIM database, the list of hereditary disease genes is described in the OMIM morbid map. The CTD database describes about the effects of environmental chemicals on human health by integrating data from curated scientific literature related to chemical interactions with genes and proteins, and associations between diseases and chemicals, and diseases and genes/proteins.

For the autism-related miRNAs, we extracted from the HMDD database, the MIR2disease database, the dataset published in [7], and the data manually curated from GEO array [2]. The HMDD database retrieves the associations of miRNA and disease from literatures [21]. The MIR2disease database is manually curated and aims at providing a comprehensive resource of miRNA deregulation in various human diseases [14].

The extracted gene data was processed and identified in gene symbols. The combination of the above-mentioned data sources provided us the comprehensive and reliable data for better predicting autism genes.

2.2 Constructing the Integrated Network of Protein Interactions and miRNA-Target Interactions

To construct the integrated network of the PPIs and the miRNA-target interactions, we firstly considered the large-scale PPI network with nodes as proteins and undirected edges as interactions, called Net_{PPI}. Secondly, we built the network of miRNA-target interactions represented by a bipartite graph, called $Net_{miRNA-target}$. Figure 1 shows a toy model for illustrating the procedure of constructing the integrated network. By merging Net_{PPI} and $Net_{miRNA-target}$ we obtained the graph with two types of nodes: grey nodes representing the normal proteins, and green nodes representing the miRNA proteins (which have interactions with miRNAs). Thirdly, the autism-related proteins (autism proteins in short) are labeled as red nodes. Similarly, the autism proteins that have interactions with miRNAs are called miRNA-autism proteins represented as purple nodes. In the final integrated network contained four types of proteins: normal proteins, miRNA proteins, autism proteins, miRNA-autism proteins (in the most right block in Figure 1).

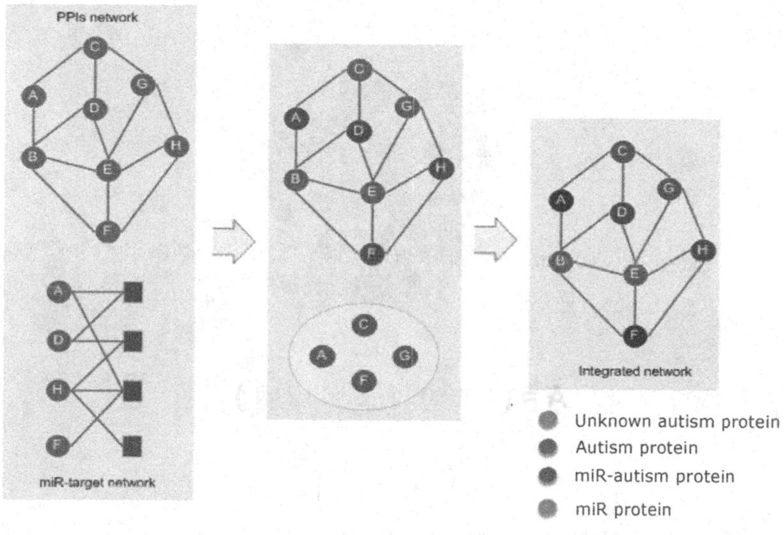

Fig. 1 A toy model for illustrating the integrated network of PPIs and miRNA-target interactions

2.3 Representing Topological Features of Proteins in the Integrated Network

We propose a new method for representing the topological features of proteins. The topological features shows the topological relationships of one proteins with its neighbors, its miRNA, and autism genes in the PPI network. For each protein p in the integrated network, we measure four topological data features as follows:

- N_{np}^{l} is the number of normal protein having the maximal distance l from protein p
- N_{mp}^{l} is the number of miRNA proteins having the maximal distance l from protein p.
- N_{dp}^{l} is the number of autism proteins having the maximal distance l from protein p.
- N_{dmp}^{l} is the number of miRNA-autism proteins having the maximal distance l from protein p.

where l ranging over $[1..L_{max}]$, L_{max} being the maximum length of the shortest path $path_{ij}$ between any two proteins (p_i, p_j) in the network.

Finally, the topological features of a protein p are represented in term of a feature vector that has $4 * L_{max}$ dimensions. Note that the order of features in the feature vector is $(N_{np}^{l}, N_{mp}^{l}, N_{dp}^{l}, N_{dmp}^{l})$. By running experiments in various PPI networks, we found that usually $L_{max} \leq 20$. For testing several models, we customized L_{max} as an input parameter. Figure 2 shows an example of how to quantify the topological features of protein A, with $L_{max} = 2$.

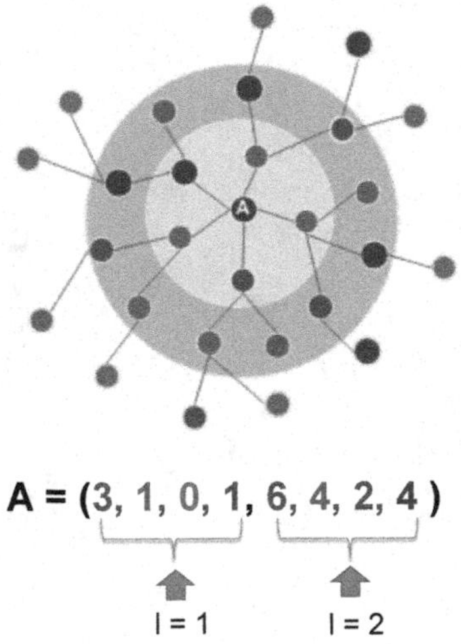

Fig. 2 An example of the feature vector of protein A with $L_{max} = 2$

2.4 Predicting Autism-Related Genes Using Supervised Learning

We carried out three supervised learning methods, i.e., the Bayesian network, the random tree and th randon forest (RF). The performance of the three methods were compared by some measures. The Bayesian network is a probabilistic graphical model (a type of statistical model) that represents a set of random variables and their conditional dependencies via a directed acyclic graph. Random tree algorithm constructs multiple decision trees randomly. The RF algorithm [6] started with a

Algorithm 2: NetAnalyzer: Representing the feature vectors and label the values for all proteins in the integrated network.

Input: List of proteins
Integrated network of PPI and miRNA-target interactions
the parameter L_{max}
Output: Feature vectors representing the topological features of proteins

```
foreach (p ∈ Protein_List) do
    Call NodeAnalyzer(p, L_max);
    foreach (p ∈ Protein_List) do
        p.visitedstatus ← true;
        switch p.type do
            case normal_protein and microRNA_protein
                p.label = 'non − autism'
            case autism_protein and miRNA − autism_protein
                p.label = 'autism'
```

standard machine learning technique called a decision tree. It consists of an arbitrary number of simple trees, which are used to determine the final outcome. For classification problems, the ensemble of simple trees votes for the most popular class. In the regression problem, their responses are averaged to obtain an estimate of the dependent variable. Using tree ensembles can lead to significant improvement in prediction accuracy (i.e., better ability to predict new data cases). There are a lot of advantages of the RF algorithm, such as running efficiently on large data bases, achieving one of highest accuracy among current algorithms, etc. After processing extracted data and representing in the feature vectors, those data feature vectors were input into the the three algorithms.

3 Results

The PPI data, the miRNA-target data and the autism genes were extracted from different sources. Those processed data were then combined and represented using the topological feature vectors. Table 1 shows the data statistics corresponding to the three considered PPI networks from the BioGRID, the HINT and the HPRD databases.

We used the machine learning workbench package Weka [28] to run the three algorithms, the RF, the Bayesian network, and the random tree. We run 10-fold cross validation to compare the performance of the three algorithms. Moreover, the experiments were run in three different PPI networks extracted from the BioGRID database, the HINT database and the HPRD database (with the same datasets of autism-related genes, miRNA-target interactions, and autism-related miRNAs). For each network, the parameter L_{max} were set with different values from 1 to 5 for showing the feasibility of the method with various scales of the PPI networks.

Algorithm 3: NodeAnalyzer: Calculate the topological feature vector values for all proteins in the integrated network

Input: *root*; L_{max}

Output: Topological feature vector values for all proteins in the network

```
vector < int > Idx;
vector < p > Q;
r ← 0;
l ← 0;
level ← −1;
Idx ← −1;
Q ← root;
root.visitedstatus ← false;
while (level < lmax) and (l ≤ r) do
    currpro ← Q.pop();
    if (l = 0) then
        level++;
    else
        if (Idx[l − 1] ≠ Idx[l]) then
            level++;
    foreach (p ∈ currgene.neibourhood) do
        if (p.visitedstatus then
            r++;
            Q.push(p);
            Idx.push(level);
            gen.visitedstatus ← false;
            switch p.type do
                case normal_protein
                    root.count[level].np++
                case microRNA_protein
                    root.count[level].mp++
                case autism_protein
                    root.count[level].dp++
                case microRNA − autism_protein
                    root.count[level].mdp++
```

The three measures of prediction quality are as follows.

$$Precision = TP/(TP+FP) \tag{1}$$

$$Recall = TP/(TP+FN) \tag{2}$$

$$F1 = 2*(Precision*Recall)/(Precision+Recall) \tag{3}$$

Table 1 Statistics of the extracted data corresponding to the BioGRID, the HINT and the HPRD databases

Statistics	BioGRID	HINT	HPRD
#protein interactions	196,866	27,297	39,240
#miRNA-target interactions	1,880	1,880	1,880
#autism genes	612	612	612
#normal gene nodes	13,793	7,548	8,797
#microRNA gene nodes	777	524	603
#autism gene nodes	231	151	202

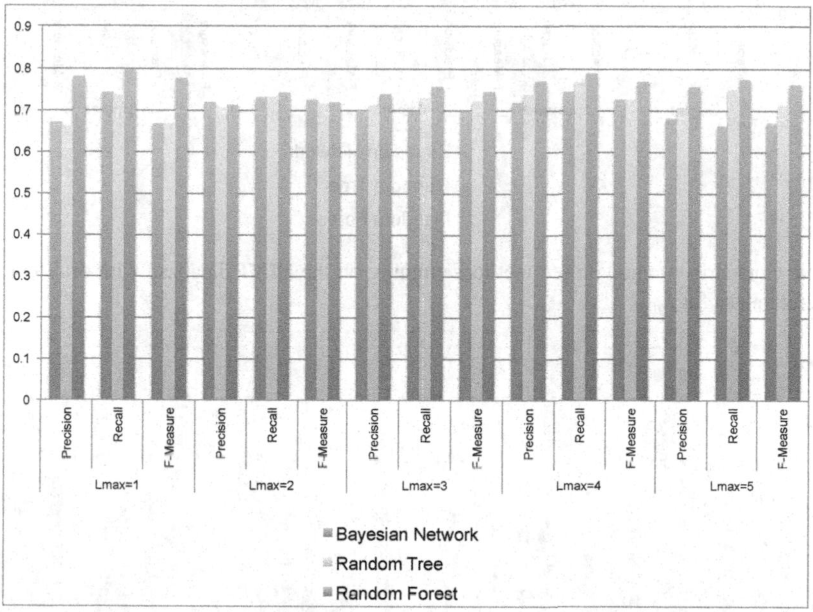

Fig. 3 Comparison among three methods running on the BioGRID database with different parameter values for L_{max}

where TP, FN, FP, and F1 denote true positive, false negative, false positive, and F-measure respectively.

Figures 3, 4, and 5 demonstrate the performance of the three methods, the Bayesian network and the random tree and the random forest when considering the BioGRID database, the HINT database, the HPRD database respectively. In almost all the experiments, the RF method achieved better performance in terms of recall, precision, and F-measure. The RF performed most efficiently on the BioGRID databases. Furthermore, with the different scales of the network from the small one ($L_{max} = 1$) to the large one ($L_{max} = 5$), the RF method achieves the high performance stably. For all above, the RF method is more suitable for resolving the problem of autism gene prediction.

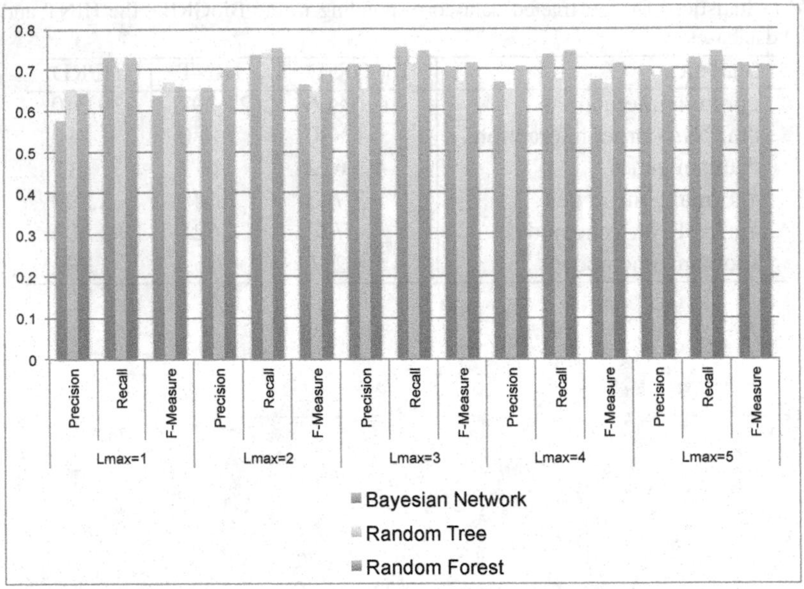

Fig. 4 Comparison among three methods running on the HINT database with different parameter values for L_{max}

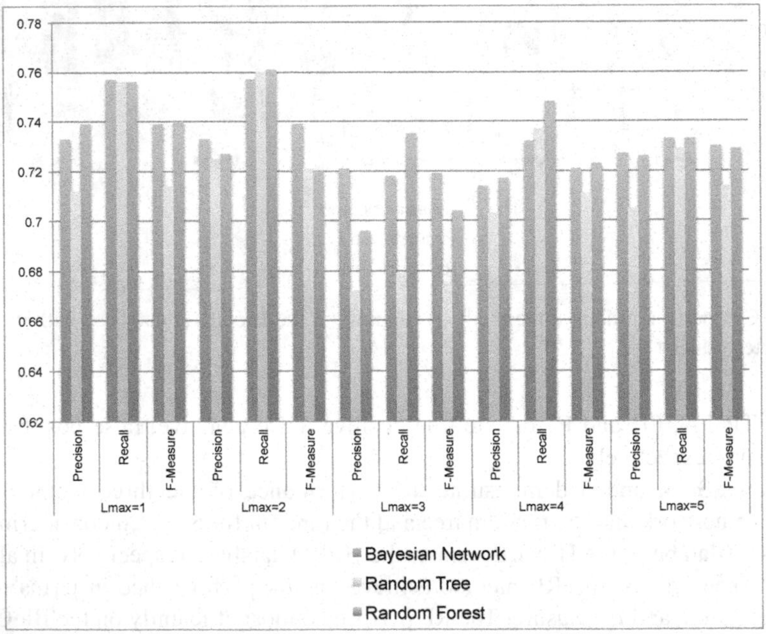

Fig. 5 Comparison among three methods running on the HPRD database with different parameter values for L_{max}

4 Conclusion

In this paper, we have introduced a method integrating multiple data features for the prediction of genes which could be potentially involved in the pathophysiology of ASD. PPIs and miRNA-target interactions were extracted and they were effectively combined in the proposed method. The experimental results demonstrated that the RF method performed well with high accuracy in comparison with other supervised learning methods. Performing the experiments with different shortest path lengths calculated, the results demonstrated the advance of the method in studying autism genes based on the large scale network. Future studies will be focused on the integration of ASD-relevant data from multiple sources (e.g. genomic, trascriptomic, clinical). The prediction could suggest protein or pathway perturbed in autism, potential targets for therapeutical intervention.

Acknowledgments. This work was supported by Vietnam National Foundation for Science and Technology Development (NAFOSTED Project No.102.01-2011.05).

References

1. Adie, E.J., Adams, R.R., Evans, K.L., Porteous, D.J., Picard, B.S.: Speeding disease gene discovery by sequence based candidate prioritization. BMC Bioinformatics 6, 55 (2005)
2. Barrett, T., Edgar, R.: Mining microarray data at NCBI's Gene Expression Omnibus (GEO). Methods in Molecular Biology (Clifton, N.J.) 338, 175–190 (2006)
3. Bartel, D.P.: MicroRNAs: genomics, biogenesis, mechanism, and function. Cell 116(2), 281–297 (2004)
4. Benjamin, S.B., Alex, B.: Protein interactions in human genetic diseases. Genome Biology 9(1), R9.1–R9.12 (2008)
5. Borgwardt, K.M., Kriegel, H.: Graph kernels for disease outcome prediction from protein-protein interaction networks. In: Pacific Symposium on Biocomputing, vol. 12, pp. 4–15. World Scientific Publishing Company, Singapore (2007)
6. Breiman, L.: Random forests. Machine Learning 45, 5–32 (2001)
7. Chan, A.W.S., Kocerha, J.: The path to microrna therapeutics in psychiatric and neurodegenerative disorders. Frontiers in Genetics 82(3), 1–10 (2012)
8. Chatr-aryamontri, A., Breitkreutz, B.-J., Heinicke, S., Boucher, L., Winter, A., Stark, C., Nixon, J., Ramage, L., Kolas, N., ODonnell, L., Reguly, T., Breitkreutz, A., Sellam, A., Chen, D., Chang, C., Rust, J., Livstone, M., Oughtred, R., Dolinski, K., Tyers, M.: The biogrid interaction database: 2013 update. Nucleic Acids Research (2012)
9. Das, J., Yu, H.: HINT: High-quality protein interactomes and their applications in understanding human disease. BMC Systems Biology 6(1), 92 (2012)
10. Davis, A.P., Murphy, C.G., Johnson, R., Lay, J.M., Lennon-Hopkins, K., Saraceni-Richards, C., Sciaky, D., King, B.L., Rosenstein, M.C., Wiegers, T.C., Mattingly, C.J.: The Comparative Toxicogenomics Database: update 2013. Nucleic Acids Research 41(D1), D1104–D1114 (2013)

11. Goh, K.I., Cusick, M.E., Valle, D., Childs, B., Vidal, M., Barabasi, A.L.: The human disease network. Proceedings of the National Academy of Sciences 104(21), 8685–8690 (2007)
12. Hamosh, A., Scott, A.F., Amberger, J.S., Bocchini, C.A., McKusick, V.A.: Online Mendelian Inheritance in Man (OMIM), a knowledgebase of human genes and genetic disorders. Nucleic Acids Res. 33(Database Issue) (2005)
13. Ideker, T., Sharan, R.: Protein networks in disease. Genome Research 18(4), 644–652 (2008)
14. Jiang, Q., Wang, Y., Hao, Y., Juan, L., Teng, M., Zhang, X., Li, M., Wang, G., Liu, Y.: Mir2disease: a manually curated database for microrna deregulation in human disease. Nucleic Acids Research 37(suppl. 1), D98–D104 (2009)
15. Kann, M.G.: Protein interactions and disease: computational approaches to uncover the etiology of diseases. Briefings in Bioinformatics 8(5), 333–346 (2007)
16. Karni, S., Soreq, H., Sharan, R.: A network-based method for predicting disease-causing genes. Journal of Computational Biology 16(2), 181–189 (2009)
17. Keshava Prasad, T.S., Goel, R., Kandasamy, K., Keerthikumar, S., Kumar, S., Mathivanan, S., Telikicherla, D., Raju, R., Shafreen, B., Venugopal, A., Balakrishnan, L., Marimuthu, A., Banerjee, S., Somanathan, D.S., Sebastian, A., Rani, S., Ray, S., Harrys Kishore, C.J., Kanth, S., Ahmed, M., Kashyap, M.K., Mohmood, R., Ramachandra, Y.L., Krishna, V., Abdul Rahiman, B., Mohan, S., Ranganathan, P., Ramabadran, S., Chaerkady, R., Pandey, A.: Human Protein Reference Database–2009 update. Nucleic Acids Research 37(Database Issue), D767–D772 (2009)
18. Krauthammer, M., Kaufmann, C.A., Gilliam, T.C., Rzhetsky, A.: Molecular triangulation: Bridging linkage and molecular-network information for identifying candidate genes in Alzheimer's disease. PNAS 101(42), 15148–15153 (2004)
19. Krol, J., Loedige, I., Filipowicz, W.: The widespread regulation of microRNA biogenesis, function and decay. Nature Reviews. Genetics 11(9), 597–610 (2010)
20. Lage, K., Karlberg, E.O., Størling, Z.M., Ólason, P.Í., Pedersen, A.G., Rigina, O., Hinsby, A.M., Tümer, Z., Pociot, F., Tommerup, N., Moreau, Y., Brunak, S.: A human phenome-interactome network of protein complexes implicated in genetic disorders. Nature Biotechnology 25(3), 309–316 (2007)
21. Lu, M., Zhang, Q., Deng, M., Miao, J., Guo, Y., Gao, W., Cui, Q.: An Analysis of Human MicroRNA and Disease Associations. PLoS One 3(10), e3420+ (2008)
22. Nguyen, T.-P., Ho, T.-B.: Detecting disease genes based on semi-supervised learning and protein-protein interaction networks. Artif. Intell. Med. 54(1), 63–71 (2012)
23. Oti, M., Snel, B., Huynen, M.A., Brunner, H.G.: Predicting disease genes using protein-protein interactions. Journal of Medical Genetics 43, 691–698 (2006)
24. Radivojac, P., Peng, K., Clark, W.T., Peters, B.J., Mohan, A., Boyle, S.M., Mooney, S.D.: An integrated approach to inferring gene-disease associations in humans. Proteins: Structure, Function, and Bioinformatics 72(3), 1030–1037 (2008)
25. Sethupathy, P., Corda, B., Hatzigeorgiou, A.G.: TarBase: A comprehensive database of experimentally supported animal microRNA targets. RNA (New York, N.Y.) 12(2), 192–197 (2006)
26. Tu, Z., Wang, L., Xu, M., Zhou, X., Chen, T., Sun, F.: Further understanding human disease genes by comparing with housekeeping genes and other genes. BMC Genomics 7, 31 (2006)
27. Turner, F.S., Clutterbuck, D.R., Semple, C.A.M.: Pocus: mining genomic sequence annotation to predict disease genes. Genome Biology 4, R75 (2003)

28. Witten, I.H., Eibe, F.: Data Mining: Practical machine learning tools and techniques. Morgan Kaufmann Publishers Inc., San Fransisco (2005)
29. Wu, X., Jiang, R., Zhang, M.Q., Li, S.: Network-based global inference of human disease genes. Molecular Systems Biology 4 (May 2008)
30. Xu, J., Li, Y.: Discovering disease-genes by topological features in human protein-protein interaction network. Bioinformatics 22(22), 2800–2805 (2006)

Modeling and Verifying Imprecise Requirements of Systems Using Event-B

Hong Anh Le, Loan Dinh Thi, and Ninh Thuan Truong

Abstract. Formal methods are mathematical techniques for describing system model properties. Such methods providing frameworks to specify and verify the correctness of systems which are usually described by precise requirements. In fact, system requirements are sometimes described with vague, imprecise, uncertain, ambiguous, or probabilistic terms. In this paper, we propose an approach to model and verify software systems with imprecise requirements using a formal method, e.g. Event-B. In the first step, we generalize our approach by representing some fuzzy concepts in the classical set theory. We then use such definitions to formalize the fuzzy requirements in Event-B and finally verify its properties such as safety, inconsistency and redundancy by using the Rodin tool. We also take a case study to illustrate the approach in detail.

1 Introduction

Requirement specification is one of the most significant steps in software development. System requirements aim to include all customers' needs. Formal methods are one of used techniques to specify and verify software system properties formally. They have been mainly addressed to verification of the system with precise requirements.

In fact, customers sometimes describe the system with ambiguous, vague or fuzzy terms such as very good, very far, low important, etc.. because they do not know how to describe precisely the system. With such requirements, one always use the fuzzy set and fuzzy logic to formally specify the system. However, we

Hong Anh Le
Hanoi University of Mining and Geology, Dong Ngac, Tu Liem, Hanoi, Vietnam

Loan Dinh Thi · Ninh Thuan Truong
VNU - University of Engineering and Technology, 144 Xuan Thuy, Cau Giay, Hanoi, Vietnam
e-mail: {anhlh.di10,loandt_54,thuantn}@vnu.edu.vn

V.-N. Huynh et al. (eds.), *Knowledge and Systems Engineering, Volume 1*, Advances in Intelligent Systems and Computing 244,
DOI: 10.1007/978-3-319-02741-8_27,

cannot prove the correctness of the model system as we do not have a support tool which used to prove fuzzy logic predicates.

In order to analyze the fuzzy requirements of a model, several works have been proposed to modeling and verifying them using formal methods. Matthews C. *et al* [13, 14] presented an approach to formalize fuzzy concepts by Z specification. However, the approach does not mention or show how we can prove system correctness using these mapping specifically. Researchers have been proposing the use of fuzzy Petri Net to model fuzzy rules-base or uncertain information systems [16]. The paper [9] introduced the use of model checking with extended Murphi verifier to verify fuzzy control systems.

However, in our thought, these results are not only insufficient for modeling and verifying fuzzy systems but also are complex to bring to real fuzzy systems development. In this paper, we propose an efficient approach to use a formal method, e.g Event-B, to formally model fuzzy requirements of systems and verify their properties.

The B method [1] is a formal software development method, originally created by J.-R. Abrial. The B notations are based on the set theory, generalized substitutions and the first order logic. Event-B [2] is an evolution of the B method that is more suitable for developing large reactive and distributed systems. Software development in Event-B begins by abstractly specifying the requirements of the whole system and then refining them through several steps to reach a description of the system in such a detail that can be translated into code. The consistency of each model and the relationship between an abstract model and its refinements are obtained by formal proofs. Support tools have been provided for Event-B specification and proof in the Rodin platform.

The novel of our approach is that we use the classical set theory to represent fuzzy terms in requirements. Since Event-B language is based on the classical set theory and the structure of an imprecise requirement is similar to an Event-B event, a collection of imprecise requirements is formalized by an Event-B model. From the target Event-B model, the correctness of requirements can be achieved by formal proofs that support to detect inconsistency in a model such as the preservation of safety, termination, deadlock properties, etc.

The rest of the paper is structured as follows. Section 2 provides some background of fuzzy systems and Event-B. In Section 3, we generalize the approach by presenting some representation of fuzzy terms in classical sets. Using such representation, we introduce rules to model fuzzy requirements using the Event-B notation. Section 4 presents a case study of Crane controller to illustrate how we model and verify imprecise requirements of the system. Section 5 summarizes some related works. We give some conclusion and present future works in Section 6.

2 Backgrounds

In this section, we briefly introduce the overview of fuzzy logics, fuzzy sets which are usually used to describe imprecise requirements. We also summarize basic knowledge of the Event-B notation.

2.1 Fuzzy Logic and Fuzzy Sets

Fuzzy logic has been applied to many fields, from control theory to artificial intelligence. Fuzzy logic concentrates on the partially true and partially false situations which accounts for almost human reasoning in everyday life. In contrast with traditional logic theory, where binary sets have two-valued logic, true or false, fuzzy logic variables may have a truth value that ranges in degree between 0 and 1. Furthermore, when linguistic variables are used, there degrees may be managed by specific functions.

Fuzzy sets are actually functions that map a value that might be a member of the set to a number between zero and one indicating its actual degree of membership. A fuzzy set F defined on an universal set X is a set, each element of which is a pair of values $(x, \mu_F(x))$, where $x \in X$ and $\mu_F(x) : X[0,1]$.

Linguistic variables are words of English to present a fuzzy set and depend on time and circumstances. It is a key part of the fuzzy controller system. For example, to represent temperature, we can tell hot, very hot, warm, cold and very cold; they are realistic concepts but cannot be represented as real number variables but linguistic variables. A linguistic variable encapsulates the properties of approximate or imprecise concepts in a systematic and computationally useful way. It reduces the apparent complexity of describing a system by matching a semantic tag to underlying concept.

Hedges are linguistic variables which carry with it the concept of fuzzy set qualifiers. Hedges allow us to model closely the semantics of the underlying knowledge as well as maintain a symmetry between fuzzy sets and numbers. In terms of fuzzy logic, hedges are operators that are applied on the membership function.

However, the weakness of fuzzy logic and fuzzy sets in description of imprecise requirements is that deduction of fuzzy predicates is approximate rather than fixed and exact. Therefore we cannot use the fuzzy logic based specification to analyse properties and the correctness of system models.

2.2 Event-B

Event-B is a formal method for system-level modeling and analysis. Key features of Event-B are the use of set theory as a modeling notation, the use of refinement to represent systems at different abstraction levels and the use of mathematical proof to verify consistency between refinement levels [2]. An Event B model encodes a state transition system where the variables represent the state and the events represent the transitions from one state to another. A basic structure of an Event-B model consists of a MACHINE and a CONTEXT.

A MACHINE is defined by a set of clauses which is able to refine another MACHINE. We briefly introduce main concepts in Event-B as follows:

VARIABLES represent the state variables of the model of the specification.

INVARIANTS described by first order logic expressions, the properties of the attributes defined in the VARIABLES clause. Typing information, functional and safety properties are described in this clause. These properties are true in the whole model. Invariants need to be preserved by events execution.

THEOREMS define a set of logical expressions that can be deduced from the invariants. Unlike invariants, they do not need to be preserved by events.

EVENTS define all the events that occur in a given model. Each event is characterized by its guard (i.e. a first order logic expression involving variables). An event is fired when its guard evaluates to true. If several guards evaluate to true, only one is fired with a non deterministic choice. The events occurring in an Event B model affect the state described in VARIABLES clause.

An Event B model may refer to a CONTEXT describing a static part where all the relevant properties and hypotheses are defined. A CONTEXT consists of the following items:

SETS describe a set of abstract and enumerated types.

CONSTANTS represent the constants used by the model.

AXIOMS describe with first order logic expressions, the properties of the attributes defined in the CONSTANTS clause. Types and constraints are described in this clause.

THEOREMS are logical expressions that can be deduced from the axioms.

After having the system modeled in Event-B, we need to reason about the model to understand it. To reason about a model, we use its proof obligation which show its soundness or verify some properties. As we mention in the first part of this Subsection, behaviors of the system are represented by machines. Variable v of a machine defines state of a machine which are constrained by invariant $I(v)$. Events E_m that describes possible changes of state consisting of guards $G_m(v)$ and actions $S_m(v,v')$ and they are denoted by

when $G_m(v)$ then $v :| \, S_m(v,v')$ end

Properties of an Event-B model are proved by using proof obligations (PO) which are generated automatically by the proof obligation generator of Rodin platform. The outcome of the proof obligation generator are transmitted to the prover of the Rodin tool to perform automatic or interactive proofs. In this paper, we mainly consider proof obligation rules which preserve all invariants of a machine.

3 Event-B for Fuzzy Requirements Modeling and Verification

In this section, we introduce an approach to model and verify imprecise requirements by Event-B. Before modeling, we present an idea to represent fuzzy terms in traditional set theory. Based upon these representation, we use an Event-B model to formalize requirements by giving transformation rules.

3.1 Representation of Fuzzy Terms in Classical Sets

As stated above, one always uses fuzzy sets and fuzzy logic to formalize fuzzy requirements. In a fuzzy reasoning system there are several different classes of hedge operations, each represented by a linguistic-like construct. There are hedges that intensify the characteristic of a fuzzy set (very, extremely), dilute the membership curve (somewhat, rather, quite), form the complement of a set (not), and those that approximate a fuzzy region or convert a scalar to a fuzzy set (about, near, close to, approximately). Hedges play the same role in fuzzy production rules that adjectives and adverbs play in English sentences, and they are processed in a similar manner. The application of a hedge changes the behavior of the logic, in the same way that adjectives change the meaning and intention of an English sentence.

Fuzzy logic usually uses IF-THEN rules, or constructs that are equivalent, such as fuzzy associative matrices. Rules are usually expressed in the form: IF variable IS property THEN action.

Example: IF costs are very high THEN margins are small.

Then the general form *FR*, also called well-defined form, of a fuzzy requirement can be represented as:

$$\text{IF } x \text{ is } \delta Y \text{ THEN } m \text{ is } P$$

Recall that, in classical set theory, sets can be combined in a number of different ways to produce another set such as Union, Intersection, Difference, Cartesian product. Below we recall some definitions related to Cartesian product operation, the definition of an ordered pair and Cartesian product of two sets using it. Then the Cartesian product of multiple sets is also defined using the concept of n-tuple.

Definition 0.1 (ordered pair). An ordered pair is a pair of objects with an order associated with them. If objects are represented by x and y, then we write the ordered pair as $\langle x, y \rangle$.

Definition 0.2 (Cartesian product). The set of all ordered pairs $\langle a, b \rangle$, where a is an element of A and b is an element of B, is called the Cartesian product of A and B and is denoted by $A \times B$.

Example 1: Let A = {1, 2, 3} and B = {a, b}.

Then $A \times B = \{\langle 1,a \rangle, \langle 1,b \rangle, \langle 2,a \rangle, \langle 2,b \rangle, \langle 3,a \rangle, \langle 3,b \rangle\}$.

The concept of Cartesian product can be extended to that of more than two sets. First we are going to define the concept of ordered n-tuple.

Definition 0.3 (ordered n-tuple). An ordered n-tuple is a set of n objects with an order associated with them (rigorous definition to be filled in). If n objects are represented by $x_1, x_2, ..., x_n$, then we write the ordered n-tuple as $\langle x_1, x_2, ..., x_n \rangle$.

Definition 0.4 (Cartesian product). Let $A_1, ..., A_n$ be n sets. Then the set of all ordered n-tuples $\langle x_1, ..., x_n \rangle$, where $x_i \in A_i$, $\forall\, i = \overline{1,n}$, is called the Cartesian product of $A_1, ..., A_n$, and is denoted by $A_1 \times ... \times A_n$.

Corollary 0.1. *A collection of well-defined fuzzy requirements can be specified by classical sets.*

Proof. Suppose that, fuzzy requirements of a system are specified by $FR = \{FR_i\}$, $FR_i = \{x_i, \delta_i Y_i, m_i, P_i\}$, $i = \overline{1,n}$. Clearly that, x_i, m_i are considered as variables in the specification, P_i is obviously described by a set of values. We consider if $\delta_i Y_i$ can be specified by a classical set in which δ_i is a hedge, Y_i is a generator of a fuzzy clause. As FR is a finite set of rules so hedges and generators in the system can be established two different finite sets, δ and Y, respectively. According to the Definition 4, $\delta_i Y_i$ is a membership of the Cartesian product of 2 sets $\delta \times Y$. Consequently, every elements in FR can be specified by classical sets. ■

3.2 Modeling Fuzzy Systems Using Event-B

Suppose that, a system is specified by a collection of requirements FR_i :

if x_i is $\delta_i Y_i$ **then** m_i is P_i

According to the Corollary 0.1, the above requirements can be represented by classical sets. Since Event-B is a language based on the classical theory set, hence in this subsection, we propose an approach to model the system using Event-B method. A system consisting a collection of requirements $FR_i, i = \overline{1,n}$ is modeled by an Event-B model $FR_B = \langle FR\_C, FR\_M \rangle$, where FR_C and FR_M are Event-B context and machine respectively. We propose below partial transformation rules to map fuzzy requirements to Event-B's elements. The important principle of the transformation process is that we can preserve the structure, represent all fuzzy requirements using the Event-B notation. Moreover, the safety properties must be preserved by actions of the system.

Transformation rules:

- Rule 1. All hedges δ_i, generators Y_i and values P_i in the collection of requirements are established as three sets δ, Y, and P respectively. They are stated in the SETS clause of FR_C.
- Rule 2. Linguistic variables x_i, m_i in each FR_i are mapped to variables x_i, m_i of the Event-B machine FR_M.
- Rule 3. Each variable x_i is described as a membership of a Cartesian product of two sets $\delta \times Y$, m_i is described as a membership of a Cartesian product of two sets $\delta \times P$ (Corollary 0.1).
- Rule 4. Each requirement FR_i is modeled by an event in Event-B machine FR_M.
- Rule 5. The safety properties of the system are modeled as invariants $\mathscr{I}$ of the machine FR_M.

Note that, these are only partial transformation rules, we need to give more additional parts to obtain the completed Event-B specification (Figure 1).

Corollary 0.2. *With the modeling proposed in transformation rules, the safety properties are preserved by all actions in fuzzy requirements of the system.*

CONTEXT FR_C
SETS
δ
P
Y
END

MACHINE FR_M
SEES FR_C
VARIABLES
x_i
m_i
INVARIANTS
inv1 : $x_i \in \delta \times Y$
inv2 : $m_i \in \delta \times P$
inv3 : $\mathscr{I}$
EVENTS
Event $FR_i \mathrel{\widehat{=}}$
when
grd1 : $x_i = \{\delta_i \mapsto Y_i\}$
then
act1 : $m_i := P_i$
end
END

Fig. 1 A part of Event-B specification using transformation rules

Proof

Suppose that, a collection of fuzzy requirements $FR = \{FR_i\}$, $i = \overline{1,n}$, is specified by corresponding event evt_i. Safety properties of the system are specified in the invariant $\mathscr{I}$. We have to prove that safety constraints are persevered though all fuzzy requirements by showing that it remains true before and after firing (executing) each event. This is abviously get through proof obligations of the Event-B machines which is used to preserve their invariants.

Without loss of the generality, we assume that the fuzzy requirement and constraint contain one variable v, hence we need to prove:

$$\mathscr{I}(v) \wedge evt_i(v,v') \vdash \mathscr{I}(v')$$

This predicate allow us to ensure the safety properties after executing the events in model, it is also the form of a proof obligation generated from Event-B machine and we can analyse them using the support tool (Rodin). Apparently, the safety properties stated in requirements are preserved. ■

3.3 Verifying System Properties

After the transformation, taking advantages of Event-B method and its support tool, we are able to verify some properties of a fuzzy system as follows:

- Safety properties: Since these properties are modeled by Event-B INVARIANTS, hence we can prove them by using proof obligations generated to prove machine invariants (see Corollary 0.2).
- Requirements consistency: Since all requirements are formalized by events, so that inconsistent and incorrect requirements can be checked through event specifications.

4 A Case Study

In this section, we first present a case study of Container Crane Control [3], we then show how we model and verify this system.

4.1 Case Study Description

Container cranes are used to load and unload containers on a off ships in most harbors. They pick up single containers with flexible cables that are mounted on the crane head. The crane head moves on a horizontal track. When a container is picked up and the crane head starts to move, the container begins to sway. While sway is no problem during transportation, a swaying container cannot be released (Figure 2).

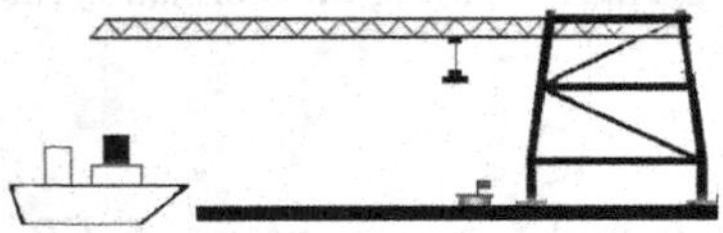

Fig. 2 Container Crane Control model

We control the system by the following strategy. Starting with the medium power, if you get started and you are still far away from target, adjust the motor power so that the container gets a little behind the crane head. If you are closer to the target, reduce speed so the container gets a little ahead of the crane head. When the container is very close to the target position, power up the motor. When the container is over the target and the sway is zero, stop the motor. *The system has a safety property such as the power can not be fast if the target is not far.*

From the description of the system, a collection of fuzzy requirements is extracted as follows:

r1. if distance is very far, then power is very high
r2. if distance is far, then power is fast
r3. if distance is medium, then power is normal
r4. if distance is near, then power is very small
r5. if distance is very near, then power is nearly stopped
r6. if distance is zero, then power is stopped

4.2 Modeling Container Crane Control System

According to the transformation rules presented in Subsection 3.2, we model the Container Crane Control system using B notation. First, all hedges, generators and values in the collection of requirements are established as the sets HEDGES, DISTANCE_VAL and POWER_VAL. They are stated in SETS clause of the CONTEXT machine. The CONTEXT machine is called C_CraneCtrl described partly as follows:

CONTEXT C_CraneCtrl
SETS

HEDGES
DISTANCE_VAL
POWER_VAL

AXIOMS

$axm1 : \quad HEDGES = \{very, quite, nearly, no\_hedge\}$
$axm2 : \quad DISTANCE\_VAL = \{far, near, zero, medium\}$
$axm3 : \quad POWER\_VAL = \{high, fast, small, normal, stopped\}$

END

In the MACHINE part, we translate linguistic variables in the requirements to Event-B variables such as *distance* and *power*. Types of these two variables are represented by the Cartesian product of the HEDGES set and their value sets: *DISTANCE_VAL* and *POWER_VAL*. Each fuzzy requirement r_i of the system is translated to an EVENT evt_i, $i = \overline{1,6}$. We represent a part of the dynamic information of the system in the machine called M_CraneCtrl as follows:

MACHINE M_CraneCtrl
SEES C_CraneCtrl
VARIABLES

power
distance

INVARIANTS

$inv1 : \quad power \in HEDGES \leftrightarrow POWER\_VAL$
$inv2 : \quad distance \in HEDGES \leftrightarrow DISTANCE\_VAL$
$inv3 : \quad distance \neq \{no\_hedge \mapsto far\} \Rightarrow power \neq \{no\_hedge \mapsto fast\}$

EVENTS
Initialisation

begin
$act1 : \quad distance := \{very \mapsto far\}$
$act2 : \quad power := \{no\_hedge \mapsto stopped\}$
end

Event *evt1* $\widehat{=}$

when
  grd1 : $distance = \{very \mapsto far\}$
then
  act1 : $power := \{very \mapsto high\}$
end

Event *evt2* $\widehat{=}$

when
  grd1 : $distance = \{no\_hedge \mapsto far\}$
then
  act1 : $power := \{no\_hedge \mapsto fast\}$
end

Event *evt3* $\widehat{=}$

when
  grd1 : $distance = \{no\_hedge \mapsto medium\}$
then
  act1 : $power := \{no\_hedge \mapsto normal\}$
end

Event *evt4* $\widehat{=}$

when
  grd1 : $distance = \{no\_hedge \mapsto near\}$
then
  act1 : $power := \{very \mapsto small\}$
end

Event *evt5* $\widehat{=}$

when
  grd1 : $distance = \{very \mapsto near\}$
then
  act1 : $power := \{nearly \mapsto stopped\}$
end

Event *evt6* $\widehat{=}$

when
  grd1 : $distance = \{no\_hedge \mapsto zero\}$
then
  act1 : $power := \{no\_hedge \mapsto stopped\}$
end

END

The system has a safety property which is stated that the power can not be fast if the target is not far. It is formalized as: $distance \neq \{no\_hedge \mapsto far\} \Rightarrow power \neq \{no\_hedge \mapsto fast\}$. This property is preserved by all events in the model by proving the proof obligations generated in Rodin tool. In addition, formalization makes requirements unambiguous, correct and complete.

5 Related Works

Several approaches have been addressed to modeling and verification of fuzzy requirements using formal methods, however, these are different with our approach. Chris Matthews has introduced a fuzzy logic toolkit for the formal specification language Z [14, 13]. This toolkit defines the operators, measure and modifiers necessary for the manipulation of fuzzy sets and relations. A series of laws are provided that establish an isomorphism between conventional Z and the extended notation when applied to boolean sets and relation. It can be modeled as a partial rather than as a total function. The focus is on the specifications of the rule base and the operations necessary for fuzzy inferencing. However, we cannot refine system at different abstraction levels and also proof to verify consistency between refinement levels. It just provides definition and manipulation of fuzzy sets and relations by using Z.

Pavliska and Knybel [15, 10] have introduced the use modified Petri Nets as a tool for fuzzy modeling. Basic concepts and relations between Fuzzy Petri Nets and Fuzzy IF-THEN rules will be described and algorithm for decomposition of fuzzy Petri net into set of linguistic descriptions will be presented and its implementation mentioned. This result just show how we model the system and does not mention how we verify the system properties.

Intrigila, B [9] have introduced the approach of verification of fuzzy control systems using model-checking technique with Murphi verifier. The authors eased the modeling phase by using finite precision real numbers and external C functions.

Yang, S.J.H. *et al.* [16] have proposed to using high-level Petri Net in order to verify fuzzy rule-based systems. This method can detect the system's errors such as redundancy, inconsistency, incompleteness, and circularity but it has to take a more step to normalize the rules into Horn clauses before transforming these rules to and use incidence matrix as fixed-value matrix for degree membership.

Jonathan Lee *et al.* [12] have extended object oriented modeling by using fuzzy objects to capture and analyze imprecise requirements. Even this approach is straight to follow and the fuzzy classes can be mapped into XML schemas and XML documents [11], we can not verify these requirements with fuzzy object model.

Recently, Goncalves, M. *et al.* [7] have presented an approach to a database application method which translates the formal specifications to implementations in the structured query language (SQL) enhanced with fuzzy logic (SQLf). This method allows to extend the tuple calculus in order to express fuzzy queries. However, the input of the conversion rules are needed to be formal speficications which are represented by fuzzy logic.

6 Conclusion and Future Works

There are few works up to date that have addressed the problem of modeling and verifying fuzzy systems. The results have focused on modeling and representing fuzzy terms. In this paper, we use classical set theory to represent fuzzy terms, after that model and verify it by an Event-B model. Since formal methods has been researched so far to dedicatedly formalize precise requirements, therefore our work provides a novel approach to model and verify imprecise requirements. The modeling can be done automatically as the structure of a fuzzy requirement matches to an Event-B event. After modeling, some system properties can be proved automatically using the supported tool (Rodin) such as safety, deadlock properties.

We are working on the tool that allows to model the input fuzzy requirements using Event-B notation. Taking consider on the membership degree of hedges in Event-B is another our future work.

Acknowledgments. This work is partly supported by the research project "Methods and tools for program analysis and their applications in education", No. QGTD.13.01, granted by Vietnam National University, Hanoi.

References

1. B method web site, `http://www.bmethod.com`
2. Event-b and the rodin platform, `http://www.event-b.org`
3. Fuzzytech home page (2012), `http://www.fuzzytech.com`
4. Prevention of load sway by a fuzzy controller (2013), `http://people.clarkson.edu`
5. Set operations (2013), `http://www.cs.odu.edu/~toida`
6. Aziz, M.H., Bohez, E.L.J., Parnichkun, M., Saha, C.: Classification of fuzzy petri nets, and their applications. Engineering and Technology, World Academy of Science 72, 394–407 (2011)
7. Goncalves, M., Rodríguez, R., Tineo, L.: Formal method to implement fuzzy requirements. RASI 9(1), 15–24 (2012)
8. Hoang, T.S., Iliasov, A., Silva, R., Wei, W.: A survey on event-b decomposition. ECE-ASST, Automated Verification of Critical Systems 46 (2011)
9. Intrigila, B., Magazzeni, D., Melatti, I., Tronci, E.: A model checking technique for the verification of fuzzy control systems. In: CIMCA 2005: Proceedings of the International Conference on Computational Intelligence for Modelling, Control and Automation and International Conference on Intelligent Agents, Web Technologies and Internet Commerce (CIMCA-IAWTIC 2006), vol. 1, pp. 536–542. IEEE Computer Society, Washington, DC (2005)
10. Knybel, J., Pavliska, V.: Representation of fuzzy if-then rules by petri nets. In: ASIS 2005, Prerov, Ostrava, pp. 121–125 (September 2005)
11. Lee, J., FanJiang, Y.-Y., Kuo, J.-Y., Lin, Y.-Y.: Modeling imprecise requirements with xml. In: Proceedings of the 2002 IEEE International Conference on Fuzzy Systems, FUZZ-IEEE 2002, vol. 2, pp. 861–866 (2002)
12. Lee, J., Xue, N.-L., Hsu, K.-H., Yang, S.J.: Modeling imprecise requirements with fuzzy objects. Inf. Sci. 118(1-4), 101–119 (1999)

13. Matthews, C., Swatman, P.A.: Fuzzy concepts and formal methods: A fuzzy logic toolkit for z. In: Bowen, J.P., Dunne, S., Galloway, A., King, S. (eds.) ZB 2000. LNCS, vol. 1878, pp. 491–510. Springer, Heidelberg (2000)
14. Matthews, C., Swatman, P.A.: Fuzzy concepts and formal methods: some illustrative examples. In: Proceedings of the Seventh Asia-Pacific Software Engineering Conference, APSEC 2000, pp. 230–238. IEEE Computer Society, Washington, DC (2000)
15. Pavliska, V.: Petri nets as fuzzy modeling tool. Technical report, University of Ostrava - Institute for Research and Applications of Fuzzy Modeling (2006)
16. Yang, S.J.H., Tsai, J.J.P., Chen, C.-C.: Fuzzy rule base systems verification using high-level petri nets. IEEE Trans. Knowl. Data Eng. 15(2), 457–473 (2003)
17. Zhong, Y.: The design of a controller in fuzzy petri net. Fuzzy Optimization and Decision Making 7, 399–408 (2008)

Resolution in Linguistic Propositional Logic Based on Linear Symmetrical Hedge Algebra

Thi-Minh-Tam Nguyen, Viet-Trung Vu, The-Vinh Doan, and Duc-Khanh Tran

Abstract. The paper introduces a propositional linguistic logic that serves as the basis for automated uncertain reasoning with linguistic information. First, we build a linguistic logic system with truth value domain based on a linear symmetrical hedge algebra. Then, we consider Gödel's t-norm and t-conorm to define the logical connectives for our logic. Next, we present a resolution inference rule, in which two clauses having contradictory linguistic truth values can be resolved. We also give the concept of reliability in order to capture the approximative nature of the resolution inference rule. Finally, we propose a resolution procedure with the maximal reliability.

1 Introduction

Automated reasoning is an approach to model human thinking. The resolution rule introduced by Robinson (1965) [5] marked an important point in studying automated reasoning. Resolution based on fuzzy set theory of Zadeh [18] has been studied to deal with uncertain information. In fuzzy logic, each clause has a membership function in $[0,1]$. Since then subtantial works [5, 17, 1, 13, 6, 14] have been done on the fuzzy resolution.

In two-valued logic, each clause has a truth value True or False. Therefore, the logical inference is absolutely accurate. However, in linguistic logic, each literal has a linguistic truth value such as MoreTrue, MoreFalse, PossibleVeryTrue,

Thi-Minh-Tam Nguyen
Faculty of Information Technology - Vinh University
e-mail: nmtam@vinhuni.edu.vn

Viet-Trung Vu · The-Vinh Doan · Duc-Khanh Tran
School of Information and Communication Technology,
Hanoi University of Science and Technology, Vietnam
e-mail: {trungvv91,doanthevinh1991}@gmail.com,
khanhtd@soict.hut.edu.vn

V.-N. Huynh et al. (eds.), *Knowledge and Systems Engineering, Volume 1*, Advances in Intelligent Systems and Computing 244,
DOI: 10.1007/978-3-319-02741-8_28, © Springer International Publishing Switzerland 2014

LessTrue,..., where True, False are *generators* and More, PossibleVery, Less,... are strings of hedges which increase or decrease the semantic of generators. Thus the accuracy of logical inference is approximate. For instance the two clauses $A^{\mathsf{True}} \vee B^{\mathsf{MoreTrue}}$ and $B^{\mathsf{LessFalse}} \vee C^{\mathsf{True}}$ can be resolved to obtain $A^{\mathsf{True}} \vee C^{\mathsf{True}}$. However the literals B^{MoreTrue} and $B^{\mathsf{LessFalse}}$ are not totally contradictory, they are only contradictory at a certain degree. Consequently the resolution inference is only reliable at a certain degree. Therefore, when the inference is performed, the infered formula should be associated with a certain reliability. Automated reasoning in linguistic logic has been attracting many researchers. Many works presented resolution algorithms in linguistic logics with truth value domain based on the implication lattice algebraic structures [2, 3, 15, 16, 19] or based on hedge algebra [4, 8, 10, 11].

Along the line of these research directions, we study automated reasoning based on resolution for linguistic propositional logic with truth value domain is taken from linear symmetrical hedge algebra. The syntax of linguistic propositional logic is constructed. To define the semantics of logical connectives we consider t-norm and t-conorm operators in fuzzy logic, specially t-norm and t-conorm operators of Gödel and Łukasiewicz. We show that logical connectives based on Gödel connectives are more appropriate to construct logical connectives for our linguistic logic. A resolution rule and resolution procedure are given. The concept of reliability of inference is introduced in such a way that the reliability of the conclusion is smaller than or equal to the reliabilities of the premises. We also present a resolution procedure with maximal reliability and prove the soundness and completeness of the resolution procedure.

The paper is structured as follows: Section 2 introduces basic notions and results on linear symmetrical hedge algebras. Section 3 presents the syntax and semantics of our linguistic propositional logic with truth value domain based on linear symmetrical hedge algebra. The resolution rule and resolution procedure are introduced in Section 4. Section 4 concludes and draws possible future work. Proofs of theorems, lemmas and proposition are presented in the Appendix.

2 Preliminaries

We recall only the most important definitions of hedge algebra for our work and refer the reader to [9, 7, 8] for further details.

We will be working with a class of abstract algebras of the form $AX = (X, G, H, \leq)$ where X is a term set, G is a set of generators, H is a set of linguistic hedges or modifiers, and $\leq$ is a partial order on X. AX is called a *hedge algebra* (HA) if it satisfies the following:

- Each hedge h is said to be positive w.r.t k, i.e. either kx $\geq$ x implies hkx $\geq$ kx or kx $\leq$ x implies hkx $\leq$ kx; similarly h is said to be negative w.r.t k, i.e. either kx $\geq$ x implies hkx $\leq$ kx or kx $\leq$ x implies hkx $\geq$ kx (for x $\in$ X);
- If terms u and v are independent, then, for all $x \in H(u)$, we have $x \notin H(v)$. If u and v are incomparable, i.e. $u \not< v$ and $v \not< u$, then so are x and y, for every $x \in H(u)$ and $y \in H(v)$;

- If $x \neq hx$, then $x \notin H(hx)$, and if $h \neq k$ and $hx \leq kx$, then $h'hx \leq k'kx$, for all $h,k,h',k' \in H$ and $x \in X$. Moreover, if $hx \neq kx$, then hx and kx are independent;
- If $u \notin H(v)$ and $u \leq v (u \geq v)$, then $u \leq hv (u \geq hv)$ for any $h \in H$.

Let $AX = (X, G, H, \leq)$ where the set of generators G contains exactly two comparable ones, denoted by $c^- < c^+$. For the variable Truth, we have $c^+ = True > c^- = False$. Such HAs are called *symmetrical* ones. For symmetrical HAs, the set of hedges H is decomposed into two disjoint subsets $H^+ = \{h \in H | hc^+ > c^+\}$ and $H^- = \{h \in H | hc^+ < c^+\}$. Two hedges h and k are said to be converse if $\forall x \in X, hx \leq x$ iff $kx \geq x$, i.e., they are in different subsets; h and k are said to be compatible if $\forall x \in X, hx \leq x$ iff $kx \leq x$, i.e. they are in the same subset. Two hedges in each of the sets H^+ and H^- may be comparable. Thus, H^+ and H^- become posets.

A symmetrical HA $AX = (X, G = \{c^-, c^+\}, H, \leq)$ is called a *linear symmetrical HA* (*LSHA*, for short) if the set of hedges H can be decomposed into $H^+ = \{h \in H | hc^+ > c^+\}$ and $H^- = \{h \in H | hc^+ < c^+\}$, and H^+ and H^- are linearly ordered.

Let $h_n h_{n-1} \dots h_1 u$, $k_m k_{m-1} \dots k_1 u$ be the canonical presentations of values x, y respectively. $x = y$ iff $m = n$ and $h_j = k_j$ for every $j \leq n$. If $x \neq y$ then there exists an $j \leq min\{m,n\} + 1$ (there is one convention is understood that if $j = min\{m,n\} + 1$, then $h_j = I$ where $j = n + 1 \leq m$ or $k_j = I$ where $j = m + 1 \leq n$) such that $h_{j'} = k_{j'}$ with all $j < j$. Denote that $x_{<j} = h_{j-1} h_{j-2} \dots h_1 u = k_{j-1} k_{j-2} \dots k_1 u$, we have: $x < y$ iff $h_j x_{<j} < k_j x_{<j}$, $x > y$ iff $h_j x_{<j} > k_j x_{<j}$.

Let x be an element of the hedge algebra AX and the canonical representation of x is $x = h_n ... h_1 a$ where $a \in \{c^+, c^-\}$. The contradictory element of x is an element y such that $y = h_n ... h_1 a'$ where $a' \in \{c^+, c^-\}$ and $a' \neq a$, denoted by $\bar{x}$. In LSHA, every element $x \in X$ has a unique contradictory element in X.

It is useful to limit the set of values X only consists of finite length elements. This is entirely suitable with the practical application in natural language, which does not consider infinite number of hedge of string.

From now on, we consider a LSHA $AX = (X, G, H, \leq, \neg, \vee, \wedge, \rightarrow)$ where $G = \{\bot, \mathsf{False}, \mathsf{W}, \mathsf{True}, \top\}$; $\bot, \top$ and W are the least, the neutral, and the greatest elements of X, respectively; $\bot < \mathsf{False} < \mathsf{W} < \mathsf{True} < \top$.

3 Propositional Logic with Truth Value Domain Based on Symmetrical Linear Hedge Algebra

Below, we define the syntax and semantics of the linguistic propositional logic.

Definition 0.1. An alphabet consists of:

- constant symbols: $\mathsf{MoreTrue}, \mathsf{VeryFalse}, \bot, \top, \dots$
- propositional variables: $A, B, C, \dots$
- logical connectives: $\vee, \wedge, \rightarrow, \neg, \equiv$, and
- auxiliary symbols: $\Box, (,), \dots$

Definition 0.2. An atom is either a propositional variable or a constant symbol.

Definition 0.3. Let A be an atom and α be a constant symbol. Then A^{α} is called a literal.

Definition 0.4. Formulae are defined recursively as follows:

- either a literal or a constant is a formula,
- if P is a formula, then $\neg P$ is a formula, and
- if P, Q are formulae, then $P \vee Q$, $P \wedge Q$, $P \rightarrow Q, P \leftrightarrow Q$ are formulae.

Definition 0.5. A clause is a finite disjunction of literals, which is written as $l_1 \vee l_2 \vee ... \vee l_n$, where l_i is a literal. An empty clause is denoted by $\square$.

Definition 0.6. A formula F is said to be in conjunctive normal form (CNF) if it is a conjunction of clauses.

In many-valued logic, sets of connectives called Łukasiewicz, Gödel, and product logic ones are often used. Each of the sets has a pair of residual t-norm and implicator. However, we cannot use the product logic connectives when our truth values are linguistic.

We recall the operators t-norm(T) and t-conorm(S) on fuzzy logic. It is presented detailed in [12, 14].

T-norm is a dyadic operator on the interval $[0,1]$: $T : [0,1]^2 \longrightarrow [0,1]$, satisfying the following conditions:

- Commutativity: $T(x,y) = T(y,x)$,
- Associativity: $T(x,T(y,z)) = T(T(x,y),z)$,
- Monotonicity: $T(x,y) \leq T(x,z)$ where $y \leq z$, and
- Boundary condition: $T(x,1) = x$.

If T is a t-norm, then its dual t-conorm S is given by $S(x,y) = 1 - T(1-x, 1-y)$.

Let $K = \{n | n \in \mathbb{N}, n \leq N_0\}$. Extended T-norm is a dyadic operator $T_E : K^2 \longrightarrow K$ and satisfies the following conditions:

- Commutativity: $T_E(m,n) = T_E(n,m)$,
- Associativity: $T_E(m, T_E(n,p)) = T_E(T_E(m,n),p)$,
- Monotonicity: $T_E(m,n) \leq T_E(m,p)$ where $n \leq p$, and
- Boundary condition: $T_E(n, N_0) = n$.

The Extended T-conorm is given by: $S_E(m,n) = N_0 - T_E(N_0 - n, N_0 - m)$. It is easy to prove that S_E is commutative, associate, monotonous. The boundary condition of S_E is: $S_E(0,n) = n$.

Two common pairs (T,S) in fuzzy logic: Gödel's(T,S) and Łukasiewicz's(T,S) are defined as following:

- Gödel:
  - $T_G(m,n) = \min(m,n)$
  - $S_G(m,n) = \max(m,n)$
- Łukasiewicz:

- $T_L(m,n) = \max(0, m+n-N_0)$
- $S_L(m,n) = \min(m+n, N_0)$

Given a SLHA AX , since all the values in AX are linearly ordered, we assume that they are $\bot = v_0 \leq v_1 \leq v_2 \leq \ldots \leq v_n = \top$.

Clearly, the pair (T,S) is determined only depending on max and min operators. Commonly, the truth functions for conjunctions and disjunctions are t-norms and t-conorms respectively.

Example 3.1. *Consider a SLHA* $AX = (X, \{\mathsf{True}, \mathsf{False}\}, \{\mathsf{More}, \mathsf{Less}\}, \leq)$ *with* $\mathsf{Less} < \mathsf{More}$. *We assume the length of hedge string is limited at 1. Then* $AX = \{v_0 = \mathsf{MoreFalse}, v_1 = \mathsf{False}, v_2 = \mathsf{LessFalse}, v_3 = \mathsf{LessTrue}, v_4 = \mathsf{True}, v_5 = \mathsf{MoreTrue}\}$. *We determine the truth value of logical connectives based on t-norm and t-conorm operators of Gödel and Łukasiewicz:*

- *Gödel:*
  - $\mathsf{LessFalse} \vee \mathsf{False} = max\{\mathsf{LessFalse}, \mathsf{False}\} = \mathsf{LessFalse}$
  - $\mathsf{MoreTrue} \wedge \mathsf{True} = min\{\mathsf{MoreTrue}, \mathsf{True}\} = \mathsf{True}$
- *Łukasiewicz:*
  - $\mathsf{LessFalse} \vee \mathsf{False} = \mathsf{LessTrue}$
  - $\mathsf{MoreTrue} \wedge \mathsf{True} = \mathsf{LessFalse}$

In fact, if the same clause has two truth values LessFalse or False, then it should get the value LessFalse. In the case of two truth values are MoreTrue and True then it should get the value True. We can see that the logical connectives based on Gödel's t-norm and t-conorm operators are more suitable in the resolution framework than those based on Łukasiewicz's. In this paper we will define logical connectives using Gödel's t-norm and t-conorm operators.

Definition 0.7. Let S be a linguistic truth domain, which is a SLHA $AX = (X, G, H, \leq)$, where $G = \{\top, \mathsf{True}, \mathsf{W}, \mathsf{False}, \bot\}$. The logical connectives $\wedge$ (respectively $\vee$) over the set S are defined to be Gödel's t-norm (respectively t-conorm), and furthermore to satisfy the following:

- $\neg\alpha = \overline{\alpha}$.
- $\alpha \to \beta = (\neg\alpha) \vee \beta$.

where $\alpha, \beta \in S$.

Proposition 3.1. *Let S be a linguistic truth domain, which is a SLHA* $AX = (X, \{\top, \mathsf{True}, \mathsf{W}, \mathsf{False}, \bot\}, H, \leq)$; $\alpha, \beta, \gamma \in X$, *we have:*

- *Double negation:*
  - $\neg(\neg\alpha) = \alpha$
- *Commutative:*
  - $\alpha \wedge \beta = \beta \wedge \alpha$

– $\alpha \vee \beta = \beta \vee \alpha$

- *Associative:*

– $(\alpha \wedge \beta) \wedge \gamma = \alpha \wedge (\beta \wedge \gamma)$
– $(\alpha \vee \beta) \vee \gamma = \alpha \vee (\beta \vee \gamma)$

- *Distributive:*

– $\alpha \wedge (\beta \vee \gamma) = (\alpha \wedge \beta) \vee (\alpha \wedge \gamma)$
– $\alpha \vee (\beta \wedge \gamma) = (\alpha \vee \beta) \wedge (\alpha \vee \gamma)$

Definition 0.8. An interpretation consists of the followings:

- a linguistic truth domain, which is a SLHA $AX = (X, G, H, \leq)$, where the set of generators $G = \{\top, \mathsf{True}, \mathsf{W}, \mathsf{False}, \bot\}$,
- for each constant in the alphabet, the assignment of an element in X,
- for each formula, the assignment of a mapping from X to X.

Definition 0.9. Let I be an interpretation and A be an atom such that $I(A) = \alpha_1$. Then the truth value of a literal A^{α_2} under the interpretation I is determined uniquely as follows:

- $I(A^{\alpha_2}) = \alpha_1 \wedge \alpha_2$ if $\alpha_1, \alpha_2 > \mathsf{W}$,
- $I(A^{\alpha_2}) = \neg(\alpha_1 \vee \alpha_2)$ if $\alpha_1, \alpha_2 \leq \mathsf{W}$,
- $I(A^{\alpha_2}) = (\neg\alpha_1) \vee \alpha_2$ if $\alpha_1 > \mathsf{W}, \alpha_2 \leq \mathsf{W}$, and
- $I(A^{\alpha_2}) = \alpha_1 \vee (\neg\alpha_2)$ if $\alpha_1 \leq \mathsf{W}, \alpha_2 > \mathsf{W}$.

Definition 0.10. The truth value of formulae under an interpretation is determined recursively as follows:

- $I(P \vee Q) = I(P) \vee I(Q)$,
- $I(P \wedge Q) = I(P) \wedge I(Q)$,
- $I(\neg P) = \neg I(P)$,
- $I(P \rightarrow Q) = I(P) \rightarrow I(Q)$

The following result follows from the properties of the $\wedge$ and $\vee$ operators.

Proposition 3.2. *Let A, B and C are formulae, and I be an arbitrary interpretation. Then,*

- *Commutative:*

– $I(A \vee B) = I(B \vee A)$
– $I(A \wedge B) = I(B \wedge A)$

- *Associative:*

– $I((A \vee B) \vee C) = I(A \vee (B \vee C))$
– $I((A \wedge B) \wedge C) = I(A \wedge (B \wedge C))$

- *Distributive:*

– $I(A \vee (B \wedge C)) = I((A \vee B) \wedge (A \vee C))$

– $I(A \wedge (B \vee C)) = I((A \wedge B) \vee (A \wedge C))$

Definition 0.11. Let F be a formula and I be an interpretation. Then

- F is said to be true under interpretation I iff $I(F) \geq \mathsf{W}$, I is also said to satisfy formula F, F is said to be satisfiable iff there is an interpretation I such that I satisfies F, F is said to be tautology iff it is satisfied by all interpretations;
- F is said to be false under interpretation I iff $I(F) \leq \mathsf{W}$, I is also said to falsify formula F, F is said to be unsatisfiable iff it is falsified by all interpretations.

Definition 0.12. Formula B is said to be a logical consequence of formula A, denoted by $A \models B$, if for all interpretation I, $I(A) > \mathsf{W}$ implies that $I(B) > \mathsf{W}$.

Proposition 3.3. *Let A and B be formulae. Then, $A \models B$ iff $\models (A \to B)$.*

Definition 0.13. Two formulae A and B are logically equivalent, denoted by $A \equiv B$, if and only if $A \models B$ and $B \models A$.

Proposition 3.4. *Let A, B and C be formulae. Then the following properties hold:*

- *Idempotency:*
  – $A \vee A \equiv A$
  – $A \wedge A \equiv A$
- *Implication:*
  – $A \to B \equiv (\neg A) \vee B$
  – $(A \equiv B) \equiv (A \to B) \wedge (B \to A)$
- *Double negation:*
  – $\neg\neg A \equiv A$
- *De Morgan:*
  – $\neg(A \vee B) \equiv (\neg A) \wedge (\neg B)$
  – $\neg(A \wedge B) \equiv (\neg A) \vee (\neg B)$
- *Commutativity:*
  – $A \vee B \equiv B \vee A$
  – $A \wedge B \equiv B \wedge A$
- *Associativity:*
  – $A \vee (B \vee C) \equiv (A \vee B) \vee C$
  – $A \wedge (B \wedge C) \equiv (A \wedge B) \wedge C$
- *Distributivity:*
  – $A \vee (B \wedge C) \equiv (A \vee B) \wedge (A \vee C)$
  – $A \wedge (B \vee C) \equiv (A \wedge B) \vee (A \wedge C)$

We will be working with resolution as the inference system of our logic. Therefore formulae need to be converted into conjunctive normal form. The equivalence properties in Proposition 3.4 ensure that the transformation is always feasible.

Theorem 0.1. *Let F be a formula of arbitrary form. Then F can be converted into an equivalent formula in conjunctive normal form.*

4 Resolution

In the previous section, we have described the syntax and semantics of our linguistic logic. In this section, we present the resolution inference rule and the resolution procedure for our logic.

Definition 0.14. The clause C with reliability α is the pair (C,α) where C is a clause and α is an element of SLHA AX such that $\alpha > \mathsf{W}$. The same clauses with different reliabilities are called variants. That is (C,α) and (C,α') are called variants of each other.

For a set of n clauses $S = \{C_1, C_2, ..., C_n\}$, where each C_i has a reliability α_i, then the reliability α of S is defined as: $\alpha = \alpha_1 \wedge \alpha_2 \wedge ... \wedge \alpha_n$.

An inference rule R with the reliability α is represented as:

$$\frac{(C_1,\alpha_1),(C_2,\alpha_2),\ldots,(C_n,\alpha_n)}{(C,\alpha)}$$

We call α the reliability of R, provided that $\alpha \le \alpha_i$ for $i = 1..n$.

Definition 0.15. The fuzzy linguistic resolution rule is defined as follows:

$$\frac{(A^a \vee B^{b_1},\alpha_1),(B^{b_2} \vee C^c,\alpha_2)}{(A^a \vee C^c,\alpha_3)}$$

where b_1, b_2 and α_3 satisfy the following conditions:

$$\begin{cases} b_1 \wedge b_2 \le \mathsf{W}, \\ b_1 \vee b_2 > \mathsf{W}, \\ \alpha_3 = f(\alpha_1,\alpha_2,b_1,b_2) \end{cases}$$

with f is a function ensuring that $\alpha_3 \le \alpha_1$ and $\alpha_3 \le \alpha_2$.

α_3 is defined so as to be smaller or equal to both α_1 and α_2. In fact, the obtained clause is less reliable than original clauses. The function f is defined as following:

$$\alpha_3 = f(\alpha_1,\alpha_2,b_1,b_2) = \alpha_1 \wedge \alpha_2 \wedge (\neg(b_1 \wedge b_2)) \wedge (b_1 \vee b_2) \tag{1}$$

Obviously, $\alpha_1, \alpha_2 \ge W$, and α_3 depends on b_1, b_2. Additionally, $b_1 \wedge b_2 \le \mathsf{W}$ implies $\neg(b_1 \wedge b_2) > \mathsf{W}$. Moreover, $(b_1 \vee b_2) > \mathsf{W}$. Then, by Formula (1), we have $\alpha_3 > \mathsf{W}$.

Lemma 0.1. *The fuzzy linguistic resolution rule 0.15 is sound.*

We define a fuzzy linguistic resolution *derivation* as a sequence of the form $S_0, \ldots, S_i, \ldots$, where:

- each S_i is a set of clauses with a reliability, and
- S_{i+1} is obtained by adding the conclusion of a fuzzy linguistic resolution inference with premises from S_i, that is $S_{i+1} = S_i \cup \{(C, \alpha)\}$, where (C, α) is the conclusion of the fuzzy linguistic resolution

$$\frac{(C_1, \alpha_1), (C_2, \alpha_2)}{(C, \alpha)},$$

and $(C_1, \alpha_1), (C_2, \alpha_2) \in S_i$.

A *resolution proof* of a clause C from a set of clauses S consists of repeated application of the resolution rule to derive the clause C from the set S. If C is the empty clause then the proof is called a *resolution refutation*. We will represent resolution proofs as *resolution trees*. Each tree node is labeled with a clause. There must be a single node that has no child node, labeled with the conclusion clause, we call it the root node. All nodes with no parent node are labeled with clauses from the initial set S. All other nodes must have two parents and are labeled with a clause C such that

$$\frac{C_1, C_2}{C}$$

where C_1, C_2 are the labels of the two parent nodes. If RT is a resolution tree representing the proof of a clause with reliability (C, α), then we say that RT has the reliability α.

Different resolution proofs may give the same the conclusion clause with different reliabilities. The following example illustrate this.

Example 4.1. *Let* $AX = (X, G, H, \leq, \neg, \vee, \wedge, \rightarrow)$ *be a SRHA where* $G = \{\bot, \mathsf{False}, \mathsf{W}, \mathsf{True}, \top\}$, $\bot, \mathsf{W}, \top$ *are the smallest, neutral,biggest elements respectively, and* $\bot < \mathsf{False} < \mathsf{W} < \mathsf{True} < \top$;$H^+ = \{\mathsf{V}, \mathsf{M}\}$ *and* $H^- = \{\mathsf{P}, \mathsf{L}\}$ *(V=Very, M=More, P=Possible, L=Less); Consider the following set of clauses:*

1. $A^{\mathsf{MFalse}} \vee B^{\mathsf{False}} \vee C^{\mathsf{VMTrue}}$
2. $B^{\mathsf{LTrue}} \vee C^{\mathsf{PTrue}}$
3. A^{PTrue}
4. B^{VTrue}
5. C^{VFalse}

At the beginning, each clause is assigned to the highest reliability $\top$. *We have:*

$$\cfrac{\cfrac{\cfrac{(A^{\mathsf{PTrue}}, \top)(A^{\mathsf{MFalse}} \vee B^{\mathsf{False}} \vee C^{\mathsf{VMTrue}}, \top)}{(B^{\mathsf{False}} \vee C^{\mathsf{VMTrue}}, \mathsf{PTrue})(B^{\mathsf{VTrue}}, \top)}}{(C^{\mathsf{VMTrue}}, \mathsf{PTrue}) \qquad (C^{\mathsf{VFalse}}, \top)}}{(\Box, \mathsf{PTrue})}$$

$$\cfrac{\cfrac{\cfrac{(A^{\mathsf{MFalse}} \vee B^{\mathsf{False}} \vee C^{\mathsf{VMTrue}}, \top)(B^{\mathsf{LTrue}} \vee C^{\mathsf{PTrue}}, \top)}{(A^{\mathsf{MFalse}} \vee C^{\mathsf{VMTrue}}, \mathsf{LTrue})\ (A^{\mathsf{PTrue}}, \top)}}{(C^{\mathsf{VMTrue}}, \mathsf{LTrue}) \qquad (C^{\mathsf{VFalse}}, \top)}}{(\Box, \mathsf{LTrue})}$$

Since different proofs of the same clause may have different reliabilities, it is natural to study how to design a resolution procedure with the best reliability. Below we present such a procedure.

We say that a set of clauses S is *saturated* iff for every fuzzy linguistic resolution inference with premises in S, the conclusion of this inference is a variant with smaller or equal reliability of some clauses in S. That is for every fuzzy linguistic resolution inference

$$\frac{(C_1,\alpha_1),(C_2,\alpha_2)}{(C,\alpha)}$$

where $(C_1,\alpha_1),(C_2,\alpha_2) \in S$, there is some clause $(C,\alpha') \in S$ such that $\alpha \leq \alpha'$.

We introduce a resolution strategy, called *α-strategy*, which guarantees that the resolution proof of each clause has the maximal reliability. An α-strategy derivation is a sequence of the form $S_0,\ldots,S_i,\ldots$, where

- each S_i is a set of clauses with reliability, and
- S_{i+1} is obtained by adding the conclusion of a fuzzy linguistic resolution inference with premises with maximal reliabilities from S_i, that is $S_{i+1} = S_i \cup \{(C,\alpha)\}$, where (C,α) is the conclusion of the fuzzy linguistic resolution inference
$$\frac{(C_1,\alpha_1),(C_2,\alpha_2)}{(C,\alpha)}$$
$(C_1,\alpha_1),(C_2,\alpha_2) \in S_i$ and there are not any clauses with reliability $(C_1,\alpha_1'),$ $(C_2,\alpha_2') \in S_i$ such that $\alpha_1' > \alpha_1$ and $\alpha_2' > \alpha_2$, or
- S_{i+1} is obtained by removing a variant with smaller reliability, that is $S_{i+1} = S_i \setminus \{(C,\alpha)\}$ where $(C,\alpha) \in S_i$ and there is some $(C,\alpha') \in S_i$ such that $\alpha < \alpha'$.

Define *the limit of a derivation* $S_0,\ldots,S_i,\ldots$

$$S_\infty = \bigcup_{i\geq 0}\bigcap_{j\geq i} S_j$$

The following result establishes the soundness and completeness of the resolution procedure.

Theorem 0.2. *Let $S_0,\ldots,S_i,\ldots$ be a fuzzy linguistic resolution α-strategy derivation. S_n contains the empty clause iff S_0 is unsatisfiable (for some $n = 0,1,\ldots$).*

Lemma 0.2. *Consider the following resolution inferences:*

$$\frac{(A^a \vee B^{b_1},\alpha),(B^{b_2} \vee C^c,\beta)}{(A^a \vee C^c,\gamma)}$$

$$\frac{(A^a \vee B^{b_1}, \alpha), (B^{b_2} \vee C^c, \beta')}{(A^a \vee C^c, \gamma')}$$

Then, $\beta' > \beta$ implies $\gamma' \geq \gamma$.

Lemma 0.3. *Let $S_0, \ldots, S_i, \ldots$ be a fuzzy linguistic resolution α-strategy derivation, and S_∞ be the the limit of the derivation. Then S_∞ is saturated.*

Theorem 0.3. *Let $S_0, \ldots, S_i, \ldots$ be a fuzzy linguistic resolution α-strategy derivation, and S_∞ be the the limit of the derivation. Then for each clause (C, α) in S_∞, there is not any other resolution proof of the clause (C, α') from S_0 such that $\alpha' > \alpha$.*

Example 4.2. *Consider again Example 4.1. Applying the α-strategy we get the following saturated set of clauses*

1. $(A^{\mathsf{MFalse}} \vee B^{\mathsf{False}} \vee C^{\mathsf{VMTrue}}, \top)$
2. $(B^{\mathsf{LTrue}} \vee C^{\mathsf{PTrue}}, \top)$
3. $(A^{\mathsf{PTrue}}, \top)$
4. $(B^{\mathsf{VTrue}}, \top)$
5. $(C^{\mathsf{VFalse}}, \top)$
6. $(B^{\mathsf{False}} \vee C^{\mathsf{VMTrue}}, \mathsf{PTrue})$
7. $(A^{\mathsf{MFalse}} \vee C^{\mathsf{VMTrue}}, \mathsf{True})$
8. $(A^{\mathsf{MFalse}} \vee C^{\mathsf{VMTrue}}, \mathsf{LTrue})$
9. $(A^{\mathsf{MFalse}} \vee B^{\mathsf{False}}, \mathsf{VTrue})$
10. $(C^{\mathsf{VMTrue}}, \mathsf{PTrue})$
11. $(A^{\mathsf{MFalse}}, \mathsf{True})$
12. $(\square, \mathsf{PTrue})$

The initial set of clauses is unsatisfiable, and the resolution futation is the following

$$\cfrac{\cfrac{\cfrac{(A^{\mathsf{MFalse}} \vee B^{\mathsf{False}} \vee C^{\mathsf{VMTrue}}, \top)(C^{\mathsf{VFalse}}, \top)}{(A^{\mathsf{MFalse}} \vee B^{\mathsf{False}}, \mathsf{VTrue})(B^{\mathsf{VTrue}}, \top)}}{(A^{\mathsf{MFalse}}, \mathsf{True}) \qquad (A^{\mathsf{PTrue}}, \top)}}{(\square, \mathsf{PTrue})}$$

5 Conclusion

We have presented a linguistic logic system with the basic components: syntax, semantics and inference. The syntax have been defined as usual. To define the semantics, the truth value domain have been taken from linear symmetrical hedge algebra and logical connectives have been defined based on Gödel's t-norm and t-conorm. We have also introduced an inference rule associated with a reliability which guarantees that the reliability of the inferred clause is less than or equal to reliaility of the premise clauses. Moreover, we have given a resolution procedure which ensures that the proof of clauses has the maximal reliability. The soundness and completeness of the resolution procedure are also proved. The proofs of the theorems, propositions and lemmas are omitted due to lack of space. They can be found in the full paper at http://arxiv.org/pdf/1307.7661v2.pdf. There are several lines of future works. It would be natural to consider the linguistic first order logic in the same settings as our logic here. It would be worth investivating how to extend our result to other hedge algebra structures and to other automated reasong methods.

References

1. Ebrahim, R.: Fuzzy logic programming. Fuzzy Sets and Systems 117(2), 215–230 (2001)
2. He, X., Liu, J., Xu, Y., Martínez, L., Ruan, D.: On -satisfiability and its -lock resolution in a finite lattice-valued propositional logic. Logic Journal of the IGPL 20(3), 579–588 (2012)
3. Lai, J., Xu, Y.: Linguistic truth-valued lattice-valued propositional logic system *l*p(x) based on linguistic truth-valued lattice implication algebra. Inf. Sci. 180(10), 1990–2002 (2010)
4. Le, V.H., Liu, F., Tran, D.K.: Fuzzy linguistic logic programming and its applications. TPLP 9(3), 309–341 (2009)
5. Lee, R.C.T.: Fuzzy logic and the resolution principle. In: IJCAI, pp. 560–567 (1971)
6. Mondal, B., Raha, S.: Approximate reasoning in fuzzy resolution. In: 2012 Annual Meeting of the North American Fuzzy Information Processing Society (NAFIPS), pp. 1–6 (2012)
7. Nguyenand, C.-H., Huynh, V.-N.: An algebraic approach to linguistic hedges in zadeh's fuzzy logic. Fuzzy Sets and Systems 129(2), 229–254 (2002)
8. Nguyen, C.-H., Tran, D.-K., Huynh, V.-N., Nguyen, H.-C.: Hedge algebras, linguistic-valued logic and their application to fuzzy reasoning. International Journal of Uncertainty, Fuzziness and Knowledge-Based Systems 7(4), 347–361 (1999)
9. Nguyen, C.H., Wechler, W.: Hedge Algebras: An Algebraic Approach in Struture of Sets of Linguistic Truth Values. Fuzzy Sets and Syst. 35, 281–293 (1990)
10. Phuong, L.A., Khang, T.D.: A deductive method in linguistic reasoning. In: 2012 2nd International Conference on Uncertainty Reasoning and Knowledge Engineering (URKE), pp. 137–140 (2012)
11. Phuong, L.A., Khang, T.D.: Linguistic reasoning based on generalized modus ponens with linguistic modifiers and hedge moving rules. In: 2012 International Conference on Fuzzy Theory and it's Applications (iFUZZY), pp. 82–86 (2012)
12. Smutná, D., Vojtás, P.: Graded many-valued resolution with aggregation. Fuzzy Sets and Systems 143(1), 157–168 (2004)
13. Vojtás, P.: Fuzzy logic programming. Fuzzy Sets and Systems 124(3), 361–370 (2001)
14. Weigert, T.J., Tsai, J.J.P., Liu, X.: Fuzzy operator logic and fuzzy resolution. J. Autom. Reasoning 10(1), 59–78 (1993)
15. Xu, Y., Ruan, D., Kerre, E.E., Liu, J.: Alpha-resolution principle based on lattice-valued propositional logic lp(x). Inf. Sci. 130(1-4), 195–223 (2000)
16. Xu, Y., Ruan, D., Kerre, E.E., Liu, J.: Alpha-resolution principle based on first-order lattice-valued logic lf(x). Inf. Sci. 132(1-4), 221–239 (2001)
17. Mukaidono, M., Shen, Z., Ding, L.: Fuzzy resolution principle. In: Proc. 18th Internat. Symp. on Multiple-valued Logic, pp. 210–215 (1989)
18. Zadeh, L.A.: Fuzzy sets. Information and Control 8(3), 338–353 (1965)
19. Zhong, X., Xu, Y., Liu, J., Chen, S.: General form of -resolution principle for linguistic truth-valued lattice-valued logic. Soft Comput. 16(10), 1767–1781 (2012)

A Subgradient Method to Improve Approximation Ratio in the Minimum Latency Problem

Bang Ban Ha and Nghia Nguyen Duc

Abstract. The Minimum Latency Problem (MLP) is a combinatorial optimization problem which has many practical applications. Recently, several approximation algorithms with guaranteed approximation ratio have been proposed to solve the MLP problem. These algorithms start with a set of solutions of the $k-$MST or $k-$troll problem, then convert the solutions into Eulerian tours, and finally, concatenate these Eulerian tours to obtain a MLP tour. In this paper, we propose an algorithm based on the principles of the subgradient method. It still uses the set of solutions of the $k-$MST or $k-$troll problem as an input, then modifies each solution into a tour with cost smaller than that of Eulerian tour and finally, uses obtained tours to construct a MLP tour. Since the low cost tours are used to build a MLP tour, we can expect the approximation ratio of obtained algorithm will be improved. In order to illustrate this intuition, we have evaluated the algorithm on five benchmark datasets. The experimental results show that approximation ratio of our algorithm is improved compared to the best well-known approximation algorithms.

1 Introduction

The minimum latency problem is also known in the literature as the delivery man problem or the traveling repairman problem. In the general case, the problem is described as NP-hard, and unless P = NP, a polynomial time approximation scheme is unlikely to exist [10]. However, the reduction from the general MLP to the problem in a metric case can be done by a simple transformation as in [15]. The metric case reflects a condition in which a complete graph with distances between vertices satisfying the triangle inequality. In this paper, we consider the problem in the metric case, and formulate the MLP as follows:
Given a complete graph K_n with the vertex set $V = \{1, 2, \ldots, n\}$ and a symmetric

Bang Ban Ha · Nghia Nguyen Duc
Hanoi University of Science and Technology, No. 1. Hai Ba Trung District, Ha Noi, Vietnam
e-mail: {BangBH,NghiaND}@soict.hut.edu.vn

V.-N. Huynh et al. (eds.), *Knowledge and Systems Engineering, Volume 1*,
Advances in Intelligent Systems and Computing 244,
DOI: 10.1007/978-3-319-02741-8_29,

distance matrix $C = \{c_{ij} \mid i, j = 1, 2, \ldots, n\}$, where c_{ij} is the distance between two vertices i and j. Suppose that $T = \{v_1, \ldots, v_k, \ldots, v_n\}$ is a tour in K_n (i.e, the path that visits each vertex of the graph exactly once). Denote by $P(v_1, v_k)$ the path from v_1 to v_k on this tour and by $l(P(v_1, v_k))$ its length. The latency of a vertex $v_k (1 < k \leq n)$ on T is the length of the path from starting vertex v_1 to v_k:

$$l(P(v_1, v_k)) = \sum_{i=1}^{k-1} c_{v_i v_{i+1}}.$$

The total latency of the tour T is defined as the sum of latencies of all vertices

$$L(T) = \sum_{k=2}^{n} l(P(v_1, v_k))$$

The minimum latency problem asks for a minimum latency tour, which starts at a given starting vertex s (which will be denoted by v_1 in the rest of the paper, for convenient) and visits each vertex in the graph exactly once.
Minimizing $L(T)$ arises in many practical situations whenever a server (for example, a repairman or a disk head) has to accommodate a set of requests with their minimal total (or average) waiting time [4], [10].

For NP-hard problem, such as the MLP problem, there are three common ways to develope algorithms for its solving. Firstly, exact algorithms guarantee to find the optimal solution and take exponential time in the worst case, but they often run much faster in practice. However, the exact algorithms in [3], [15] only solve the MLP problem with small sizes (up to 40 vertices). Secondly, the heuristic algorithms in [11], [12] perform well in practice and validate their empirical performance on an experimental benchmark of interesting instances. However, the efficiency of the heuristic algorithms cannot be evaluated in the general case. Thirdly, approximation algorithms that always produce a solution whose value at most $\alpha * f^*$, where f^* is the optimal value (the factor α is called the approximation ratio). Our algorithm belongs to the last approach.

In the thirs approach, recently, several approximation algorithms have proposed to solve the MLP problem [1], [2], [4], [5], [7]. These algorithms start with a set of solutions of the $k-$MST or $k-$troll problems. The $k-$MST problem [5] and the $k-$troll problem [6]) requires to find the minimum tree spanning k vertices and the minimum path visiting k vertices, respectively. The algorithms in [1], [2], [4], [7] use the set of solutions of the $k-$MST problem. Meanwhile, K. Chaudhuri et al.'s algorithm in [5] uses the set of solutions of the $k-$troll problem. In [5], K. Chaudhuri et al. show that using the solutions of the $k-$troll problem are better than one of the $k-$MST problem for building a MLP solution. In fact, K. Chaudhuri et al.'s algorithm gives the best approximation ratio for the MLP problem. Their algorithm consists of three steps. In first step, it finds the set of solutions of the $k-$troll problem $(k = 2, ..., n)$, each of which simultaneously is a $k-$tree. In the second step, these $k-$tree $T_k (k = 2, ..., n)$ are converted into the Eulerian tours $C_k (k = 2, ..., n)$ such that $C(C_k) \leq \beta C(T_k)$ with $\beta = 2$, where $C(C_k)$ and $C(T_k)$ denote the cost of C_k and T_k, respectively. In final step, Geomans et al.'s technique [7] is applied to construct a tour of latency at most $3.59(\beta/2)\sum_k C(C_k)$. Therefore, Chaudhuri et al.'s algorithm is a $3.59(\beta/2)-$approximation algorithm for the MLP problem. Note that in the second step Chaudhuri et al.'s algorithm use Eulerian tours whose costs are far from the optimal values of the $k-$troll problem. Therefore, if we replace C_k by any tour $C_k^{'}$ whose cost is less than $C(C_k)$, we can expect an improvement in the approximation ratio for the MLP problem and that is our main purpose.

In this paper, we propose an algorithm based on the principles of the subgradient method. It uses the set of K. Chaudhuri et al.'s $k-$trees $T_k (k = 2, 3, ..., n)$ as an input, then modifies each $k-$tree into a tour of cost less than one of the Eulerian tour and finally, the obtained tours are used to built a MLP tour. Intuitively, since we use low cost tours to build a MLP tour, we can expect an improvement on approximation ratio. In order to verify this intuition, we have evaluated the algorithm on five benchmark datasets. The experimental results show that our algorithm is better than the best well-known approximation algorithm in terms of the approximation ratio.

The rest of this paper is organized as follows. In section 2, we present the proposed algorithm. Experimental results are reported in section 3. Section 4 concludes the paper.

2 The Proposed Algorithm

We suppose that we have a set of $k-$trees $(k = 2, ..., n)$ from K. Chaudhuri et al.'s algorithm. For each $k-$tree, the subgradient method is used to modify it. Intuitively, our subgradient method tries to force it to being a tour whose cost is less than the cost of the Eulerian tour. In order to support an iterative improvement scheme, we give three Lemmas which indicate the direction for each iteration, the limits on appropriate step sizes and the best choice of step size such that our subgradient method converges. That means our subgradient method produce a tour whose cost is less than one of the Eulerian tour. This theoretical result then is used to construct our algorithm. Our algorithm includes two steps. In the first step, it uses the set of K. Chaudhuri et al.'s $k-$trees as an input and then modifies each $k-$tree into a tour by the subgradient method. In the second step, we concatenate obtained tours to obtain a MLP tour.

At first, we present the theoretical framework for the subgradient method. Without loss of generality, we pick arbitrarily a $k-$tree in the set of $k-$trees $(k = 2, ..., n)$ and then describe our theoretical framework to modify it. Similar arguments still hold for modifying any $p-$tree with $p \neq k$ and $2 \leq p \leq n$. We describe several definitions to construct our theoretical framework. Remember that $k-$tree is a tree spanning k vertices. It has a root vertex v_1 and an end-vertex v_{end}. We denote the set of the vertices of this $k-$tree by $V^{'} = \{v_1, v_2, ..., v_k = v_{end}\}$. Let U be the set of all $k-$trees with the given vertices $V^{'}$ and T^* be the minimum tree in U. Moreover, we call $k-$tour as a path visiting k vertices in $V^{'}$ from v_1 to v_k. Let I be the set of all these $k-$tours and T^*_{tour} be the minimum tour among all tour of I. For each $k-$tree $T \in U$, let $C(c_{ij}, T)$ be its cost and $d(v_i, T)$ be the degree of v_i in T. Introduce a real $k-$vector $\tau = (\tau_1, \tau_2, ..., \tau_k)$, let $C(\bar{c}_{ij}, T)$ be the cost of $k-$tree or $k-$tour with respect to $\bar{c}_{ij} = c_{ij} + \tau_i + \tau_j$. Thus, by the definition, if T is a k-tree, then

$$C(\bar{c}_{ij}, T) = L(c_{ij}, T) + \sum_{i=1}^{k} \tau_i d(v_i, T), \tag{1}$$

and if T is a $k-$tour, then its cost is

$$C(\bar{c}_{ij}, T) = C(c_{ij}, T) + 2\sum_{i=2}^{k-1} \tau_i + \tau_1 + \tau_k. \tag{2}$$

Since each $k-$tour is simultaneously is a k-tree, we have

$$\min_{T \in U} C(c_{ij}, T) \leq \min_{T \in I} C(c_{ij}, T). \tag{3}$$

Furthermore, since T^*_{tour} denotes the minimum tour, from (2) and (3) it follows that

$$\min_{T \in U} C(\bar{c}_{ij}, T) \leq C(c_{ij}, T^*_{tour}) + 2\sum_{i=2}^{k-1} \tau_i + \tau_1 + \tau_k. \tag{4}$$

Now, for short, we write $C(T)$ and $C(T_k)$ instead $C(c_{ij}, T)$ and $C(c_{ij}, T_k)$, respectively. From (1) and (4), we have

$$\min_{T \in U}\{(C(T) + \sum_{i=1}^{k} \tau_i d(v_i, T))\} \leq C(T^*_{tour}) + 2\sum_{i=2}^{k-1} \tau_i + \tau_1 + \tau_k \tag{5}$$

which implies

$$\min_{T \in U}\{C(T) + \sum_{i=2}^{k-1} \tau_i(d(v_i, T) - 2) + \tau_1(d(v_1, T) - 1) + \tau_k(d(v_k, T) - 1)\} \leq C(T^*_{tour}). \tag{6}$$

Introduce function $w(\tau)$ as follows

$$w(\tau) = \min_{T \in U}\{C(T) + \tau_1(d(v_1, T) - 1) + \tau_k(d(v_k, T) - 1) + 2\sum_{i=2}^{k-1} \tau_i(d(v_i, T) - 2)\}. \tag{7}$$

For a given tree T, let v_T be the vector $(d(v_1, T) - 1, d(v_2, T) - 2, d(v_3, T) - 2, ..., d(v_k, T) - 1)$. Then, we can rewrite $w(\tau)$ as

$$w(\tau) = \min_{T \in U}\{C(T) + \tau v_T\},$$

where τv_T stands for the scalar product of two vectors τ and v_T.

Now, from inequality (6), it is easy to see that each value of $w(\tau)$ is a lower bound for $C(T^*_{tour})$. Hence, we are interested in finding a vector τ which gives the maximum value of $w(\tau)$. So, we need to solve the following optimization problem:

$$\max_{\tau} w(\tau).$$

Intuitively, we try to find a vector τ such that the correspondent $k-$tree is as close as possible to being a tour, keeping its cost from exceeding the cost of the optimal tour T^*_{tour}. Thus by finding $\max_{\tau} w(\tau)$, we force the degree of all vertices v_i ($v_i \neq v_1$ and v_k) to be 2 and the degree of v_1 and v_k to be 1.

M. Held et al. [8] gave a theoretical framework which supported an iterative improvement scheme for $w(\tau)$. The main idea of their method consisted in trying to convert the minimum tree into a cycle in which each vertex has degree 2. However, in this paper, we are interested in a tour rather than a cycle. Therefore, we need to slightly modify their theory. We use the subgradient method based on iterative procedure, which starting with an arbitrary vector $\tau^{(0)}$, compute a sequence of vectors $\{\tau^{(m)}, m = 0, 1, 2, ...\}$ by the following formula:

$$\tau^{(m+1)} = \tau^{(m)} + t_m d^{(m)}, \tag{8}$$

where $d^{(m)}$ is a moving direction from $\tau^{(m)}$ and t_m is the stepsize. Polyak [14] showed that $w(\tau^{(m)})$ converges to $\max_{\tau} w(\tau)$ when $d^{(m)}$ is a subgradient of the function w at $\tau^{(m)}$ and the sequence of stepsizes $\{t_m\}$ satisfies the following conditions

$$\sum_{m=1}^{\infty} t_m = \infty, t_m \to 0, t_m > 0, \forall m.$$

Now, for a given τ, denote the tree that provides minimum value of right side hand of (7) by $T^*(\tau)$. Then, we have

$$w(\tau) = C(T^*(\tau)) + \tau v_{T^*(\tau)}. \tag{9}$$

For an arbitrary vector $\overline{\tau}$, we have

$$C(T^*(\tau)) + \overline{\tau} v_{T^*(\tau)} \geq \min_{T \in U}\{C(T) + \overline{\tau} v_T\} = w(\overline{\tau}). \tag{10}$$

By subtracting (9) from (10), we obtain

$$\overline{\tau} v_{T^*(\tau)} - \tau v_{T^*(\tau)} \geq w(\overline{\tau}) - w(\tau).$$

Hence, we have the following lemma:

Lemma 1. *For any vector $\overline{\tau}$*

$$(\overline{\tau} - \tau) v_{T^*(\tau)} \geq w(\overline{\tau}) - w(\tau).$$

The hyperplane through τ having $v_{T^*(\tau)}$ as normal determines a closed halfspace $\{\overline{\tau} : (\overline{\tau} - \tau) v_{T^*(\tau)} \geq 0\}$ which contains all points $\overline{\tau}$ such that $w(\overline{\tau}) \geq w(\tau)$. Thus, Lemma 1 shows that vector $v_{T^*(\tau)}$ is a subgradient of the function w at τ.

Lemma 2. *Let $\overline{\tau}$ be a vector such that $w(\overline{\tau}) \geq w(\tau)$ and*

$$0 < t < \frac{2(w(\overline{\tau}) - w(\tau))}{\|v_{T^*(\tau)}\|^2},$$

where $\|\|$ *denotes the Euclidean norm, then*

$$\| (\overline{\tau} - (\tau + tv_{T^*(\tau)}) \| < \| \overline{\tau} - \tau \|,$$

i.e. the point $\tau + tv_{T^*(\tau)}$ *is closer to* $\overline{\tau}$ *than* τ.

Proof. We have

$$\begin{aligned}
\| \overline{\tau} - (\tau + tv_{T^*(\tau)}) \|^2 &= \| \overline{\tau} - \tau \|^2 - 2t(\overline{\tau} - \tau)v_{T^*(\tau)} + t^2 \| v_{T^*(\tau)} \|^2 \\
&= \| \overline{\tau} - \tau \|^2 - t[2(\overline{\tau} - \tau)v_{T^*(\tau)} - t \| v_{T^*(\tau)} \|^2] \\
&\leq \| \overline{\tau} - \tau \|^2 - t[2(w(\overline{\tau}) - w(\tau)) - t \| v_{T^*(\tau)} \|^2] \\
&\leq \| \overline{\tau} - \tau \|^2 .
\end{aligned}$$

Lemma 2 gives the limits on appropriate stepsizes t_m. Lemma 3 further guides the choice of the values of stepsize t_m to guarantee the convergence of the sequence $\{w(\tau^{(m)})\}$.

Lemma 3. *Assume that* $\overline{w} = \max_{\tau} w(\tau)$. *Let* $\{\tau^{(m)}\}$ *be a sequence of points obtained by the following iterative process*

$$\tau^{(m+1)} = \tau^{(m)} + t_m v_{T^*(\tau^{(m)})} \tag{11}$$

where

$$t_m = \lambda_m \left(\frac{\overline{w} - w(\tau^{(m)})}{\| v_{T^*(\tau^{(m)})} \|^2} \right) \tag{12}$$

If, for some $\varepsilon, 0 < \varepsilon < \lambda_m \leq 2$ *for all m, then sequence* $\{\tau^{(m)}\}$ *either contains a some point* $\tau^{(h)} \in P(\overline{w})$, *or converges to a point on the boundary of* $P(\overline{w})$.

Proof. We use the results in [9] in order to proof Lemma 3. Let $P(\overline{w})$ denote the set of feasible solutions to the system of inequalities $\overline{w} \leq C(T^*(\tau^{(m)})) + \tau v_{T^*(\tau^{(m)})} \forall m$, i.e.

$$P(\overline{w}) = \{\tau : C(T^*(\tau^{(m)})) + \tau v_{T^*(\tau^{(m)})} \geq \overline{w} \; \forall m\}.$$

Let $\{y^m\}$ be a sequence of points which are out of $P(\overline{w})$. Sequence $\{y^m\}$ is called Fejer-monotone relative to $P(\overline{w})$ if, for every $x \in P(\overline{w})$: $y^m \neq y^{m+1}$ and $\| y^m - x \| \geq \| y^{m+1} - x \|$ $(m = 1, ..., l)$. In [9], T. Motzkin et al. showed that any sequence which is Fejer-monotone relative to $P(\overline{w})$ converges. In the present case, if the sequence $\{\tau^{(m)}\}$ does not contain any point of $P(\overline{w})$, then by Lemma 2, the sequence $\{\tau^{(m)}\}$ is Fejer-monotone relative to $P(\overline{w})$ and thus it converges according to [9]. Therefore, $\| \tau^{(m+1)} - \tau^{(m)} \| \rightarrow 0$. From (11), it follows that

$$\| \tau^{(m+1)} - \tau^{(m)} \| = \lambda_m \left(\frac{\overline{w} - w(\tau^{(m)})}{\| v_{T^*(\tau^{(m)})} \|} \right) \rightarrow 0.$$

Hence, $\overline{w} - w(\tau^{(m)}) \rightarrow 0$ and from the continuity of function $w(.)$, we have $w(\lim_{m \to \infty} \tau^{(m)}) = \overline{w}$. Therefore, $\tau^{(m)}$ converges to a point on the boundary of $P(\overline{w})$.

Algorithm 4: The Proposed Algorithm

Input: K_n, c_{ij} and v_1 is the complete graph, matrix of cost, root, respectively.
Output: The MLP solution T.

for $k = 2$ **to** n **do**
  $(T_k, v_{end}) = k$-troll(K_n, v_1, c_{ij}) according to [5]; //$v_{end} \in T_k$;
  (T_{tour}, UB_k) =SGM(T_k, v_1, v_{end}, V');
  $L.add(T_{tour})$;
end for
T = Concatenate $T_{tour} \in L$ by using Geomans et al.'s algorithm [7];
return (T);

Obviously, if we known the value $\overline{w} = \max\limits_{\tau} w(\tau)$ in Lemma 3 then we ensure convergence of $w(\tau^{(m)})$ to $\overline{w}$. However, the exact value of $\overline{w}$ is unknown. Therefore, we use either an underestimate or overestimate as an alternative value. Here we use an overestimate of $\overline{w}$. Since this overestimate is used, the Polyak's conditions cannot be satisfied. In order to satisfy the Polyak's conditions, we set $\lambda_m = 2$ and successively halves both the value of λ_m after each iteration. Our aim is to tend λ_m towards zero.

Now, we use the above theory to construct our algorithm. The pseudo-code of our algorithm is described in Algorithm 1. At first, K. Chauhuri et al.'s algorithm is used to obtain the set of $k-$trees $(k = 2, 3, ..., n)$. Each $k-$tree consists of a root v_1 and end-vertex v_{end}. If all $k-$trees are tours, our algorithm stops (since, if a minimum $k-$tree is a tour, then it must be the minimum tour from v_1 to v_{end}). When we find any $k-$tree (T_k) which is not a tour, an effort is made by the SGM Procedure. In this Procedure, we calculate a lower bound (LB) according to (7) and an upper bound (UB) is the cost of any tour from v_1 to v_{end}. We then use the algorithm in [13] for the set of vertices of T_k to achieve a new minimum spanning tree (T_k') and convert it into an Eulerian tour (T_{tour}') by the algorithm in [13]. The cost of T_{tour}' (NUB) is compared to one of the current upper bound and the upper bound is updated only if the cost of T_{tour}' is lower than one of the current upper bound. Similarly, the new value of lower bound (NLB) is calculated and if its value is higher than one of the current lower bound, then the lower bound is updated. At the end of each iteration, we update τ, t according to (11) and (12), respectively. The SGM procedure stops if ξ is less than the predefined threshold value. Finally, the solutions of SGM Procedure are concatenated according to Geoman et al.'s algorithm in [7] to obtain a MLP tour.

3 Computation Evaluation

We have implemented the algorithm in C language to evaluate its performance. The experiments were conducted on a personal computer, which is equipped with an Intel Pentium core 2 duo 2.13Ghz CPU and 4GB bytes memory.

Algorithm 5: The SGM

Input: T_k, v_1, v_{end}, V' are the $k-$troll's solution of K. Chaudhuri et al.'s algorithm, a root and end-vertex, the set of the vertices in T_k, respectively.
Output: (T_{tour}, UB).

$\tau^{(0)} = (0,0,...,0)$;
Initiate λ^0;
$UB = double.max$;
$LB = 0$;
$T_{tour} = \emptyset$;
$m = 0$;
if T_k is a tour **then**
  return $(T_k, L(T_k))$;
else
  while ($\xi <$ threshold value) **do**
    $T_k' =$ Minimum-spanning-tree-algorithm (V') according to [13];
    $NLB = w(\tau^{(m)})$ according to (7);
    $(T_{tour}', UB) =$ Eulerian-tour-algorithm (T_k') according to [13];
    if $NUB < UB$ **then**
      $UB = NUB$;
      $T_{tour} = T_{tour}'$;
    end if
    if $NLB > LB$ **then**
      $LB = NLB$;
    end if
    Update $\tau^{(m+1)}$ according to (11);
    Update $t^{(m+1)}$ according to (12);
    $\lambda = \lambda/2$;
    $m++$;
    $\xi = \frac{UB-LB}{LB} 100\%$
  end while
  return (T_{tour}, UB);
end if

3.1 Datasets

The experimental data includes three random datasets and two real datasets. Each dataset consists of several instances. For all instances, the triangle inequality is satisfied. Each instance contains the coordinate of n vertices. Based on the value of n, we divide the instances into two types: The small instances ($n \leq 40$) and the large instances ($n > 40$).

The first two random datasets comprise non-Euclidean and Euclidean instances, and were named as random dataset 1 and 2, respectively. In the former one, the instances were generated artificially with an arc cost drawn from a uniform distribution. The values of the arc costs were integers between 1 and 100. In latter one, the Euclidean distance between two vertices was calculated. The coordinates of the vertices were randomly generated according to a uniform distribution in a 200×200 square. We chose the number of vertices as 40 and generated ten different instances for each dataset. The last random dataset supported by A. Salehipour et al. [11] was named as random dataset 3.

A real dataset 1 consists of the well-known real instances from TSPLIB [16]. Besides, we added more real instances by randomly choosing partial data from the larger instances in TSPLIB. The number of vertices of each partial instance is forty. We divided the partial instances into three groups based on the following method: Suppose that X_{max}, X_{min} is the max, min abscissa of an instance, and Y_{max}, Y_{min} is the max, min ordinate of an instance, respectively. We denote $\triangle x = \frac{X_{max}-X_{min}}{n}$ and $\triangle y = \frac{Y_{max}-Y_{min}}{n}$. We have analyzed the data of TSPLIB and found that instances mostly belong to one of the three following groups. Group one with $\triangle x, \triangle y \leq 3$ where vertices are concentrated;group two, $\triangle x, \triangle y \geq 9$ where vertices are scattered; or group three where vertices are spaced in a special way such as along a line or evenly distributed. Specifically, group one includes instances extracted from Eil51, St70, and Eil76 (where 51, 70, 76 are the number of vertices). In group two, the instances are chosen from KroA100, KroB100, KroC100 and Berlin52. In the last group, the instances are from Tsp225, Tss225, Pr76, and Lin105. We gathered the partial instances into a group and named them as real dataset 2.

3.2 Results and Discussion

Two experiments are conducted on the datasets. The datasets in the first experiment include the small instances of random dataset from 1 to 3 and real dataset 2. Meanwhile, the large instances of real dataset 1 from [16] and random dataset 3 are used for the second. For the small instances, their optimal solutions from the exact algorithms in [3] let us evaluate exactly the efficiency of our algorithm. For the large instances, since their optimal solutions have been unknown, the efficiency of our algorithm is only evaluated relatively. In two experiments, the threshold value in our algorithm is 3%.

We denote KA, AA, SGA by K. Chaudhuri et al's [5], A. Archer et al.'s [1], and our algorithm, respectively. For each instance, let us OPT be the optimal solution; $L_i(i = 1, ..., 3)$ be the solutions' latency of KA, SGA, and AA, respectively; $p_i(= \frac{L_i}{OPT}), T_i(i = 1, 2)$ be the approximation ratios and running times by minute of KA, SGA, respectively; $gap_1[\%](= \sum_{k=2}^{n} \frac{UB_k^1 - UB_k^2}{UB_k^2})$ be the difference between $\sum_{k=2}^{n} UB_k^1$ and $\sum_{k=2}^{n} UB_k^2$ (UB_k^1, UB_k^2 are the cost of Eulerian tour and our tour for a certain value of k, respectively). For small instances, since the optimal solutions are known, $gap_2[\%]$ $(= \frac{p_2-p_1}{p_1})$ is the difference between p_1 and p_2. Otherwise, for large instances, we have unknown the optimal solutions. Therefore, $gap_2[\%]$ $(= \frac{L_2-L_1}{L_1})$ is the difference between L_1 and L_2. In the small datasets, random dataset 3 includes twenty instances while the others include ten instances. Otherwise, in the

Table 1 The average $\overline{gap}_1, \overline{gap}_2, \overline{p}_2, \overline{p}_1$ of the algorithms for the small instances

dataset	SGA			KA
	$\overline{gap}_1[\%]$	$\overline{gap}_2[\%]$	$\overline{p}_2$	$\overline{p}_1$
Real dataset 2	15.2	6.31	1.18	1.26
Random dataset 1	17.6	7.57	1.13	1.22
Random dataset 2	14.9	5.70	1.16	1.24
Random dataset 3	16.26	7.44	1.16	1.30
Aver.	16.0	6.8	1.15	1.26

Table 2 The average running time of the algorithms for the small instances

dataset	KA	SGA
	$\overline{T}_1$	$\overline{T}_2$
Real dataset 2	0.13	0.22
Random dataset 1	0.14	0.23
Random dataset 2	0.13	0.23
Random dataset 3	0.16	0.26
Aver.	0.14	0.24

large datasets, real dataset 1 and random dataset 3 include twenty five and twenty instances, respectively. In Table 1, .., 4, we denote $\overline{gap}_i, \overline{p}_i, \overline{T}_i$ by the average values of gap_i, p_i, and T_i $(i = 1, 2)$ for each dataset. The experimental detail of each instance can be seen in [17].

3.2.1 Experiment for Small Dataset

Each instance was tested ten times and the values in Table 1 and 2 are the average values calculated from each small dataset.

The experimental results in Table 1 indicate that the tours' costs of our algorithm are less than those of the Eulerian tours since $\overline{gap}_1$ is 16.0%.

In Table 1, $\overline{gap}_2$, $\overline{p}_2$ are 6.8% and 1.15, respectively. These mean our algorithm gives nearly optimal solutions. Moreover, its average approximation ratio is less than that of KA. However, only in a small number of cases, do our approximation ratios remain unchanged although our tours are used. Therefore, there is no guarantee that every concatenation of our tours will be always better than one of the Eulerian tours. The experimental results also indicate that the real approximation ratio of our algorithm is much better than the theoretical one.

The experimental results in Table 6 indicate that the running time of our algorithm is worse than the one of KA because our algorithm needs the extra time for running the SGM procedure.

3.2.2 Experiment for Large Dataset

Each instance was tested ten times, and the values in Table 3 and 4 are the average values from each large dataset.

Table 3 The average $\overline{gap}_1$ and $\overline{gap}_2$ of the algorithms for the large instances

dataset	SGA	
	$\overline{gap}_1[\%]$	$\overline{gap}_2[\%]$
Real dataset 1	15.46	11.82
Random dataset 3	17.07	6.30
Aver.	16.27	9.06

Table 4 The average running time of the algorithms for the large instances

dataset	KA	SGA
	$\overline{T_1}$	$\overline{T_2}$
Real dataset 1	3.00	6.69
Random dataset 3	6.24	13.53
Aver.	4.6	10.1

In Table 3, $\overline{gap}_1$ is 16.27%. It indicates that the tours' costs of our algorithm are still are less than those of the Eulerian tours.

In Table 3, $\overline{gap}_2$ is 9.06%. Therefore, our algorithm produces the average approximation ratio which is less than those of KA and AA. In most of the instances, we obtain the better approximation ratios and only in small number of cases, does our algorithm not improve approximation ratio. The experimental results also indicate that the real approximation ratio of our algorithm is much better than the one of Chaudhuri's algorithm.

The experimental results in Table 4 show that our algorithm still consumes much time than KA for all instances.

4 Conclusions

In this paper, we propose the algorithm based on the subgradient method for the MLP problem. It uses the set of K. Chaudhuri et al.'s solutions as an input, then modifies each solution into a tour of cost less than one of the Eulerian tour by using the proposed subgradient procedure. Finally, the obtained tours are concatenated to built MLP tour. Intuitively, since we use low cost tours to build a MLP tour, we can expect an improvement on approximation ratio of the proposed algorithm. The experimental results on benchmark datasets show that our algorithm have better approximation ratio than those of the other approximation algorithms. However, the approximation ratio and running time of our algorithm must be further improved. This is our research in the future.

Acknowledgements. This work is partly supported by a Ministry of Education and Training research grant for fundamental sciences, grant number: B2012 - 01 - 28, Department of Computer Science School of Information and Communication Technology provided research facilities for the study.

References

1. Archer, A., Levin, A., Williamson, D.: A Faster, Better Approximation Algorithm for the Minimum Latency Problem. J. SIAM 37(1), 1472–1498 (2007)
2. Arora, S., Karakostas, G.: Approximation schemes for minimum latency problems. In: Proc. ACM STOC, pp. 688–693 (1999)
3. Ban, H.B., Nguyen, K., Ngo, M.C., Nguyen, D.N.: An efficient exact algorithm for Minimum Latency Problem. J. Progress of Informatics (10), 1–8 (2013)
4. Blum, A., Chalasani, P., Coppersmith, D., Pulleyblank, W., Raghavan, P., Sudan, M.: The minimum latency problem. In: Proc. ACM STOC, pp. 163–171 (1994)
5. Chaudhuri, K., Goldfrey, B., Rao, S., Talwar, K.: Path, Tree and minimum latency tour. In: Proc. IEEE FOCS, pp. 36–45 (2003)
6. Garg, N.: Saving an Epsilon: A 2-approximation for the $k-$MST Problem in Graphs. In: Proc. STOC, pp. 396–402 (2005)
7. Goemans, M., Kleinberg, J.: An improved approximation ratio for the minimum latency problem. In: Proc. ACM-SIAM SODA, pp. 152–158 (1996)
8. Held, M., Karp, R.M.: The travelling salesman problem and minimum spanning tree: part II. J. Mathematical Programming 1, 5–25 (1971)
9. Motzkin, T., Schoenberg, I.J.: The relaxation method for linear inequalities. J. Mathematics, 393–404 (1954)
10. Sahni, S., Gonzalez, T.: P-complete approximation problem. J. ACM 23(3), 555–565 (1976)
11. Salehipour, A., Sorensen, K., Goos, P., Braysy, O.: Efficient GRASP+VND and GRASP+VNS metaheuristics for the traveling repairman problem. J. Operations Research 9(2), 189–209 (2011)
12. Silva, M., Subramanian, A., Vidal, T., Ochi, L.: A simple and effective metaheuristic for the Minimum Latency Problem. J. Operations Research 221(3), 513–520 (2012)
13. Rosenkrantz, D.J., Stearns, R.E., Lewis, P.M.: An analysis of several heuristics for the traveling salesman problem. SIAM J. Comput. (6), 563–581 (1977)
14. Polyak, B.T.: Minimization of unsmooth functionals. U.S.S.R. Computational Mathematics and Mathematical Physis 9(3), 14–29 (1969)
15. Wu, B.Y., Huang, Z.-N., Zhan, F.-J.: Exact algorithms for the minimum latency problem. Inform. Proc. Letters 92(6), 303–309 (2004)
16. `http://www.iwr.uniheidelberg.de/groups/comopt/software/TSPLIB96`
17. `https://sites.google.com/a/soict.hut.edu.vn/the-proposed-algorithm-the-gradient-method/`

Particulate Matter Concentration Estimation from Satellite Aerosol and Meteorological Parameters: Data-Driven Approaches

Thi Nhat Thanh Nguyen, Viet Cuong Ta, Thanh Ha Le, and Simone Mantovani

Abstract. Estimation of Particulate Matter concentration (PM_1, $PM_{2.5}$ and PM_{10}) from aerosol product derived from satellite images and meteorological parameters brings a great advantage in air pollution monitoring since observation range is no longer limited around ground stations and estimation accuracy will be increased significantly. In this article, we investigate the application of Multiple Linear Regression (MLR) and Support Vector Regression (SVR) to make empirical data models for $PM_{1/2.5/10}$ estimation from satellite- and ground-based data. Experiments, which are carried out on data recorded in two year over Hanoi - Vietnam, not only indicate a case study of regional modeling but also present comparison of performance between a widely used technique (MLR) and an advanced method (SVR).

1 Introduction

Aerosol Optical Thickness/Aerosol Optical Depth (AOT/AOD) is considered as one of the Essential Climate Variables (ECV) [1] that influences climate, visibility and quality of the air. AOT is representative for the amount of particulates present in a vertical column of the Earth's atmosphere. Aerosol concentration can be measured directly by ground-based sensors or estimated from data recorded by sensors onboard polar and geostationary satellites observing the Earth. Ground measurements have usually high accuracy and temporal frequency (hourly) but they are representative of a limited spatial range around ground sites [2]. Conversely, satellite

Thi Nhat Thanh Nguyen · Viet Cuong Ta · Thanh Ha Le
University of Engineering and Technology, VNU, 144 XuanThuy, Hanoi, Vietnam
e-mail: {thanhntn,cuongtv,ltha}@vnu.edu.vn

Simone Mantovani
MEEO S.r.l., via Saragat 9, 44122, Ferrara, Italy, and SISTEMA GmbH,
Waehringerstrasse 61, Vienna, Austria
e-mail: mantovani@meeo.it

V.-N. Huynh et al. (eds.), *Knowledge and Systems Engineering, Volume 1*,
Advances in Intelligent Systems and Computing 244,
DOI: 10.1007/978-3-319-02741-8_30,

observation provides information at global scale with moderate quality and lower measurement frequency (daily).

MODerate resolution Imaging Spectrometer (MODIS) is a multispectral sensor on-board the two polar orbiting satellites Terra and Aqua, launched in 1999 and 2002, respectively and operated by the National Aeronautic and Space Administration (NASA). Using MODIS-measured spectral radiances, physical algorithms based on Look-Up Table (LUT) approaches have used since 90s to generate the aerosol products for Land and Ocean areas in Collection 004 [2] and following improved releases (Collection 005 [4], Collection 051 and the newest Collection 006 issued in 2006, 2008 and 2012, respectively). The aerosol product provided by NASA (MOD04L2) is trusted and largely used in many studies. However, its spatial resolution (10 km x 10 km) is appropriate for applications at the global scale but adequate for monitoring at regional scale. Therefore, an augmented AOT product, with spatial resolution of 1 km x 1 km, is obtained by PM MAPPER software package [3], of which quality has been validated over Europe using three year data [4].

The usage of satellite technology for air pollution monitoring applications has been recently increasing especially to provide global distribution of aerosol and its properties for deriving Particulate Matter concentration (PM), one of the major pollutants that affect air quality. PM is a complex mixture of solid and liquid particles that vary in size and composition and remain suspended in the air. PM is classified into PM_1, $PM_{2.5}$ or PM_{10} by their aerodynamic diameters. $PM_{1/2.5/10}$ obtained by ground-based instruments has high quality and frequency but limited range, which makes the use of $PM_{1/2.5/10}$ for monitoring air pollution at the global or regional scale become challenging. Motivated from early studies of relations between $PM_{1/2.5/10}$ and AOT and the fact that satellite AOT nowadays has acceptable quality in comparison with ground-based AOT, thanks to technology achievements, deriving PM from satellite AOT is recently a promising approach.

In literature, relation between AOT and PM was considered over different areas (Italy, French, Netherland, United State, Peninsular Malaysia, Hong Kong, Sydney, Switzerland, Delhi, New York) in different experimental conditions [8, 10, 9, 10, 11, 12, 13, 14, 15, 16, 17, 18, 19]. The general methodology is applied using three main steps: (i) collecting satellite/ground-based AOT and ground-based $PM_{2.5/10}$; (ii) matching data following time and spatial constrains; (iii) investigating their relationship in different conditions. Experimental results showed that the relations are site-dependant, therefore they are not easy to be extrapolated to other locations. Besides, there are many factors effecting to PM estimation such as data collections, atmospheric conditions during studies, aerosol types and size fractionation, PM sampling techniques in which meteorological parameters are especially important. Ambient relative humidity, fractional cloud cover, mixing layer height, wind conditions, height of planetary boundary layer, temperature, pressure and wind velocity [12, 13, 14, 15, 16, 17, 18, 19] were considered seriously and used together with AOT in order to improve PM estimation accuracy.

Regarding estimation methodologies, Linear Regression (LR) or Multiple Linear Regression (MLR) are widely used to establish the AOD and $PM_{2.5/10}$ relationship, and therefore is regarded as a common and valid methodology to predict particulate matters of different sizes (10 μm and 2.5 μm in diameter) [8, 10, 9, 10][15]. These techniques also applied to create mixed effect models in which meteorological parameters and land use information were used as input to PM prediction [18]. In a further improvement of the previous work, a robust calibration approach by integrating a simple adjustment technique into a mixed effect models is applied to estimate PM concentration using satellite AOD monthly average datasets [19].

Taking advantages of machine learning techniques, researchers have recently applied them to improve PM prediction. Exploited Self Organizing Map (SOM), Yahi et al. clustered integrated data by meteorological situations and then found high relationship between AOT and PM_{10}. Their experiments were applied to Lille region (France) for summer of the years 2003-2007 [20]. On a study using three years of MODIS AOT and meteorological data over southeast United State to estimate hourly or daily $PM_{2.5}$, Artificial Neural Network (ANN) was able to improve regression coefficient R from 0.68 to 0.74 or 0.78, respectively in comparison with MLR [16]. Following the same approach, SVR is applied to estimate PM_{10} from satellite data, meteorological parameters and ancillary data over Austria in site domain (i.e. pixels around location of Austrian Air Quality ground-stations) and in model domain (i.e. the geographical domain of the Austrian Air Quality model). SVR implemented on a monthly basis in the period from 2008 to 2010 shows promising result in site domain (coefficient regression is bigger than 0.75) while further investigations should be done in model domain [21]. Originated from other sciences (Environment, Physics, Chemistry), the problem of PM estimation, which is started around 2002, has not considered the use of machine learning techniques seriously. The number of studies following the approach is limited although most of work done have shown promising results [16][20, 21].

In this article, we investigate the application of multiple linear regression and support vector regression to make empirical data models for $PM_{1/2.5/10}$ estimation from satellite aerosol product and ground-based meteorological parameters. Experiments over Hanoi, Vietnam, not only indicate a case study of regional modeling for PM_1, $PM_{2.5}$, PM_{10} but also present performance of MLR, a widely used technique, and SVR, an advanced but investigated method.

The article is organized as follows. The methodologies including data collection, data integration and modeling techniques will be presented in Section 2. Experiments and results on data modeling of MLR and SVR will be described and discussed in Section 3. Finally, conclusions are given in Section 4, together with future works.

2 Methodologies

2.1 Data Collection

2.1.1 Satellite-Based Aerosol

The aerosol product provided by NASA, namely MOD04 L2, is derived from MODIS images using aerosol retrieval algorithms over land and ocean areas. These methods match satellite observations to simulated values in LUT to derive aerosol concentration and its properties. Both algorithms perform on cloud-free pixels whose covered regions (land/water) are determined by their geographical information [2].

In an effort to improve the spatial resolution of the MODIS aerosol products, a software package called PM MAPPER was developed to increase spatial resolution from 10x10 km^2 to 3x3 km^2 and then 1x1 km^2. The PM MAPPER aerosol product at 3 km resolution was validated by direct comparison with MODIS retrievals and showed higher ability to retrieve pixels over land and coastlines [3]. The validation of the 1 km was carried out over Europe for the years 2007-2009. Comparison with both ground sun-photometers of the AERONET network and the MODIS 10 km AOT products have shown a high correlation [4]. In the article, PM MAPPER aerosol product (1 km spatial resolution) over Hanoi in two years is used in modeling and validation processes of $PM_{1/2.5/10}$ estimation.

2.1.2 Ground-Based Aerosol

AERONET is the global system of ground based Remote Sensing aerosol network established by NASA and PHOTONS (University of Lille 1, CNES, and CNRS-INSU) (NASA, 2011). At AERONET stations, Aerosol Optical Thickness is measured by CIMEL Electronique 318A spectral radiometers, sun and sky scanning sun photometers in various wavelengths: 0.340, 0.380, 0.440, 0.500, 0.675, 0.870, 1.020, and 1.640 μm in intervals of 15 minutes in average. After data processing steps, cloud-screened and quality-assured data are stored and provided as Level 2.0. In our work, AERONET aerosol product at level 2.0 are collocated in space and synchronized in time with satellite-based aerosols in order to validate the PM MAPPER AOT products over Hanoi, Vietnam.

2.1.3 Particulate Matter Concentration and Meteorological Data

Particulate matter suspended in the air is the main factor affecting to air quality and leading premature deaths. These fine particles have either anthropogenic sources (plants, burning of fossil fuels, spray can ...) or natural sources (dust storms, volcanoes, fires ...). Particles can be classified by size, referred to as fractions: PM_{10}, $PM_{2.5}$, PM_1 represent particles with aerodynamic diameters smaller than 10 μm, 2.5 μm and 1 μm, respectively. Traditionally, particulate matter is measured directly at ground stations on hourly or daily basis. The ground measurements are highly trustworthy but not representative for areas far from ground stations.

In order to obtain PM measurements for modeling in Hanoi, we have used ground data provide by a station managed by Center for Environmental Monitoring (CEM) in Vietnam Environment Administration at geospatial coordinates (21°02'56.3", 105°22'58.8"). Data include particulate matter concentration at different sizes (1, 2.5 and 10 μm) averaged by 24 hours and meteorological data (wind speed, wind direction, temperature, relative humidity, pressure and sun radiation) averaged by an hour in a period from August 2010 to July 2012.

2.2 Data Integration

Since data sets are collected from different sources, they have different temporal and spatial resolutions which can only be solved by an integration process. Satellite data include aerosol products at 1 km provided by the PM MAPPER. Ground-based measurements are obtained from the AERONET and CEM.

Satellite aerosol maps at 1 km of spatial resolution are obtained daily while $PM_{1/2.5/10}$, meteorological data and AERONET aerosol are measured at the ground stations and averaged in twenty-four hours, an hour and fifteen minutes in average, respectively. Time and location constrains are applied for data integration following the methodology proposed in [22]. Satellite data are considered if their pixels are cloudy-free and have distances to ground station within radius R. Meanwhile, contemporaneous measurements of AERONET and CEM instruments are selected and averaged within a temporal window T minutes around the satellite overpasses as illustrated in Fig. 1. The optimal thresholds of R and T will be selected by experiments presented in next sections.

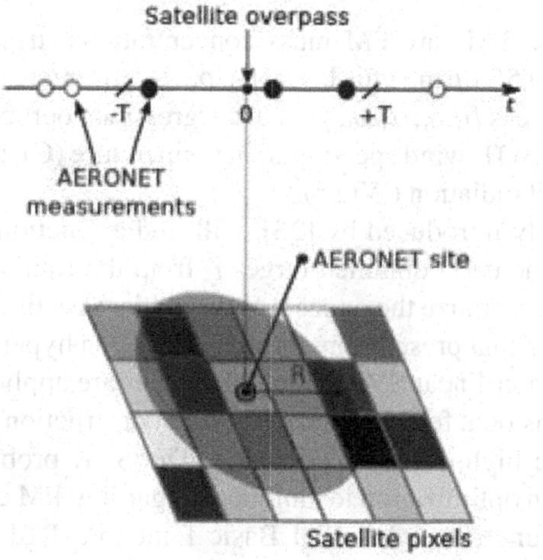

Fig. 1 Spatial-temporal window for extracting satellite and ground-based measurements [4]

2.3 Modeling Techniques

The modeling process is stated as follows. Given a training dataset including l samples:

$$\{(x_1,y_1),(x_2,y_2),\ldots,(x_l,y_l)\} \in X \times Y \tag{1}$$

where X,Y denotes the space of the input and output patterns (i.e. $X \subset R^n, Y \subset R$). The modeling process would investigate an appropriate function f presented relationship between X_i and Y_i with the minimal error ε. The general form for the model would be as follows:

$$Y = f(X) + \varepsilon \tag{2}$$

In particular, the modeling process for PM estimation is to find an appropriate function f by applying a modeling technique on integrated datasets consisting of $PM_{1/2.5/10}$, AOT, Wind Speed (Wsp), Temperature (Temp), Relative Humidity (Rel_-H), Pressure (Bar) and sun Radiation (Rad). Multiple linear regression and support vector regression are considered in our work.

Multiple linear regression technique will assume f have a linear form and the problem become to estimate weighting parameters instead of the complicated infinite dimensional f. Based on MLR techniques, particulate matter concentration in different diameters is able to be calculated using equations as follows:

$$PM_{10} = \beta_0 + \beta_1 AOT + \beta_2 Wsp + \beta_3 Temp + \beta_4 Rel\_H + \beta_5 Bar + \beta_6 Rad \tag{3}$$

$$PM_{2.5} = \alpha_0 + \alpha_1 AOT + \alpha_2 Wsp + \alpha_3 Temp + \alpha_4 Rel\_H + \alpha_5 Bar + \alpha_6 Rad \tag{4}$$

$$PM_1 = \gamma_0 + \gamma_1 AOT + \gamma_2 Wsp + \gamma_3 Temp + \gamma_4 Rel\_H + \gamma_5 Bar + \gamma_6 Rad \tag{5}$$

where PM_{10}, $PM_{2.5}$, PM_1 are PM mass concentrations (μgm^{-3}) and AOT is PM MAPPER AOT at 0.553 μm (unit less). $\beta_0, \alpha_0, \gamma_0$ are intercepts for PM_{10}, $PM_{2.5}$, PM_1 equations whereas $\beta_{1-6}, \alpha_{1-6}, \gamma_{1-6}$ are regression coefficients for the predictor variables including AOT, wind speed (ms^{-1}), temperature (C), relative humidity (%), barometer (hPa) and radiation (Wm^{-2}).

The ε-SVR, firstly introduced by [23], will find a function f that has at most ε deviation from the actually obtained target Y from the training data in order to as 'flat' as possible to minimize the expected risk. In the case of linear SVR, flatness is to find the function f that presents an optimal regression hyper-plane with minimum slope w. In case of non-linear SVR, kernel functions are applied to map input space X into a high dimensional feature space F before construction of optimal separating hyper-planes in the high-dimensional space. The SVR problem is solved by the classical Lagrangian optimization techniques. Regarding PM estimation, the ε-SVR with epsilon loss function and Radial Basic Function (RBF) kernel provided by LIBSVM [24] is applied.

3 Experiments and Results

3.1 Satellite-Based Aerosol Validation

PM MAPPER provides AOT product at 1 km of spatial resolution using the improved MODIS aerosol algorithms. A validation over Europe was done on three years data and presented a high correlation in comparison to both ground sunphotometers of the AERONET network and the MOD04L2 AOT products [4]. Although PM MAPPER provides AOT maps at the global scale, this experiment is still carried out in order to validate its aerosol product at 1km over Hanoi before the modeling step.

In fact, as the CEM station hasn't provided any aerosol measurement, we used AOT collected from NghiaDo, the unique AERONET station in Hanoi. The NghiaDo station is far from the CEM station about 10 km in west, which is close enough to make an assumption that obtained aerosol is able to representative for both locations.

The PM MAPPER AOT maps and AERONET AOT measurements are collected in a year, from December 2010 to November 2011. A satellite map obtained daily is presented by three layer matrixes including latitude, longitude and AOT at 0.553 μm. Each pixel is corresponding to 1km^2 area on the ground. Meanwhile, AERONET measurements provide AOT at various wavelengths: 0.340, 0.380, 0.440, 0.500, 0.675, 0.870, 1.020, and 1.640 μm, in intervals of 15 minutes in average. All data are integrated following the method mentioned in Section 2.2 with various spatial window R and temporal window T. Different thresholds of R (10, 15, 20, 25, 30, 35, 40 and 50 km) and T (30, 60, 120 and 1440 minutes ($\sim$24 hours)) are considered in order to investigate satellite/ground-based aerosols behaviours. The validation results are presented in Fig. 2. CORrelation coefficient (COR) increases by spatial and temporal windows (Fig. 2(a)) while Root Mean Square Error (RMSE) decreases by distance but has the same behaviours if temporal windows at 60, 120 minutes or all day are applied (Fig.2(b)). The best match should be established between satellite AOT collected in areas of 25 km around the AERONET station and ground-based AOT averaged in a day when the satellite image is recorded. In this case, we obtained PM MAPPER AOT and AERONET AOT's COR = 0.648 and RMSE = 0.421, which is good enough for using PM MAPPER AOT in $PM_{1/2.5/10}$ estimation.

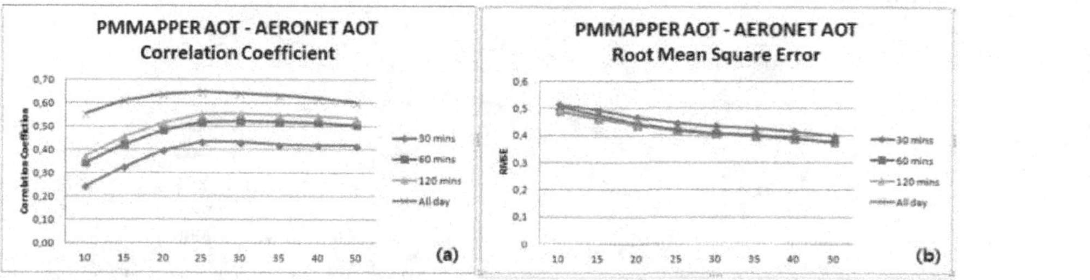

Fig. 2 (a) Correlation Coefficient and (b) Root Mean Square Error between PM MAPPER AOT and AERONET AOT

3.2 Threshold Selection

In this section, we present an experiment to identify spatial and temporal thresholds for integration data in order to obtain samples for the $PM_{1/2.5/10}$ modeling step. Data collected from August 2010 to July 2012 consist of daily AOT maps at $1km^2$, daily particulate matter concentration (PM_1, $PM_{2.5}$, PM_{10}) and hourly meteorological parameters (wind speed, temperature, relative humidity, pressure and sun radiation). The experiment is carried out to integrate data with various spatial windows for AOT maps (R=5, 10, 15, 20, 25, 30, 35, 40, 45 and 50km) and different temporal windows for meteorological data (the nearest time - T1, average of two nearest times - T2 and average of four nearest times - T3). Correlation coefficient matrixes for all parameter are calculated to investigate relations among them. However, no much difference can be found with various temporal windows, and therefore we selected the temporal threshold T1 and only present its results.

The Fig. 3 presents correlation coefficient between AOT collected at many distance thresholds and other variables on experimental datasets. AOT-PM_{10}, AOT-$PM_{2.5}$ and AOT-PM_1 correlation are optimal at 35, 30, 30 km, respectively. However, the threshold of 30 km should be selected because it almost maximizes all considered correlations (COR of AOT-PM_{10} = 0.301, COR of AOT-$PM_{2.5}$ = 0.255 and COR of AOT-PM_1 = 0.192). Regarding meteorological parameters, AOT relationship is strong with relative humidity (Rel_H.), weak with wind speed, sun radiation (Wsp., Rad.) and almost not with pressure and temperature (Bar., Temp.) (see Fig. 3).

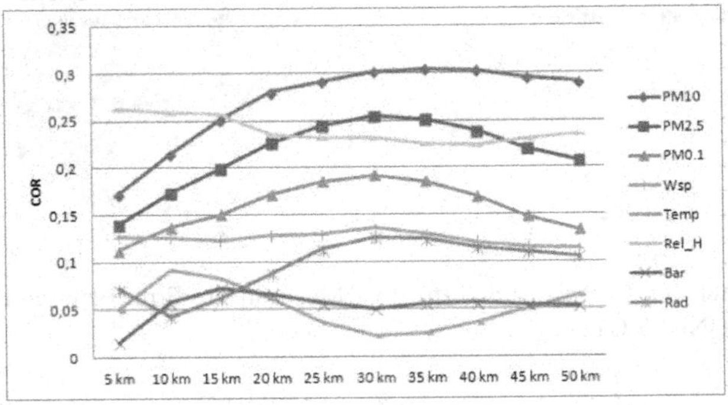

Fig. 3 Correlation Coefficients in distance between satellite AOT and other factors

From data sets selected by proposed spatial and temporal thresholds (R= 30km, T=T1), we carried out a further experiment to investigate which factors will be important to PM_{10}, $PM_{2.5}$, PM_1 estimation in the modeling step. CORs are calculated between $PM_{1/2.5/10}$ and other variables. Results presented in Fig. 4 show that dependence of PM on AOT increases in the order of their aerodynamic diameters (i.e. 1, 2.5 and then 10 μm) whereas their relations of PM with meteorological variables (Wsp., Temp., Rel_H., Bar. and Rad.) decreases with their size (i.e. 10, 2.5 and then 1 μm). The obtained results confirmed that satellite AOT is a key factor. Besides, temperature, radiation and wind speed should affect to data models more strongly than relative humidity and pressure. However, considering together with CORs between AOT and other factors (Fig. 3), we decided to use all meteorological parameters in the next modeling step.

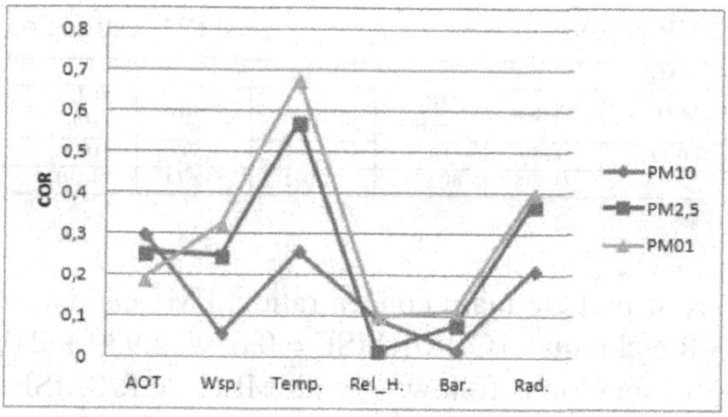

Fig. 4 Correlation coefficients between $PM_{1/2.5/10}$ and other factors in the selected dataset

3.3 *MLP and SVR Comparison*

MLP and SVR techniques presented in Section 2.3 are applied to estimate $PM_{1/2.5/10}$ values. The experiment focuses on investigating estimators of different types of particle mass concentration (PM_1, $PM_{2.5}$ and PM_{10}), role of satellite AOT and performance of two regression methodologies. Data for each type of PM are obtained by using thresholds proposed in the previous section. They are grouped into two years (August 2010 - July 2011 and August 2011 - July 2012) with statistics shown in Fig. 5. In two years, the totals of samples are comparable (55,131 and 49,908) but data distributions over many months are too much different (Aug, Oct, Nov, Jan, Feb, Mar and Apr). Therefore, we considered data in year basis instead of month or season basis as mentioned in previous studies. Moreover, datasets with (w) and without (w/o) satellite AOT are created and considered in our experiment. One year data is used for training whereas another year data is used to testing and vice versa. The final COR and RMSE are averaged on two year results for each PM_1, $PM_{2.5}$ and PM_{10} (see Table 1).

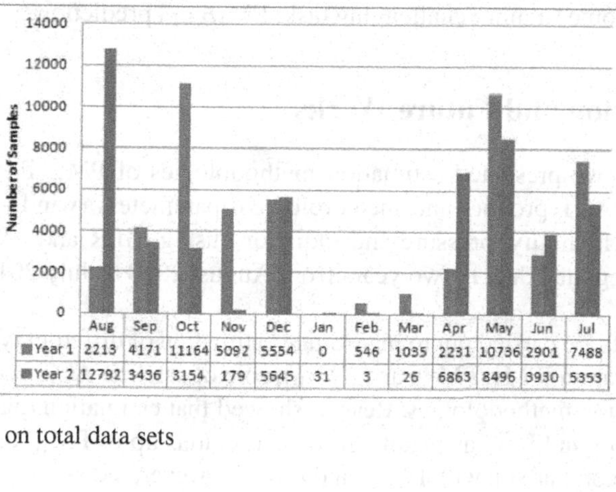

Fig. 5 Statistics on total data sets

Table 1 MLR and SVR performance on PM_{10}, $PM_{2.5}$ and PM_1 estimation

	PM_{10}			$PM_{2.5}$			PM_1		
	MRL-w/o	MRL-w	SVR-w	MRL-w/o	MRL-w	SVR-w	MRL-w/o	MRL-w	SVR-w
COR	0.038	0.174	0.239	0.429	0.598	0.593	0.608	0.659	0.694
RMSE	109.225	96.656	74.935	40.836	31.071	31.674	24.591	22.939	22.349

Regarding types of particle mass concentration, PM_1 can be estimated best by both MLR and SVR techniques (COR/RMSE = 0.659/22.939 and 0.694/22.349, respectively). $PM_{2.5}$ estimation is following with MLR COR/RMSE at 0.598/31.071 and SVR COR/RMSE at 0.593/31.674. The worst case is for PM_{10} estimation (MLR COR/RMSE = 0.174/96.656 and SVR COR/RMSE = 0.239/74.935). Based on experimental results, PM_1 and $PM_{2.5}$ estimation seem good enough while PM_{10} estimation need more data for modeling and further investigation.

The use of satellite AOT in $PM_{1/2.5/10}$ prediction is able to improve regression correlation and accuracy significantly. In case of PM_{10} estimation, regression correlation increases from 0.038 to 0.174 and 0.239 when MLR and SVR are applied. COR of $PM_{2.5}$ estimation increases from 0.429 to 0.598 and 0.593 while the same trend is also seen on PM_1 estimation (from 0.608 to 0.659 and 0.694, respectively). The strong improvement is happened to PM_{10} than $PM_{2.5}$ or PM_1 estimators. It can be explained by the different levels of relation between $PM_{1/2.5/10}$ and AOT as shown in Fig. 4.

In comparison of modeling techniques performance using datasets with satellite AOT, SVR is better than MRL for PM_{10} and PM_1 estimation. The regression correlation increases 36.9% and 5.5% while error decreases 22.4% and 2.6%, respectively. Meanwhile, MRL and SVR perform in nearly same way for $PM_{2.5}$ estimation, which is presented by a slight difference of COR (0.78%) and RMSE (1.9%). In general, SVR outperforms MLR in our experiments although its improvements are not impressive as shown in other studies [16][21]. It could be due to different datasets and data splitting methods. In the experiment, our data are limited on a CEM station and so, divided by years instead of locations. Therefore, the problem of $PM_{1/2.5/10}$ estimation becomes a more challenging task, $PM_{1/2.5/10}$ prediction.

4 Conclusion and Future Works

In this article, we presented estimation methodologies of PM_1, $PM_{2.5}$ and PM_{10} from satellite AOT product and meteorological parameters (wind speed, temperature, relative humidity, pressure and radiation) using MLR and SVR techniques applied on integrated data in two years from August 2010 to July 2012 over Hanoi, Vietnam.

Experiments are carried out to investigate estimation of different types of particle mass concentration (PM_1, $PM_{2.5}$ and PM_{10}), role of satellite AOT and performance of two regression methodologies. Results showed that estimation quality decreases by PM_{10}, $PM_{2.5}$ and PM_1 as results of loose relationship of PM_{10} on meteorology parameters in comparison with $PM_{2.5}$ and PM_1. However, the use of satellite AOT in modeling is able to improve all PM estimators accuracy. For regression techniques, SVR outperforms MLR but more data collection, ground station extension and further investigation should be done. The presented work can be considered as a case study for regional $PM_{1/2.5/10}$ models over Hanoi, Vietnam.

Acknowledgements. The authors would like to thank the research project "Analytical methods of satellite image processing for air pollution estimation in Vietnam" of the Asia Research Center, VNU for financial support, MEEO S.r.l. (Ferrara, Italy) for providing satellite aerosol maps, Center for Environmental Monitoring, Vietnam Environment Administration for supporting PM mass concentration, meteorological data and AERONET PIs for collecting aerosol data in Hanoi.

References

1. Global Climate Observing System Essential Climate Variables, `http://gosic.org/ios/MATRICESECVECV-matrix.htm`
2. Balaguer, N.C.: Combining models and monitoring. A survey to elicit expert opinion on the spatial representativeness of ground based monitoring data. Fairmode activity for WG2-SG1 (2012)
3. Kaufman, Y.J., Tanre, D.: Algorithm for remote sensing of tropospheric aerosol from modis. In: MODIS ATBD (1997)
4. Remer, L.A., Tanré, D., Kaufman, Y.J.: Algorithmfor remote sensing of tropospheric aerosol from MODIS: Collection 5. In: MODIS ATBD (2004)
5. Nguyen, T., Mantovani, S., Bottoni, M.: Estimation of Aerosol and Air Quality Fields with PMMAPPER, An Optical Multispectral Data Processing Package. In: ISPRS TC VII Symposium 100 Years ISPRS-Advancing Remote Sensing Science, vol. XXXVIII(7A), pp. 257–261 (2010)
6. Campalani, P., Nguyen, T.N.T., Mantovani, S., Bottoni, M., Mazzini, G.: Validation of PM MAPPER aerosol optical thickness retrievals at 1x1 km2 of spatial resolution. In: The 19th International Conference on Proceeding of Software, Telecommunications and Computer Networks (SoftCOM), pp. 1–5 (2011)
7. Chu, D.A., Kaufman, Y.J., Zibordi, G., Chern, J.D., Mao, J., Li, C., Holben, B.N.: Global monitoring of air pollution over land from the Earth Observing System-Terra Moderate Resolution Imaging Spectroradiometer (MODIS). Journal of Geophysical Research Atmospheres 108(D21), 4661 (2003)
8. Wang, J., Chirstopher, S.A.: Intercomparison between satellite-derived aerosol optical thickness and PM2.5 mass: Implication for air quality studies. Geophysical Research Letter 30(21), 2095 (2003)
9. Engel-Cox, J.A., Holloman, C.H., Coutant, B.W., Hoff, R.M.: Qualitative and quantitative evaluation of MODIS satellite sensor data for regional and urban scale air quality. Atmospheric Environment 38, 2495–2509 (2004)
10. Kacenelenbogen, M., Leon, J.F., Chiapello, I., Tanre, D.: Characterization of aerosol pollution events in France using ground-based and POLDER-2 satellite data. Atmospheric Chemistry and Physics 6, 4843–4849 (2006)
11. Pelletier, B., Santer, R., Vidot, J.: Retrieving of particulate matter from optical measurements: A semiparametric approach. Journal of Geophysical Research: Atmospheres 112(D6208) (2007)

12. Schaap, M., Apituley, A., Timmermans, R.M.A., Koelemeijer, R.B.A., Leeuw, G.D.: Exploring the relation between aerosol optical depth and PM2.5 at Cabauw, the Netherlands. Atmospheric Chemistry and Physics 9, 909–925 (2009)
13. Gupta, P., Christopher, S.A., Wang, J., Gehrig, R., Lee, Y., Kumar, N.: Satellite remote sensing of particulate matter and air quality assessment over global cities. Atmospheric Environment 40, 5880–5892 (2006)
14. Gupta, P., Christopher, S.A.: Seven year particulate matter air quality assessment from surface and satellite measurements. Atmospheric Chemistry and Physics 8, 3311–3324 (2008)
15. Gupta, P., Christopher, S.A.: Particulate matter air quality assessment using integrated surface, satellite, and meteorological products: Multiple regression approach. Journal of Geophysical Research: Atmospheres 114(D14205) (2009)
16. Gupta, P., Christopher, S.A.: Particulate matter air quality assessment using integrated surface, satellite, and meteorological products: A neural network approach. Journal of Geophysical Research: Atmospheres 114(D20205) (2009)
17. Zha, Y., Gao, J., Jiang, J., Lu, H., Huang, J.: Monitoring of urban air pollution from MODIS aerosol data: effect of meteorological parameters. Tellus B 62(2), 109–116 (2010)
18. Lee, H.J., Liu, Y., Coull, B.A., Schwartz, J., Koutrakis, P.: A novel calibration ap-proach of MODIS AOD data to predict PM2.5 concentrations. Atmospheric Chemistry and Physics 11, 7991–8002 (2011)
19. Yap, X.Q., Hashim, M.: A robust calibration approach for PM10 prediction from MODIS aerosol optical depth. Atmospheric Chemistry and Physics 13, 3517–3526 (2013)
20. Yahi, H., Santer, R., Weill, A., Crepon, M., Thiria, S.: Exploratory study for estimating atmospheric low level particle pollution based on vertical integrated optical measurements. Atmospheric Environment 45, 3891–3902 (2011)
21. Hirtl, M., Mantovani, S., Krger, B.C., Triebnig, G., Flandorfer, C.: AQA-PM: Extension of the Air-Quality model for Austria with satellite based Particulate Matter estimates. In: European Geosciences Union, General Assembly 2013, Austria (2013)
22. Ichoku, C., Chu, D.A., Mattoo, S., Kaufiman, Y.J.: A spatio-temporal approach for global validation and analysis of MODIS aerosol products. Geophysical Research Letter 29(12), 1616 (2002)
23. Vapnik, V.: The nature of statistical learning theory. Springer, Berlin (1995)
24. Chang, C., Lin, C.: LIBSVM: A Library for Support Vector Machines (2011)

A Spatio-Temporal Profiling Model for Person Identification

Nghi Pham and Tru Cao

Abstract. Mobility traces include both spatial and temporal aspects of individuals' movement processes. As a result, these traces are among the most sensitive data that could be exploited to uniquely identify an individual. In this paper, we propose a spatio-temporal mobility model that extends a purely spatial Markov mobility model to effectively tackle the identification problem. The idea is to incorporate temporal perspectives of mobility traces into that probabilistic spatial mobility model to make it more specific for an individual with respect to both space and time. Then we conduct experiments to evaluate the degree to which individuals can be uniquely identified using our spatio-temporal mobility model. The results show that the proposed model outperforms the purely spatial one on the benchmark of MIT Reality Mining project dataset.

1 Introduction

Mobile phones are becoming more and more popular in the modern life. The growing popularity of mobile phones has produced large amounts of user data that become valuable resources for studying. Learning patterns of human behavior from such observed data is an emerging domain with a wide range of applications ([1]).

Among several types of observed data recorded by mobile phones, mobility traces are figured out to be highly unique ([3]). As mobile phone users travel within a GSM network, their locations (or, to be more precise, their mobile phones' locations) can be recorded by the network providers. A GSM network is actually a cellular network that is comprised of several cells covering a certain geographical

Nghi Pham
Ho Chi Minh City University of Technology - VNUHCM, Vietnam

Tru Cao
Ho Chi Minh City University of Technology - VNUHCM, Vietnam, and
John von Neumann Institute - VNUHCM, Vietnam
e-mail: 511070462@stu.hcmut.edu.vn, tru@cse.hcmut.edu.vn

V.-N. Huynh et al. (eds.), *Knowledge and Systems Engineering, Volume 1*,
Advances in Intelligent Systems and Computing 244,
DOI: 10.1007/978-3-319-02741-8_31,

area. Each cell is served by at least one base station, also known as cell station. Every time a mobile phone carrier moves into a cell area, his location is registered with the corresponding cell station. Therefore, locations of a mobile user can be obtained as a chronological sequence of cell stations. These location sequences have the potential to give an insight into human behaviors as well as to identify an anonymous user or device [4].

An approach to that identification problem was introduced in [4] that concerned the construction of location profiles based on individuals' traces as well as the methods to link location traces back to an individual. A modeling method based on the Markov process, referred to as the Markov mobility model or simply the Markov model, was also introduced in order to represent location profiles of mobile phone users. Anonymous mobile phone users were then identified by matching their location profiles with those recorded in the past.

However, with location-based mobility modeling, it is not always possible to distinguish an individual from a group of individuals who reside and travel within somewhat the same areas. This is because these individuals' mobility patterns are likely to be very similar and the mobility model that is only based on geographical locations is not sufficient to uniquely identify an individual. In fact, different individuals may move to the same location but at different times of a day and stay there for different durations.

Therefore, in this paper, we propose a spatio-temporal mobility model that extends the purely spatial Markov mobility model in [4] to deal with the problem of person identification based on mobile phone information. That is, we incorporate temporal perspectives of mobility traces into that mobility model to make it more characteristic of individuals. It is expected that the more individual-characteristic the model is, the more accurate it is for individual identification.

The details of our model are explained in Section 2. Experimental results are presented and discussed in Section 3. Finally, in Section 4, we conclude the paper and discuss the directions for future work.

2 Proposed Model

For the paper to be self-contained, we summarize the spatial Markov model and its identification methods in Section 2.1 and Section 2.2 respectively. These serve as the basis for our proposed spatio-temporal model as presented in Section 2.3.

2.1 Mobility Profiling Using Location-Based Markov Model

As in [4], movement of a mobile phone user in a GSM network is considered a Markov process, and the location-based profile of one mobile user is modeled based on the sequence of cells recorded during his movement. This sequence is treated as an order-1 Markov chain in which each state corresponds to a cell. The location-based Markov mobility profile, also referred to as the location profile, comprises a transition probability matrix and a stationary distribution vector. Both are generated from the Markov chain.

2.1.1 Transition Probability Matrix

User movement from one cell to one of its neighboring cells corresponds to a state transition in the Markov chain. The probability distribution of next states for the Markov chain depends only on the current state, and not on how the Markov chain arrives at the current state. Probability of every possible state transition is computed and these probabilities are represented using a transition probability matrix. For example, transition probabilities of one mobile phone user who resides within m cells are represented by an $m \times m$ matrix P in which value of element $P(i,j)$ represents the probability that the user moves from cell i to an adjacent cell j.

2.1.2 Stationary Distribution Vector

For a user who resides within m cells, his cell residence distribution over these cells corresponds to the state occupancy distribution of the Markov chain and can therefore be represented by a stationary distribution vector π of length m. The value of element $\pi(i)$ (with $i = 1,2,3...m$) represents the residence probability of the mobile user in cell i.

Consider the following example. Given a user who resides within $m = 3$ cells and the observed sequence L of cells as follows:

$$L = \left[L_1, L_2, L_1, L_2, L_1, L_3, L_2\right]$$

where each L_i denotes a unique cell ($i = 1,2,3$). Each element $P(i,j)$ of the transition probability matrix P corresponds to the number of $L_i \rightarrow L_j$ transitions divided by the total number of transitions from state L_i. For example, the number of $L_1 \rightarrow L_2$ transitions occurring in sequence L is 2 while the total number of transitions out of cell L_1 is 3, resulting in the transition probability $P(1,2) = 2 \div 3 = 0.67$.

Meanwhile, each element $\pi(i)$ of the stationary distribution vector π corresponds to residence probability $Pr(L_i)$ of the considered user in cell L_i. For example, $\pi(1)$ corresponds to the number of occurrences of L_1 divided by the total number of cells in the Markov chain. That is, $\pi(1) = 3 \div 7 = 0.43$.

Therefore, constructing the location profile for the above sequence of cells will result in the transition probability matrix P and the stationary distribution vector π as follows:

$$P = \begin{bmatrix} 0 & 0.67 & 0.33 \\ 1 & 0 & 0 \\ 0 & 1 & 0 \end{bmatrix}$$

$$\pi = \left[0.43\ 0.43\ 0.14\right]$$

2.2 *Identification via Location-Based Markov Model*

As described in Section 2.1, mobile phone users' traces can be used to build a probabilistic mobility model that reflects the unique characteristics of an individual's movement process. Identification of a user is then achieved by comparing his location profile with the existing already known profiles, trying to find the best-matched profile. The more closely the two profiles resemble, the more likely the two corresponding individuals are the same person.

Basic steps of the identification process include:

1. Initiate the database of location profiles from known users' mobility traces within a specified period referred to as the identification period.
2. Observe an unknown user during the subsequent period (referred to as the evaluation period) and generate the new location profile for this user.
3. Compare the new location profile of the unknown user with all known profiles stored in the identification database to find the user whose profile is closest to that of the unknown user.

Two methods, namely, Residence Matching and Cell Sequence Matching, were introduced in [4] to measure how closely location profiles resemble. This closeness is referred to as the identification index between two location profiles. The higher the identification index is, the more likely the two users match.

2.2.1 Residence Matching

The identification index *iden* between two users U_x and U_k corresponds to the probability that they reside in the same set of cells, given by the following formula:

$$iden = \sum_{i,j}^{m} Pr_x(L_j|L_i)Pr_x(L_i) \times Pr_k(L_j|L_i)Pr_k(L_i) \qquad (1)$$

where m denotes the number of cells appearing in both cell sequences of U_x and U_k.

Let P_x and π_x denote, respectively, the transition probability matrix and stationary distribution vector of user U_x. The probabilities $Pr_x(L_j|L_i)$ and $Pr_x(L_i)$ correspond to the values of $P_x(i,j)$ and $\pi_x(i)$ respectively. So, Formula 1 can be rewritten as:

$$iden = \sum_{i,j}^{m} P_x(i,j)\pi_x(i) \times P_k(i,j)\pi_k(i)$$

2.2.2 Cell Sequence Matching

In this method, the identification index between two users U_x and U_k is calculated by evaluating the likelihood that the cell sequence of U_x is reflected by the location profile of U_k (or vice-versa). Let l be the number of cells in the cell sequence of user U_x. There are then $l-1$ cell transitions denoted by θ ($\theta = 1,2,3,...,l-1$). For every cell transition θ, the probability that this transition is generated from the location profile of U_k, denoted as p^{θ}, is computed by looking up for transition θ

in the transition probability matrix of U_k. Then, the identification index between U_x and U_k is obtained as the product of all these transition probabilities, given by the following formula:

$$iden = \prod_{\theta=1}^{l-1} p^{\theta}$$

2.3 Mobility Profiling Using Spatio-Temporal Markov Model

In both methods described in Section 2.2, the observed sequence of cells is expected to be individual-characteristic enough to help distinguish an individual from the others. However, there are certain cases where location profile of one user is very similar to that of another user while, in fact they may have different mobility patterns with respect to temporal characteristics.

In our approach, we make use of the same Markov model for constructing user mobility profile that is consisted of a transition probability matrix and a stationary distribution vector. However, we seek to improve the uniqueness of the mobility profile by incorporating time into the modeling process in order to address the issue that two different users have the same location profile due to the resemblance in moving patterns. We refer to the extended mobility model as the spatio-temporal Markov model.

To achieve this, we also take into account the time during which one user resides in a given cell and try to reflect the information in the mobility model. We therefore replace every state of the original Markov chain by a spatio-temporal state which is capable of representing both spatial and temporal aspects of the moving traces. To be more precise, we associate every state with a corresponding time interval during which the Markov chain stays in the state.

As an example, consider a user who resides within $m = 3$ cells and the observed cell sequence L as follows:

$$L = [L_1, L_2, L_1, L_2, L_1, L_3, L_2]$$

We associate every state L_i with a time interval Δt_i indicating the time during which the user resides in cell i. Values of Δt_i may be particular hours of a day, periods of day, or any time interval. For example, Δt_i may take the values of time intervals of a day as follows:

$$T = \{\Delta t_1 : morning, \Delta t_2 : afternoon, \Delta t_3 : evening, \Delta t_4 : night\}$$

This results in a new spatio-temporal cell sequence L' as follows:

$$L' = [\Delta t_1 L_1, \Delta t_1 L_2, \Delta t_1 L_1, \Delta t_2 L_2, \Delta t_2 L_1, \Delta t_3 L_3, \Delta t_4 L_2]$$

As we can see in the above sequence, a unique state is no longer determined by a cell's location itself but by both of the cell's geographical location and the corresponding time interval. As presented in Section 3 next, our experiments show that the way how time is partitioned into intervals has great effect on identification performance, and choosing an appropriate time partitioning scheme can help magnify the unique characteristics of individuals' mobility profiles.

The cell sequence L' can also be rewritten as:

$$L' = [S_1, S_2, S_1, S_3, S_4, S_5, S_6]$$

where S_i denotes a unique location and time interval pair. The maximum number of unique spatio-temporal states in L' is $m \times |T|$ where m is the number of unique cell locations and $|T|$ is the number of time intervals. As we can see, L and L' are of the same length but the number of unique spatial states in L is 3 whereas the number of unique spatio-temporal states in L' is 6, indicating that L' is more complex. In general, the more complex a model is (in terms of the number of unique states), the more adequately it reflects the reality and the more individual-characteristic it is. Notice that, the original cell sequence L is considered the simplest case of L' where the number of time intervals $|T|$ is equal to 1, indicating there is no time partitioning at all.

With an additional time interval associated to every state of the original Markov chain, temporal diversities in individuals' mobility traces are conserved during the modeling process, lessening the likelihood that two mobility profiles of two different individuals resemble.

Let us look at an example of how temporal factors help improve the uniqueness of mobility profiles. Consider two mobile users U_x and U_k who reside very closely to each other within the areas covered by m cells ($m = 3$). Given the observed cell sequences as follows:

$$L_x = [L_1, L_2, L_1, L_2, L_1, L_3, L_2]$$
$$L_k = [L_1, L_3, L_2, L_1, L_2, L_1, L_2]$$

The transition patterns include $L_1 \rightarrow L_2$, $L_1 \rightarrow L_3$, $L_2 \rightarrow L_1$, and $L_3 \rightarrow L_2$. These patterns appear in both sequences with the same frequencies. Consequently, constructing the location profiles for the above two sequences will result in the same transition probability matrix, and also the same stationary distribution vector.

Obviously, this is not the expected case, and U_x could possibly be mistaken as U_k (or vice-versa) in the identification process since their location profiles coincide. In reality, it is not likely that two individuals have exactly the same mobility profile. However, through this example, we emphasize the possibility that mobility profiles could coincide or highly resemble even with different mobility sequences.

To deal with the problem, a time interval Δt is associated to every state in the original chains, as described above. For this example, Δt takes the value of either $morning$ or $afternoon$ corresponding to the actual time interval during which a user resides in a given cell. Then the mobility sequences may look as follows:

$$L_x = [\Delta t_1 L_1, \Delta t_1 L_2, \Delta t_1 L_1, \Delta t_1 L_2, \Delta t_2 L_1, \Delta t_2 L_3, \Delta t_2 L_2]$$
$$L_k = [\Delta t_1 L_1, \Delta t_1 L_3, \Delta t_1 L_2, \Delta t_1 L_1, \Delta t_2 L_2, \Delta t_2 L_1, \Delta t_2 L_2]$$

Constructing the Markov mobility profiles with the above spatio-temporal sequences can be done following exactly the same process. And the resulting transition probability matrixes and stationary distribution vectors will no longer coincide since L_x and L_k now have their own specific transition patterns with respect to both location and time. For example, L_x is characterized by $\Delta t_2 L_1 \rightarrow \Delta t_2 L_3$ transition while L_k is characterized by $\Delta t_1 L_1 \rightarrow \Delta t_1 L_3$ transition.

In general, by embracing additional temporal factors we obtain more complex spatio-temporal mobility models. However, as a gain, more characteristic patterns of individuals' mobility traces are obtained.

3 Evaluation

In the previous section, we introduce a novel mobility profiling approach of conserving individuals' moving characteristics by extending the location-based Markov model with temporal perspectives. We now present the evaluation of our proposed model by employing the same two identification methods (described in Section 2.2) within both of the location-based model and the spatio-temporal model.

3.1 The Dataset

For our experiments, we use the dataset of real mobility traces obtained from the MIT Reality Mining project. The dataset contains several types of information recorded from mobile phone users over the course of nine months, from September 2004 to May 2005. However, for the scope of our research, only mobility logs are taken into consideration.

3.2 Experiments

Following the identification process mentioned earlier in this paper, for every experiment, we choose one month as the identification period and its succeeding month as the evaluation period. The number of correctly identified users is considered the performance measure for identification methods. To illustrate how identification performance varies, we evaluate different methods on the same dataset, namely:

1. *SpatioRes*. Identification using the Residence Matching method within the location-based model.
2. *SpatioSeq*. Identification using the Cell Sequence Matching method within the location-based model.
3. *SpatioTempoRes*. Identification using the Residence Matching method within the spatio-temporal model.
4. *SpatioTempoSeq*. Identification using the Cell Sequence Matching method within the spatio-temporal model.

3.2.1 Experiments with Common Time Intervals

In our experiments, the methods *SpatioRes* and *SpatioSeq* (referred to as the location-based methods) are not affected by how time is partitioned into intervals. On the other hand, for *SpatioTempoRes* and *SpatioTempoSeq* (referred to as the spatio-temporal methods), we first evaluate them by dividing the day into four meaningful and typical time intervals as follows:

- *Morning*, from 08:00 AM to 12:00 PM.
- *Afternoon*, from 12:00 PM to 06:00 PM.
- *Evening*, from 06:00 PM to 10:00 PM.
- *Night*, from 10:00 PM to 08:00 AM.

Table 1 Experiment results for different identification/evaluation periods. Columns (a), (b), (c) and (d) show the percentage of correctly identified users using the corresponding methods.

Identification Period	Evaluation Period	*SpatioRes* (a)	*SpatioSeq* (b)	*SpatioTempoRes* (c)	*SpatioTempoSeq* (d)
Sep 2004	Oct 2004	10%	59%	19%	65%
Oct 2004	Nov 2004	29%	62%	29%	70%
Nov 2004	Dec 2004	27%	77%	27%	84%
Dec 2004	Jan 2005	41%	74%	44%	80%
Jan 2005	Feb 2005	49%	67%	53%	81%
Feb 2005	Mar 2005	58%	80%	65%	88%
Mar 2005	Apr 2005	34%	76%	31%	86%

As shown in Table 1, the spatio-temporal methods perform better than the corresponding location-based ones. Indeed, for every period, (c) is greater than or equal to (a), and (d) is greater than (b). It is also interesting that, *SpatioTempoSeq* performs considerably better than its counterpart *SpatioSeq*, whereas *SpatioTempoRes* only shows little improvement over its counterpart *SpatioTempoRes*. That means the identification methods based on cell sequence matching can better exploit the specific characteristics of the spatio-temporal mobility profile.

We do not expect the percentage of correctly identified users to reach 100% due to the inherent noise in the dataset ([1]). Also, for certain users, the amount of mobility trace data in either the identification period or the evaluation period is not sufficient to make up a highly characteristic mobility profile. In such cases, the identification process fails to match the unknown mobility profile with the correct individual's one.

However, in reality, humans tend to follow certain repeated movement patterns. These patterns commonly dominate individuals' mobility profiles and are not likely to dramatically change over the course of months or even years. So, for the majority of cases, users could be correctly identified thanks to these dominant patterns.

3.2.2 Experiments with Arbitrary Time Intervals

In these experiments, instead of using the common four intervals corresponding to *morning*, *afternoon*, *evening* and *night*, we divide the 24 hours of a day into four arbitrary intervals of equal duration (6 hours each) as follows:

- From 12:00 AM to 06:00 AM.
- From 06:00 AM to 12:00 PM.
- From 12:00 PM to 06:00 PM.
- From 06:00 PM to 12:00 AM.

Table 2 Experiment results for different identification/evaluation periods. Columns (a), (b), (c) and (d) show the percentage of correctly identified users using the corresponding methods. Note that the results for methods *SpatioRes* and *SpatioSeq* remain unchanged, they are included in the table for reference only.

Identification Period	Evaluation Period	*SpatioRes* (a)	*SpatioSeq* (b)	*SpatioTempoRes* (c)	*SpatioTempoSeq* (d)
Sep 2004	Oct 2004	10%	59%	13%	61%
Oct 2004	Nov 2004	29%	62%	29%	62%
Nov 2004	Dec 2004	27%	77%	27%	78%
Dec 2004	Jan 2005	41%	74%	44%	75%
Jan 2005	Feb 2005	49%	67%	51%	67%
Feb 2005	Mar 2005	58%	80%	65%	78%
Mar 2005	Apr 2005	34%	76%	38%	76%

As shown in Table 2, the spatio-temporal methods still yield better results. However, the improvement is not as good as that of the first experiments in which time is partitioned into meaningful intervals of a day. This is because individuals have their own moving patterns during particular time intervals of the day that are commonly *morning*, *afternoon*, *evening* and *night*. By associating individuals' moving behaviors with these particular time intervals, we can magnify the differences in terms of time among mobility profiles. Using arbitrary time intervals, on the contrary, could blur these interval-based characteristics of mobility behaviors.

For the spatio-temporal methods, with n time intervals, the number of unique states in the Markov chain may increase by n times in the extreme case. As a result, more computation time is required for processing the mobility profiles. This is obviously a processing overhead penalty for the improvement. So, there is a problem of balancing between the actual improvement in identification performance and the processing cost required for that improvement.

It is also noticeable that our experiments are made using a particular dataset in which mobility traces are not in the form of exact locations, but rather in the form of cell stations recorded by users' mobile phones. The mobility traces are therefore coarse-grained and contain a large amount of noisy data due to cell oscillation problems in a GSM network as pointed out in [4]. Removing noise and distortion by a clustering algorithm may help yield better results. However, this is beyond the scope of our research since we only focus on evaluating the degree to which temporal factors help improve the existing identification methods.

4 Conclusion

Through this paper, we demonstrate how mobility traces are modeled in such a way that makes it possible to uniquely identify an individual. We also develop a profiling method for incorporating time as an input into the Markov mobility model, resulting in highly unique mobility profiles.

Our experiment results show that the more characteristic the mobility model is, the more possible it is to distinguish an individual from the others. We can further enhance our proposed model, making it even more characteristic and realistic, by discovering more possible underlying patterns from the mobility traces and incorporating those into the mobility model. For example, through examining the mobility traces of the dataset, we find that individuals have their own dwell-time in certain cells, and transition time between contiguous cells also differs among individuals. These differences, if appropriately modeled, are a key to the identification process.

Apart from mobility modeling, identification methods themselves also play a crucial role in the identification process. For the methods mentioned in this paper, identification is achieved by computing the average resemblance between two mobility profiles. However, the approach may not work as expected in case mobility traces contain large numbers of common patterns that are shared among several individuals. Sometimes the specific mobility patterns of an individual are overwhelmed by these common patterns. As a result, an ideal method should be able to recognize and focus on key patterns that uniquely characterize an individual, rather than treating all patterns the same.

References

1. Bayir, M.A., Demirbas, M., Eagle, N.: Discovering Spatio-Temporal Mobility Profiles of Cellphone Users. In: Proc. of the 10th IEEE International Symposium on a World of Wireless, Mobile and Multimedia Networks, pp. 1–9 (2009)
2. Cvrcek, D., Kumpost, M., Matyas, V., Danezis, G.: A Study on the Value of Location Privacy. In: Proc. of the 5th ACM Workshop on Privacy in Electronic Society, pp. 109–118 (2006)

3. De Montjoye, Y.A., Hidalgo, C.A., Verleysen, M., Blondel, V.D.: Unique in the Crowd: The Privacy Bounds of Human Mobility. Scientific Reports 3(1376) (2013)
4. De Mulder, Y., Danezis, G., Batina, L., Preneel, B.: Identification via Location-Profiling in GSM Networks. In: Proc. of the 7th ACM Workshop on Privacy in the Electronic Society, pp. 23–32 (2008)
5. Eagle, N., Pentland, A., Lazer, D.: Inferring Friendship Network Structure by Using Mobile Phone Data. Proceedings of The National Academy of Sciences 106(36), 15274–15278 (2009)
6. Eagle, N., Quinn, J.A., Clauset, A.: Methodologies for Continuous Cellular Tower Data Analysis. In: Tokuda, H., Beigl, M., Friday, A., Brush, A.J.B., Tobe, Y. (eds.) Pervasive 2009. LNCS, vol. 5538, pp. 342–353. Springer, Heidelberg (2009)
7. Ghahramani, Z.: An Introduction to Hidden Markov Models and Bayesian Networks. International Journal of Pattern Recognition and Artificial Intelligence 15(1), 9–42 (2002)
8. Gonzalez, M.C., Hidalgo, C.A., Barabasi, A.L.: Understanding Individual Human Mobility Patterns. Nature 453(7196), 779–782 (2008)
9. Lane, N.D., Xie, J., Moscibroda, T., Zhao, F.: On the Feasibility of User De-Anonymization from Shared Mobile Sensor Data. In: Proc. of the 3rd International Workshop on Sensing Applications on Mobile Phones, pp. 1–5 (2012)
10. Song, C., Qu, Z., Blumm, N., Barabasi, A.L.: Limits of Predictability in Human Mobility. Science 327(5968), 1018–1021 (2010)
11. Sundaresan, A., Chowdhury, A.K.R., Chellappa, R.: A Hidden Markov Model Based Framework for Recognition of Humans from Gait Sequences. In: Proc. of the International Conference on Image Processing, pp. 93–96 (2003)
12. Xia, J.C., Zeephongsekul, P., Packer, D.: Spatial and Temporal Modeling of Tourist Movements Using Semi-Markov Processes. Tourism Management 32(4), 844–851 (2011)

Secure Authentication for Mobile Devices Based on Acoustic Background Fingerprint

Quan Quach, Ngu Nguyen, and Tien Dinh

Abstract. In this paper, we propose a method to establish a secured authentication among the mobile devices based on the background audio. The devices will record synchronous ambient audio signals. Fingerprints from background is used to generated the shared cryptographic key between devices without sending the information from ambient audio itself. Noise in fingerprints will be detected and fixed by using Reed-Solomon error correcting code. In order to examine this approach, fuzzy commitment scheme is introduced to enable the adaption of the specific value of tolerated noise in fingerprint by altering internal parameters of error correction. Besides, we introduce the method of background classification to alter the threshold of error correction automatically. Furthermore, the silent background issue is also solved thoroughly by several approaches. The system is build and tested on iOS and indoor environment.

1 Introduction

One of the most fundamental features of smart phones is that it has the ability to access digital media such as pictures, videos and music. It is natural that people want to share their digital media with each other. By this way, secure communication plays a crucial role in pervasive environments. It is required to protect confidential information. Every transmission in the network is required to prevent security breach from the outside. With mobile devices, communication protocol may change frequently, which makes it a problem to provide a confident security method for the transmission.

We utilize background sound as a potential method for creation shared cryptographic keys among devices. In 2007, Jaap Haitsma and Ton Kalker [1] presented

Quan Quach · Ngu Nguyen · Tien Dinh
Advanced Program in Computer Science, Faculty of Information Technology,
VNU - University of Science, Ho Chi Minh City, Vietnam
e-mail: qmquan@apcs.vn, {nlnngu,dbtien}@fit.hcmus.edu.vn

V.-N. Huynh et al. (eds.), *Knowledge and Systems Engineering, Volume 1*,
Advances in Intelligent Systems and Computing 244,
DOI: 10.1007/978-3-319-02741-8_32,

the mechanism to build a fingerprint system in purpose to recognize songs. Being inspired by their mechanism, we construct the system which takes the background audio as input and establish the secure authentication among iOS devices. Indeed, the devices, which are willing to start a connection, will take a synchronized audio recording and compute unique audio fingerprints based on that audio as well. Depending on extracted fingerprints, we demonstrate the mechanism of generating cryptographic key among devices using fuzzy commitment scheme [13] and Reed-Solomon error correcting code [14]. Besides, peak detection algorithm [16] is used to generate the value of threshold corresponding to environment noise levels (silent, quite noisy and noisy background). However, the drawback of this approach occurs in silent environment when devices produce the same fingerprint although they are far away. In order to overcome this issue, we propose several approaches. In the first one, networks surround each device are collected and analyzed without requiring any Internet connection. Another one is to establish the audio communication among devices.

The rest of this paper is organized as follows. Section 2 presents the related research about generating audio fingerprint from background environment. Section 3 describes system overview as well as the detail of each step in purpose of implementation. Section 4 focuses on classifying the environment noise levels. In section 5, we propose two methods which deal with the problem of silent environment. The experimental results and conclusions are presented in the last two sections.

2 Related Work

There is a variety of methods to extract the audio fingerprint such as the audio fingerprinting system introduced by Jaap Haitsma and Ton Kalker [1] and audio fingerprint extraction method introduced by J.Deng, W.Wan, X.Yu and W.Yang [2]. The first method mentioned in "A highly robust audio fingerprinting system" by Jaap Haitsma and Ton Kalker introduces the way to extract the fingerprint of a song based on the energy difference in energy of frequency bands. A similar technique can be found in [3][4]. Another audio fingerprint extraction method was introduced by J.Deng, W.Wan, X.Yu and W.Yang. The scheme was based on spectral energy structure and non-negative matrix (NMF). Besides, this method [5] uses the acoustic background to decide mobiles' location.

One of the closest ideas to our work which has been commercialized is Bump [6]. It is a cross-platform application that allows any two mobile devices that communicate with each other to initialize a connection that allows a data transmission among them. Its general idea is similar to [7][8] which utilize the accelerometer sensor on devices which are shaken together.

In order to generate the cryptographic key, fuzzy commitment scheme is necessary because similarity of fingerprint is typically not sufficient to establish the secure channel. With fuzzy cryptography scheme, the identical fingerprint based on noisy data is possible [9]. Li et al. analyze the usage of biometric or multimedia data as

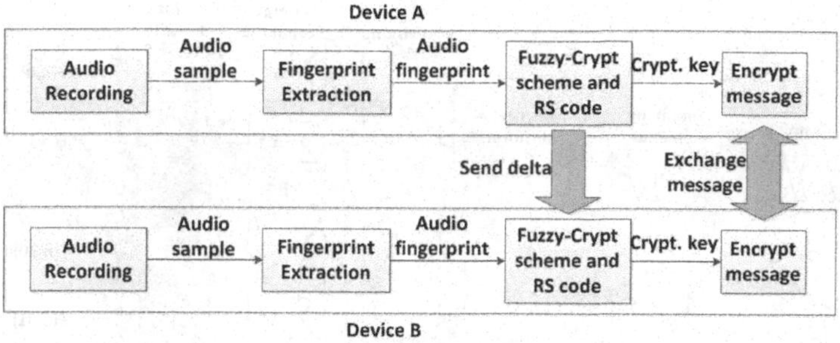

Fig. 1 Process diagram

a part of authentication process and propose the protocol [10]. The similar study is presented by Miao et al [11].

3 System Overview and Algorithm

3.1 System Overview

As mentioned previously, we divide the whole process into four big stages. Figure 1 depicts our creation process. As shown in figure 1, two testing devices will record audio of environment. Two audio recorded must be taken in close proximity for the fingerprint to be similar enough. Then the recording audio will be put into fingerprint extraction scheme in order to produce the fingerprint by calculating the energy difference between frequency bands of each frame [1]. After acquiring the fingerprint, a mechanism to exchange and correct the fingerprints between the two devices will be used based on part of the fuzzy commitment scheme [13] and Reed-Solomon error correcting code [14]. The fingerprint will then become a cryptographic (common) key to encrypt and decrypt the message (or data, photo, etc) by using advanced encryption standard (AES) [15]. The data is now transmitted by the mean of Blue-tooth, local network or through the Internet. The users can choose the similarity percentage threshold over which the fingerprints recognize to be in the same background, or they can keep the default one. Moreover, in order to make a deal with silent issue, Wi-Fi footprints at devices' location are generated while recording audio or audio communication among devices is constructed by themselves.

3.2 Fingerprint Extraction Scheme

The idea of generating the fingerprint is mainly from the method of Jaap Haitsma and Ton Kalker [1] that we mentioned previously. The audio of environment is taken with the length of approximately 6.8 seconds. The audio samples are encoded with the sample resolution of 16 bits per sample and the sample rate of 44100 Hz (44100

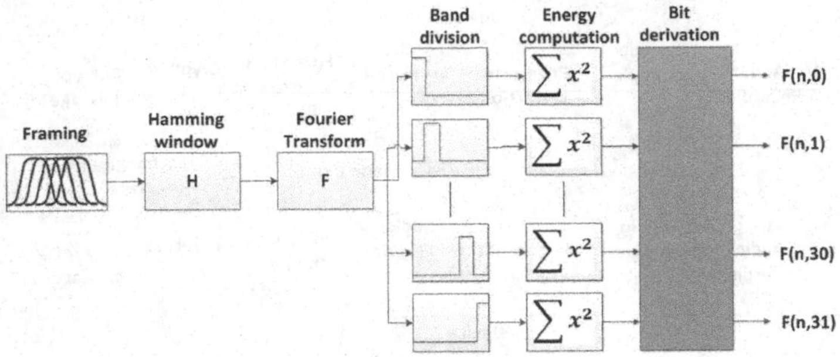

Fig. 2 Fingerprint extraction scheme overview

samples per second). The sound channel is mono. After acquiring the audio record, it provides not much information in the time domain. In order to get a better identity of the background, we need to transform them into the frequency domain of the audio with the help of a fast Fourier transform [12] (FFT) in our case.

The goal of the fingerprint extraction scheme is to generate a 512-binary-bits fingerprint. As we could see in the figure 2, at first we need to segment the record into the smaller frames. The technique will examine every two consecutive frames. As each sub-fingerprint contains 32 binary bits, we should split the record into 17 **non-overlapping** frames. Before proceeding any further, one thing that differs from some techniques in References section is that we apply a Hamming window to each frame. The reason why we apply the window to our sample is because the FFT always assumes that the computed signals are periodic, which is mostly impossible with real life signals. If we try to apply the FFT on the non-periodic signal, unwanted frequency will appear in our final result which is called spectral leakage. The window forces the samples of the signal at the beginning and the end to be zero, thus the signal will become periodic, which will reduce the leakage. After that, we apply the FFT to each frame to extract their frequency spectrum from the time domain samples [12]. The reason we need to examine the frequency domain is because it contains most of the audio features that are understandable by the machine. After obtaining the result of the frequency domain for the audio frames, each result is divided into 33 **non-overlapping** frequency bands.

Finally, we use a formula provided by [1] since authors already verified that the sign of energy differences along the time and frequency axes is a property that is very robust to many kind of processing. The formula is as follow: if we denote the energy of the frequency band m of frame n by $E(n,m)$ and the mth bit of the sub-fingerprint of frame n by $F(n,m)$, the bit of the sub-fingerprint are formally defined as:

$$F(n,m) = \begin{cases} 1 & \text{if } E(n,m) - E(n,m+1) - (E(n-1,m) - E(n-1,m+1)) > 0 \\ 0 & \text{if } E(n,m) - E(n,m+1) - (E(n-1,m) - E(n-1,m+1)) \leq 0 \end{cases}$$

After we do all of steps above, the result is a 512-binary-bits fingerprint as we can see in figure 3.

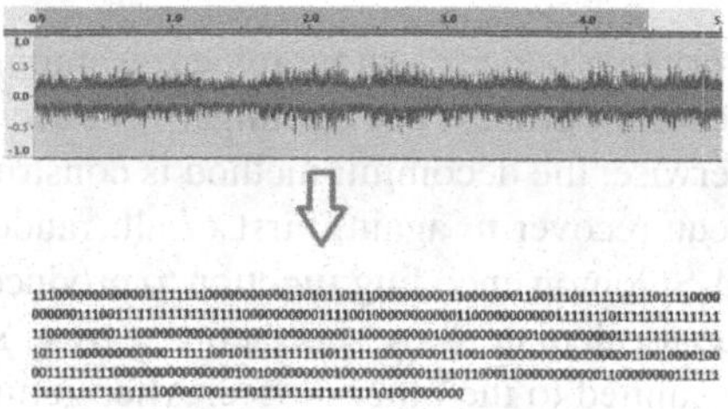

Fig. 3 An audio signal is represented by wave form and fingerprint

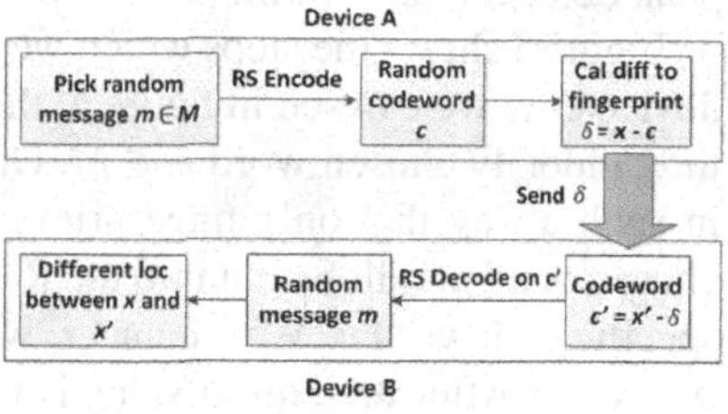

Fig. 4 Fuzzy commitment scheme diagram

3.3 *Fuzzy Commitment Scheme and Reed-Solomon Error Correcting Code*

As mentioned earlier, fuzzy commitment scheme [13] and Reed-Solomon error correcting code [14] help us generate the cryptographic key. With fuzzy commitment scheme, the secret x is used to hide the key k in the set of possible keys K in such a way that only the similar secret x' can find and recover the original k. The secrets x and x', in our case, are considered as the audio fingerprints.

Fuzzy commitment scheme is usually constructed with Reed-Solomon error correcting code. The following discussion provides a short introduction to this code. Consider the set of message M which have the m-bit representation with the length k and the set of codeword C with the length n. The Reed-Solomon code $RS(m,n,k)$ is the mapping between two sets and is initialized as follow: $m = 9$, $n = 2^m - 1$, k is calculated depending on peak detection algorithm (see section 4).

Reed-Solomon code has two main functions: encode and decode. Encode function is the mapping from a message $m \in M$ of length k uniquely to a specific codeword $c \in C$ of length n, says,

$$m \xrightarrow{encode} c$$

This step adds the redundant information at the end of message to generate the code word. Otherwise, the decode process maps the set of codeword $\{c, c', \ldots\} \in C$ to the only one single origin message.

$$c \xrightarrow{decode} m \in M$$

The value

$$t = \frac{n-k}{2} \tag{1}$$

defines the maximum number of bits that the decode process can fix to recover the origin message. In our case, t is chosen arbitrarily by the noise of fingerprints which can be obtained by classifying environment. Consequently, t is also the result of peak detection algorithm.

Figure 4 shows the steps which we work to produce cryptographic key among the devices. As we can see in figure 4, the commit process use the fingerprint x to hide the randomly chosen word $m \in M$. Otherwise, the decommit method is constructed in such a way that only fingerprint x' can recover m again. First of all, randomly chosen word m will be put into the Reed-Solomon encoding function to produce the specific codeword c. With c and x, we can calculate δ by subtracting x by c, says, $\delta = x - c$. After that, the δ value is transmitted to the other device. After getting δ value, the receiver will also subtract its fingerprint x' by δ, to produce its codeword c', says, $c' = x' - \delta$ and applies Reed-Solomon decoding method on c'

$$c' \xrightarrow{decode} m \in M$$

At the moment, m has been obtained completely. Consequently, we get the different locations between c' and m. Furthermore, the different locations between c' and m are also the different locations between two fingerprints x' and x. By that way, it is straightforward to fix bit errors of fingerprint on receiver's side with given error locations since the fingerprint only contains 0 and 1. Instead of using m to generate cryptographic key as described in [4], devices use their fingerprints (sender utilizes its own fingerprint and receivers use their fingerprints which are already corrected) as the cryptographic key. Data which was encrypted by the cryptographic key using AES [15] will be exchanged back and fourth among devices.

4 Background Classification

As we discussed previously, the environment noise affects the extracted fingerprints. Actually, by experiment, the similarity of fingerprints in noisy background are lower than in quite environment. Moreover, dependent on the noise of background, the value of threshold for the maximum number of bits between codeword (Section 3.3) that can be corrected through Reed-Solomon process [14] is chosen arbitrarily. Therefore, we need to use peak detection algorithm [16] to classify environment noise into three categories: silent, quite noise and noise. The general idea of this algorithm is to associate each sample value to a score which is determined by using peak function. We can conclude whether a point is local peak or not by evaluating its score. A data point in time-series is a *local peak* if (a) it is a large and locally maximum value within a window, which is unnecessarily large nor globally maximum value in entire time-series; and (b) it is isolated, says, not too many points in the window have similar value.

The value of *peak function* [16] can be describes as follow:

$$S(k,i,x_i,T) = \frac{\frac{x_i - x_{i-1} + x_i - x_{i-2} + \ldots + x_i - x_{i-k}}{k} + \frac{x_i - x_{i+1} + x_i - x_{i+2} + \ldots + x_i - x_{i+k}}{k}}{2}$$

where, k is the window size, x_i is the sample value at index i and T is the sample array. Peak-function S computes the average of (i) average of signed distances of x_i from its k left neighbors and (ii) average of signed distances of x_i from its k right neighbors.

The detail of peak detection algorithm can be found in [16]. The algorithm inputs the sample array, window size around a peak k and a specified value h. After that, it produces the **average** of all peak values.

In previous section, we already knew the aim of peak detection algorithm is not only classifying environment noise, but also altering the parameters of Reed-Solomon error correction [14]. Here, we provide a short discussion about how it work. Again, result of this peak detection algorithm is the **average** of all peak values which is corresponding to threshold of similarity of fingerprint. For instance, in silent background (**average** of all peak values ranges from 0 to 0.04), similarity of fingerprint between devices is always above 90 % (see experimental section for more detail). That means, we can fix **at most** 10 % of 512-bits-length of fingerprint. By that way, internal parameters of error correction in silent environment is initialized as follow:

$$m = 9, n = 2^9 - 1 = 511, t = 10\% * 511 = 51,$$
$$k = 511 - 2 * 51 = 409(\text{by equation } 1)$$

5 Silent Background Issue

A little problem arises in silent background, especially very silent environment when two devices get too far away, but still extract similar fingerprints. For instances, one device is in my house and the other is in my partner's house at the midnight. We hence propose two approaches to overcome this issue. The first method utilizes network information surrounding each device to generate the Wi-Fi footprint.We thus determine whether or not two devices are nearby. Another approach is called audio communication among devices. Each device emits a unique short sound. When two devices are nearby, they could record the other sound and the created fingerprint of each device thus contains the information of other device. Otherwise, it only extracts information of its own audio.

5.1 Wi-Fi Footprint

While recording audio of environment, devices also analyze the surrounding networks and produce Wi-Fi footprints at that location. A Wi-Fi footprint has two type of information we have a concern: media access control (MAC) address and received signal strength indicator (RSSI). Hence, by matching each pair of Wi-Fi footprints, we also compute the number of common networks which have the same MAC address, and the list of δ values which are absolutely difference of each pair of RSSI coming from common networks. After that, similarity of Wi-Fi footprints among devices is defined by the ratio between the number of common networks c'

which satisfy $\delta \leq$ *specific threshold* and the minimum of total number of networks between two Wi-Fi footprints, says, $similarity\ of\ footprint = \frac{c'}{min(network_1, network_2)}$.

5.2 Audio Communication among Devices

As mentioned previously, both two devices should emit their sounds simultaneously while recording. The acoustic emitted is hence designed carefully. The signal should be different from various background noise. Because most of background noise are located at lower frequency band (e.g., conversation between 300 Hz to 3400 Hz, music from 50 Hz to 15 kHz, which covers almost all naturally occurring sounds). 16-20 kHz is still audible to human ears [17], but much less noticeable and thus present less disturbances.

6 Experimental Results

6.1 Experiment Setup

We conducted several experiments to demonstrate the correctness of our implementation. The data (both audio fingerprint and Wi-Fi footprint) for these experiments were taken by the two devices: Apple iPhone 4 and Apple iPhone 4s in three different backgrounds which represent different background conditions. The distance between two devices varies from 0 to 4 meters. The three backgrounds are:

- In the student lab, the background is quite silent with only a few students discussing.
- In the university hallway, the background audio came from various sources and is quite noisy.
- In a mall, the background acoustic came from different places and is noisy.

6.2 Similarity of Fingerprint Experiment

We recorded approximately 20 samples for each background conditions and then applied the fingerprint extraction schemes on them. Finally, the fingerprints were compared with each other. If the two records were taken at the same place and time, the outcome fingerprints should have the similar percentage above 75% and varied depended on the background conditions. The result is demonstrated in Figure 5a.

In the same backgrounds, the extracted fingerprint results were mostly above 75% percent, which is the minimum correction percentage that the Reed Solomon can correct. As we can see from the results, the accuracy between fingerprints is highest in the Lab (above 90%), then university hallway (varies from 82.23% to 90%) and the mall (below 86%). The ambiguous between quite noisy and noisy background (similarity in 80%-85%) just occurs in few number of tests. We concludes from the result that the more noisy and chaotic the background, the less similar the

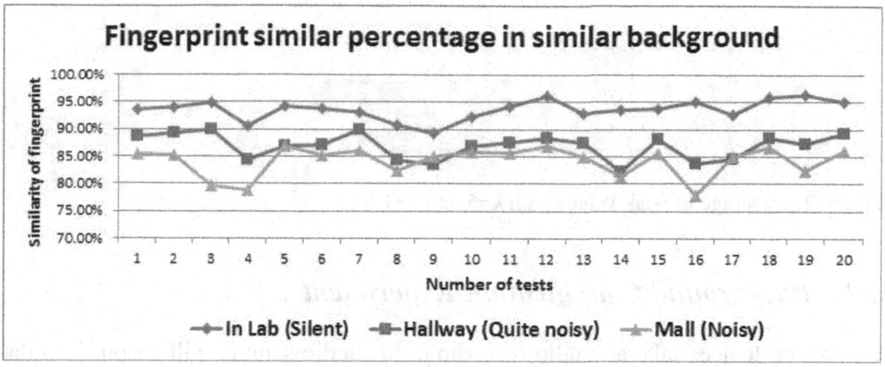

Fig. 5a

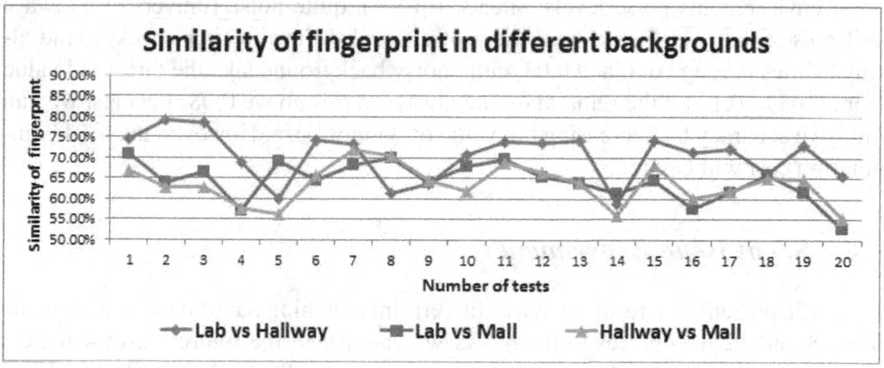

Fig. 5b

Fig. 5 Similarity of fingerprint: (a) On the same background. (b) On the different background

fingerprints are. This is understandable. The similarity percentage is also above the correction threshold which is 75%.

Besides, we take 20 recording audio samples in different environment and attempt to compare their fingerprints to each other. Figure 5b depicts the similarity of fingerprint in different background. As we can see in Figure 5b, in different environment background, the average similarity of the fingerprints drops down below the threshold 75% which is a minimum threshold values that the Reed-Solomon error correcting code can not correct. These results prove the point that the fingerprint correctly represents the corresponding background.

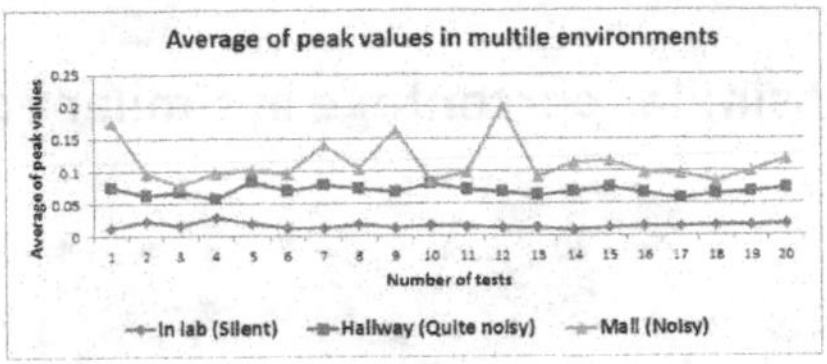

Fig. 6 The average of peak values with k=5 and h=1.5

6.3 Background Classification Experiment

Whenever devices take an audio recording, this audio sample will be put into the peak detection algorithm to determine environment noise levels. Figure 6 shows result of background classification with window size k=5 and specific value h=1.5 in three environments noise levels: silence (in lab), quite noise (university hallway) and noise (mall). In Figure 6, average of all peak values in silent background always varies in range of 0 and 0.04, quite noisy background take the threshold value from 0.05 to 0.1 and the remaining one always varies above 0.08. Besides, we can improve accuracy by decreasing the value of window size. However, the total computation cost will enlarge.

6.4 Silent Issue Experiment

Figure 7a presents the result of Wi-Fi fingerprint matching based on distances value of 4, 8 and 12 meters, respectively. As we can see in the figure 7a, in 4 meters distance, most of the similarity of fingerprints are above 80%. Meanwhile, the Wi-Fi footprint similar percentage in 8 meters distance varies from 62.5 % to 80% and the similarity of footprint in 12 meters far away are always below 60%. Consequently, whenever similarity of audio fingerprint is above 90%, however, the similarity of Wi-Fi footprint is below 60%, we conclude that this case is in silent background issue. After that, we change the threshold of similarity of fingerprint in order to prevent correcting errors successfully. For example, similarity of audio fingerprint is 90%, but the Wi-Fi footprint test fails. Threshold of similar of audio fingerprint is then modified from 90% to 85% so that Reed-Solomon error correction fails to generate cryptographic key.

Figure 7b depicts result of alternative solution for silent background problem. In the experiment design, two sounds emitted differ significantly. Basically, the similar percentage of fingerprint between two far away devices (above 4 meters) varies from 60% to 70%. Otherwise, it always take the value above 80% within 1 meter distance when the ambient audio is mixed of two emitted sounds. The reason why one device can not be apart from the other one for too long because microphone of devices are not good enough to record audio from long distance.

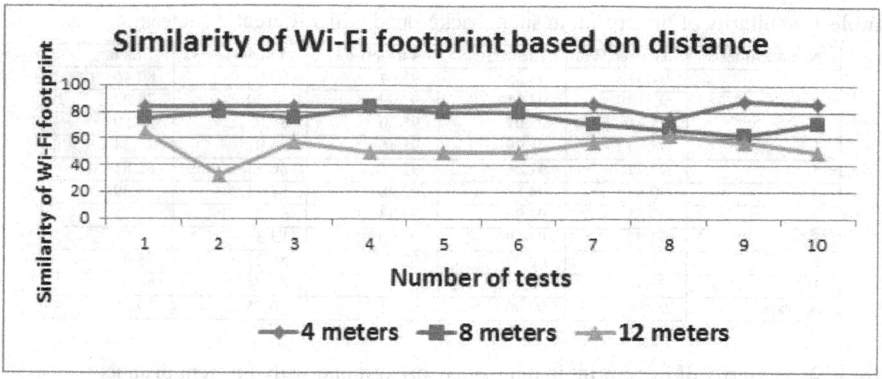

Fig. 7a

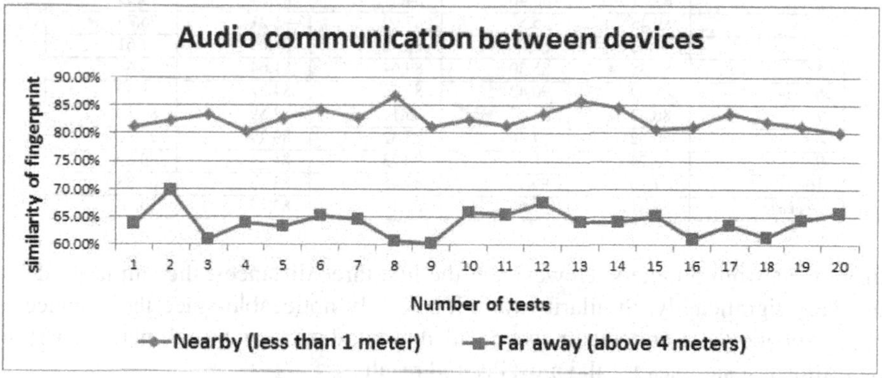

Fig. 7b

Fig. 7 Silent issue solutions: (a) Similarity of Wi-Fi fingerprint with *threshold* = 10 (b) Similarity of created fingerprints after establishing the audio communication between devices

6.5 Most Effected Distance Experiment

One of the reason why the distances from 0 to 4 meters between devices is preferred because it is the most effected distance. That means in the range of 0 to 4 meters, the audio surrounding each device efficiently help to determine each other. For this specific experiment, we collect the test data by separating two mobile devices from each other by the distance of 1, 2, 4, 5 and 7 meters with the devices microphone always pointing at the same direction. The sample tests for each distance are taken independently. Then one device is hold on, the other devices is moved circular around the first device with the radius 1, 2, 4, 5 and 7 meters, respectively. We do the test at hallway university (quite noise) and inside lab (silence). The results of similarity of fingerprints based on distance between devices are displayed in Table 1 and Table 2. From two tables, we can see that the fingerprint similarity percentage begins to

Table 1 Similarity of fingerprint in silent background with different distances

Number of tests	1 meter(%)	2 meters(%)	4 meters (%)	5 meters (%)	7 meters (%)
1	94,53	93,35	95,04	90,62	80,46
2	92,18	94,31	93,09	91,70	78,32
3	95,31	95,07	95,50	88,08	83,39
4	95,89	91,99	91,46	86,91	77,34
5	92,77	91,01	92,67	86,13	82,81
6	90,62	94,53	92,28	84,43	75,39
7	93,55	92,96	94,53	90,23	86,71
8	94,53	93,16	91,60	83,39	72,65
9	91,40	92,72	92,11	80,27	78,32
10	93,16	90,16	90,82	85,74	80,27
Average	93,39	92,96	92,91	86,75	79,56

Table 2 Similarity of fingerprint in quite noisy background with different distances

Number of tests	1 meter(%)	2 meters(%)	4 meters (%)	5 meters (%)	7 meters (%)
1	88,08	84,76	89,60	82,22	78,32
2	89,21	88,40	84,57	85,74	79,99
3	89,45	86,57	87,69	91,21	76,75
4	91,60	88,9	88,45	81,05	82,61
5	87,35	85,50	89,64	81,64	80,66
6	87,69	90,42	85,93	84,37	78,90
7	86,32	88,47	85,11	83,59	84,17
8	90,62	87,30	85,32	80,27	77,92
9	85,74	87,69	90,23	81,44	80,85
10	87,10	89,45	84,45	83,98	81,83
Average	88,31	87,53	87,1	83,45	80,20

drop down from 5 meters. However, at the first three distances, the similarity does not vary significantly. Similarity variation is only noticeable since the distance 5 meters or above. In conclusion, we decided to put devices within 4 meters apart in recording test samples for the most desired result.

7 Conclusion

In this paper, we proposed the method for iOS devices to utilize the background audio and establish secure communication channel. Audio fingerprint is generated based on fingerprint extraction scheme and then used as cryptographic key. Moreover, the proposed fuzzy commitment scheme enable the adaption of noise tolerance in fingerprint by altering the parameters of Reed-Solomon error correcting code. Several protocol to exchange data among devices are constructed such as Bluetooth, local network and through Internet. Also, we have studied how to detect a peak in a given time-series. Besides, Iphone indoor localization faces some difficult to implement because localization library in Apple library system are unavailable.

There are still some limitations we need to improve in the future research. First, most effected distance (within 4 meters) is small. Second, audio coming from unlimited source in outdoor environment is another challenge. Finally, silent background issue which was mentioned many times earlier still requires some extended features to evaluate.

Acknowledgments. This research is supported by research funding from Advanced Program in Computer Science, University of Science, Vietnam National University - Ho Chi Minh City.

References

1. Haitsma, J., Kalker, T.: A Highly robust audio fingerprint system (2002)
2. Wan, W., Yu, X., Yang, W., Deng, J.: Audio fingerprints based on spectral energy structure and NMF. In: Communication Technology (ICCT), pp. 1103–1106 (September 2011)
3. Blum, T., Keislar, D., Wheaton, J., Wold, E.: Content-Based Classification, Search, and Retrieval of Audio. IEEE Multimedia 3, 27–36 (1996)
4. Schrmann, D., Sigg, S.: Secure communication based on ambient audio (February 2013)
5. Dinda, P., Dick, R., Memik, G., Tarzia, S.: Indoor localization withouth infrastructure using the Acoustic Background Spectrum. In: Proceedings of the 9th International Conference on Mobile Systems, Applications, and Services, MobiSys 2011, pp. 155–168 (2011)
6. Wikipedia: Bump, `http://en.wikipedia.org/wiki/Bump_application` (accessed June 6, 2013)
7. Mayrhofer, R., Gellersen, H.: Spontaneous mobile device authentication based on sensor data. Information Security Technical Report 13(3) (2008)
8. Mayrhofer, R., Gellersen, H.: Shake well before use: Intuitive and secure pairing of mobile devices. IEEE Transactions on Mobile Computing 8(6) (2009)
9. Tuyls, P., Skoric, B., Kevenaar, T.: Security with Noisy Data. Springer (2007)
10. Li, Q., Chang, E.-C.: Robust, short and sensitive authentication tags using secure sketch. In: Proceedings of the 8th Workshop on Multimedia and Security, pp. 56–61. ACM (2006)
11. Miao, F., Jiang, L., Li, Y., Zhang, Y.-T.: Biometrics based novel key distribution solution for body sensor networks. In: IEEE Engineering in Medicine and Biology Society, pp. 2458–2461 (2009)
12. AlwaysLearn.com Website, A DFT and FFT TUTORIAL (2012), `http://www.alwayslearn.com/DFT%20and%20FFT%20Tutorial/DFTandFFT_BasicIdea.html`
13. Juels, A., Wattenberg, M.: A fuzzy commitment scheme. In: CCS 1999 Proceedings of the 6th ACM Conference on Computer and Communications Security, pp. 28–36 (1999)
14. Clarke, C.K.P.: Reed Solomon Error Correction. BBC Research & Development (July 2002)
15. Federal Information Processing Standards Publication 197, Announcing the Advanced Encryption Standard (AES) (November 26, 2001)
16. Palshikar, G.K.: Simple Algorithms for Peak Detection in Time-Series (2009)
17. Gelfand, S., Levitt, H.: Hearing: An Introduction to Psychological and Physiological Acoustics. Marcel Dekker, New York (2004)

Pomelo's Quality Classification Based on Combination of Color Information and Gabor Filter

Huu-Hung Huynh, Trong-Nguyen Nguyen, and Jean Meunier

Abstract. Vietnam is a country with strength in fruit trees, including many fruits well-known to the world, such as pomelo, dragon fruit, star apple, mango, durian, rambutan, longan, litchi and watermelon. However, the competitiveness and export of these fruits are low and incommensurate with the existing potential. To solve this problem, Vietnam is studying sustainable directions by investing in machinery for automation process to meet international standards. In this paper, we introduce an effective method for detecting surface defects of the pomelo automatically based on the combination of color information and Gabor filter. Our approach begins by representing the input image in HSV color space, computing the compactness based on the H channel, extracting texture parameters and using the K-nearest neighbor algorithm for quality classification. The proposed approach has been tested with high accuracy and is promising.

1 Introduction

Nowadays, some pomelo types in Viet Nam such as Nam roi and Green skin have high economic value and potential for exporting to Europe, Russia, USA, Japan, Philippines. However, to be exported to these difficult markets, they must be grown and harvested in accordance with GAP standards [1]. Thus, an automatic system for quality control during the export process is necessary. The problem of defect detection on the surface of the fruit has been studied by many researchers, especially in the countries that have high production levels of vegetables for export such as:

Huu-Hung Huynh · Trong-Nguyen Nguyen
DATIC, Department of Computer Science, University of Science and Technology, Danang, Vietnam
e-mail: hhhung@dut.udn.vn, ntnguyen.dn@gmail.com

Jean Meunier
DIRO, University of Montreal, Montreal, Canada
e-mail: meunier@iro.umontreal.ca

V.-N. Huynh et al. (eds.), *Knowledge and Systems Engineering, Volume 1*, Advances in Intelligent Systems and Computing 244,
DOI: 10.1007/978-3-319-02741-8_33,

Chile, South Korea, and India. However, Vietnam farmers cannot apply these solutions because of their high costs. In this paper, we propose an effective method to detect defects on surfaces of the pomelo via a computer combined with image processing techniques to improve productivity, reduce costs, save time, and support for packing process.

Our solution needs three basic steps to extract features for classifying a pomelo into the satisfactory class or unsatisfactory class. We represent the pomelo image in HSV color space. Then one feature is computed based on the H channel. The next five characteristics are calculated after applying a Gabor filter for the V channel. At the final step, the pomelo fruit is classified using the k-NN algorithm combined with the obtained features and some others.

2 Related Work

Some researches on fruit identification and quality evaluation have been done.

Angel Dacal-Nieto et al [2] conducted a potato quality assessment based on computer vision technology. Potatoes are segmented by converting the image to HSV color space. Then authors used gray level histogram and Gray-Level Co-occurrence Matrix (GLCM) to extract characteristics and put them in the training set. The algorithm k-NN with $k = 1$ was used, and this algorithm was optimized by combining with an ad-hoc algorithm to select some of the obtained characteristics.

The authors in [3] developed an algorithm to classify foods based on the shape and texture characteristics. They were extracted using morphological operations and GLCM, and then they were categorized by the mean-shift algorithm. Results were tested with 6 different foods: apple, banana, orange, peanut, pear and donuts, with 100 training samples for each. The average recognition accuracy is 97.6%.

Hetal N. Patel et al [4] introduced a method to recognize fruit on tree based on the characteristics: intensity, color, border, and direction. The input image is converted into HSV color space. Then the channel H and S are combined to represent the color characteristic of the input image. The channel V is used to estimate the intensity features. The direction component is extracted by using Gabor filters with four different orientations: $0^\circ, 45^\circ, 90^\circ, 135^\circ$. The boundary feature is detected based on the gradient method. Authors combined all the characteristics into a feature map, then used local thresholds to separate objects from the background. The accuracy is greater than 95%.

Panli HE [5] proposed a model to detect defects on the surface of fruits. The fruit is segmented by morphological operations and the Fourier transform is applied. The texture characteristic and gray level are calculated by GLCM. The SVM method is used to classify defects of the fruit.

Deepesh Kumar Srivastava [6] proposed a method for removing glare and detecting surface defects on fruit using Gabor filters. Input image is assumed to be dazzling, it is transformed to binary image using a threshold value of about 0.8-0.9, and then the glare is removed using some morphological operations. To detect defects, Gabor filters with the values ($\lambda, \theta, \phi, \gamma$, bandwidth)

= (8, [0°,30°,45°,60°,90°,120°,180°], 0, 0.5, 1) are applied to the input image after removing glare to segment the image.

The authors [7] introduced a method to detect defects on the surface of the fruit based on color characteristics. RGB color image is converted into HSI color space and each channel H, S, I is separated. These channels were put into a Gray Level Dependence Matrix (SGDM) to conduct texture calculation.

Md. Zahangir Alom and Hyo Jong Lee [8] proposed a segmentation method to detect disease in rice leaves. Input image is represented in the HSV color space, and then H channel is used to obtain some parameters based on statistical methods. The Gaussian model is used to calculate the average value and the difference between all the pixels in the input image. An optimal threshold is calculated based on this model; this threshold value is used to segment the image.

3 Proposed Approach

In this section, we propose our method for pomelo's quality classification. Our approach mainly uses the shape and surface texture information for characteristic extraction. The following will present the main contents in our method.

3.1 Color Space Representation

Pomelo pictures should be obtained via a high-resolution camera in order to get a clear image, reduce noise and blur, and facilitate image analysis. With each pomelo fruit, we capture two sides to be more fully analyzed.

The characteristics we use later are based on HSV color space because of its advantages. The HSV color space is quite similar to the way in which humans perceive color. The other models, except for HSL and HSI, define color in relation to the primary colors. The colors used in HSV can be clearly defined by human perception. Each channel will be computed to extract the shape and texture characteristics of the pomelo fruit. An example of pomelo image in HSV color space is shown below:

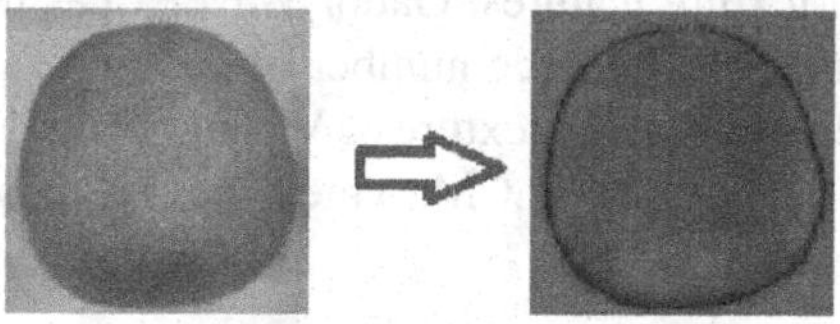

Fig. 1 A pomelo fruit in HSV color space

3.2 Shape Feature

In many real-time applications, the object boundary is often used to describe the shape which is one of the important parameters for the classification. In this approach, we use the compactness as a characteristic for pomelo's quality classification.

The HSV color image is separated into three-color channels to serve different purposes. The H-channel is selected to measure the shape parameter of the pomelo fruit. The H-channel image is converted to binary image by using Otsu thresholding method.

Fig. 2 Binary image from H channel in HSV color space

After this step, the next task is computing the area and perimeter of the object. The area is the number of pixels inside the object, and the perimeter is total pixels on the boundary of the object. The value of compactness is calculated by the following formula.

$$compactness = 4\pi \frac{area}{perimeter^2} \tag{1}$$

For a perfectly circular fruit, compactness = 1.

3.3 *Surface Texture*

The texture describes the attributes of the elements that constitute the object's surface, so the information extracted from the texture is important for object recognition.

In order to measure the texture characteristics, we propose the following model: after representing the image in HSV color space, the V channel is processed with Gabor filters to extract texture features. Gabor Wavelet has many different parameters, so we experimented with a large number of parameters to select the suitable values for the Gabor filter so that the texture of V channel image is most highlighted. The obtained image is put into the GLCM to measure texture values. The results are stored for object recognition.

3.3.1 Gabor Wavelet

The Gabor filter is a linear filter which is often used for edge detection. Frequency and orientation representations of Gabor filters are similar to those of the human visual system, and they have been found to be particularly appropriate for texture representation and discrimination. In the spatial domain, a 2D Gabor filter is a Gaussian kernel function modulated by a sinusoidal plane wave. The Gabor filters are self-similar: all filters can be generated from one mother wavelet by dilation and rotation. The Gabor function in the spatial domain has the following form [10]:

$$g_{\lambda,\theta,\phi,\sigma,\gamma}(x,y) = exp(-\frac{x'^2 + \gamma^2 y'^2}{2\sigma^2})cos(2\pi\frac{x'}{\lambda} + \phi) \quad (2)$$

$$x' = xcos(\theta) + ysin(\theta) \quad (3)$$

$$y' = -xsin(\theta) + ycos(\theta) \quad (4)$$

where λ represents the wavelength of the sinusoidal factor, θ represents the orientation of the normal to the parallel stripes of a Gabor function, ϕ is the phase offset, σ is the sigma of the Gaussian envelope and γ is the spatial aspect ratio, and specifies the ellipticity of the support of the Gabor function. Some results after applying Gabor filters to a pomelo fruit are presented below:

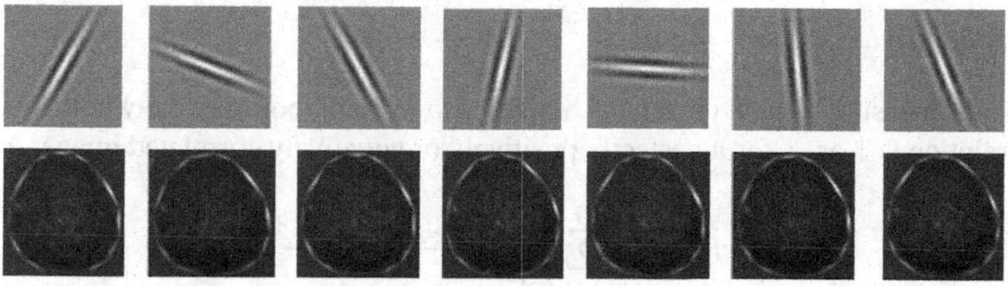

Fig. 3 Some pomelo images processed with Gabor filters with different orientations

3.3.2 Gray-Level Co-occurrence Matrix

A GLCM is a matrix or distribution that is defined over an image to be the distribution of co-occurring values at a given offset. Mathcmatically, a co-occurrence matrix *C* is defined over an *n*m* image *I*, parameterized by an offset (Δx, Δy), as:

$$C_{\Delta x,\Delta y}(i,j) = \sum_{p=1}^{n}\sum_{q=1}^{m} \begin{cases} 1, if \begin{cases} I(p,q) = i \\ I(p+\Delta x, q+\Delta y) = j \end{cases} \\ 0, otherwise \end{cases} \quad (5)$$

where *i* and *j* are the image intensity values of the image, *p* and *q* are the spatial positions in the image *I* and the offset (Δx, Δy) depends on the direction used θ and the distance at which the matrix is computed *d*. The value of the image originally referred to the grayscale value of the specified pixel. Any matrix or pair of matrices can be used to generate a co-occurrence matrix, though their main applicability has been in the measuring of texture in images.

Five features calculated from GLCM to be used in our approach are the energy, contrast, entropy, correlation and homogeneity. The calculation method is presented below, in which the GLCM *C* of the image is a *G* x *G* matrix and has *G* gray levels.

Energy

This feature represents the local homogeneity in the image. The value is high when the gray-scale pixel values are uniform. This value is in the range from 0 to 1.

$$Energy = \sum_{i=1}^{G}\sum_{j=1}^{G} C^2(i,j) \tag{6}$$

Contrast

This value is a measure of the intensity contrast between a pixel and its neighbor over the whole image. Contrast is 0 for a constant image. The contrast value is in the range $[0,(M-1)^2]$.

$$Contrast = \sum_{i=1}^{G}\sum_{j=1}^{G} (i-j)^2 C(i,j) \tag{7}$$

Correlation

This is a measure of how correlated a pixel is to its neighbor over the whole image. Correlation is 1 or -1 for a perfectly positively or negatively correlated image.

$$Correlation = \sum_{i=1}^{G}\sum_{j=1}^{G} \frac{(i-\mu_i)(j-\mu_j)C(i,j)}{\sigma_i \sigma_j} \tag{8}$$

where μ_i, μ_j and σ_i, σ_j respectively the mean values and standard deviations of the row i and column j of the matrix. The values are calculated as follows:

$$\mu_i = \sum_{i=1}^{G} i \sum_{j=1}^{G} C(i,j) \tag{9}$$

$$\mu_j = \sum_{j=1}^{G} j \sum_{i=1}^{G} C(i,j) \tag{10}$$

$$\sigma_i = \sum_{i=1}^{G} (i-\mu_i)^2 \sum_{j=1}^{G} C(i,j) \tag{11}$$

$$\sigma_j = \sum_{j=1}^{G} (j-\mu_j)^2 \sum_{i=1}^{G} C(i,j) \tag{12}$$

Entropy

Entropy is a measure of randomness. The highest entropy value is 1 when the elements of C are all equal; the lowest is 0 if all values in the matrix are different. The entropy is computed as follows:

$$Entropy = -\sum_{i=1}^{G}\sum_{j=1}^{G} C(i,j) log C(i,j) \tag{13}$$

Homogeneity

This is a value that measures the closeness of the distribution of elements in the GLCM to the GLCM diagonal. The homogeneity value is in the range [0, 1]. Homogeneity is 1 for a diagonal GLCM.

$$Homogeneity = \sum_{i=1}^{G}\sum_{j=1}^{G} \frac{C(i,j)}{1+|i-j|} \tag{14}$$

Table 1 shows the values of these features for the images presented in Fig. 3.

Table 1 Obtained features when using Gabor filter with 7 different orientations

Orientation	Entropy	Contrast	Correlation	Energy	Homogeneity
0°	0.065914	0.0414	0.9343	0.6820	0.9793
30°	0.065914	0.0470	0.9290	0.6273	0.9765
45°	0.065914	0.0402	0.9348	0.6966	0.9799
60°	0.065914	0.0408	0.9335	0.6741	0.9796
90°	0.065914	0.0397	0.9334	0.7035	0.9802
120°	0.065914	0.0429	0.9362	0.6624	0.9786
180°	0.065914	0.0399	0.9334	0.6995	0.9801

3.4 *Quality Classification*

3.4.1 Feature Vector

The feature vector used for classification is created by combining the 12 values below: three mean values of three channels in HSV color space, three standard deviations of these channels, compactness, entropy, energy, contrast, correlation and homogeneity.

3.4.2 Classification Using K-nearest Neighbor Algorithm

In pattern recognition, the k-nearest neighbor (k-NN) algorithm is a non-parametric method for classifying objects based on closest training examples in the feature space. The k-NN algorithm is simple: an object is classified by a majority vote of its neighbors, with the object being assigned to the class most common amongst its k nearest neighbors (k is a positive integer, typically small).

The classification is done with two classes (satisfactory and unsatisfactory) based on the k-NN algorithm. New samples are computed to get 12 values for the feature vector, and the distances from these samples to all stored examples are calculated.

The *k* minimum distances are chosen, and samples are classified into corresponding classes. After testing with many numbers, the value of *k* was chosen as 1.

3.4.3 Counting Surface Defects

We found that the surface defects are shown clearly on the a* channel in L*a*b* color space. So, to improve the accuracy of our approach, we developed a method that counts the number of surface defects:

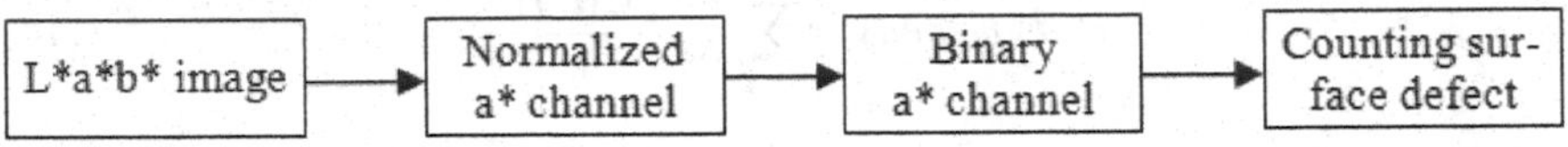

Fig. 4 Steps for counting surface defects

The pomelo fruit image is converted into L*a*b* color space, and the a* channel is used for surface defect extraction. First, the a* channel of the pomelo fruit image in L*a*b* color space is normalized. Then we use the Otsu thresholding method to get the binary image. Finally, the number of surface defects is determined by counting the connected components (regions) in the obtained image. This process is shown in the Fig. 5.

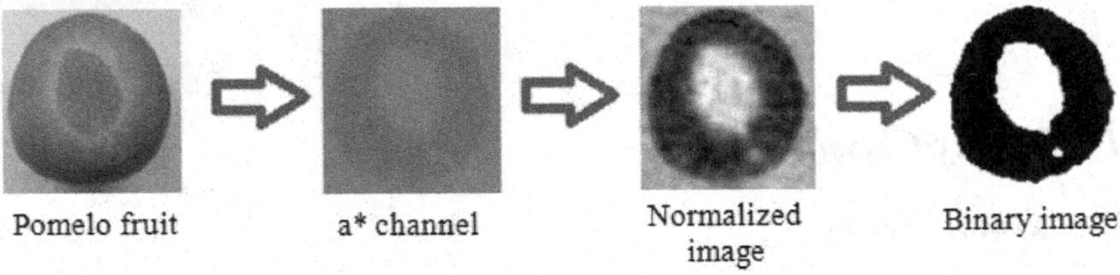

Fig. 5 Two surface defects are detected based on the a* channel

4 Experimental Results

We used the Logitech 9000 webcam for this research. In experiments, distance of the webcam to fruits was about 1.8-2.2m. Our system is implemented in Matlab and C++ using the OpenCV library. Training data consisted of 100 samples. 50 fruits had good quality and meet the GAP requirement. The remaining ones had some surface defects such as bruises, holes, rot, scratches, scars and some other unusual marks. Testing data included 75 images collected from different sources. Some images are shown in Fig 6, 7.

After testing with many different Gabor parameters, we decided to select the following values to obtain high accuracy in classification: $\lambda = 8$, $\phi = [0, \pi/2]$,

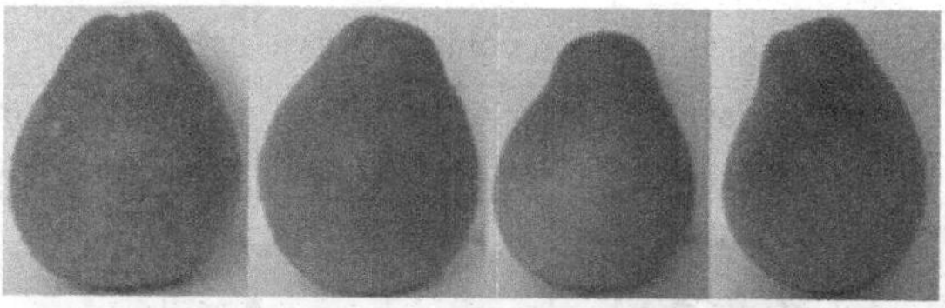

Fig. 6 Some good-quality pomelo fruits

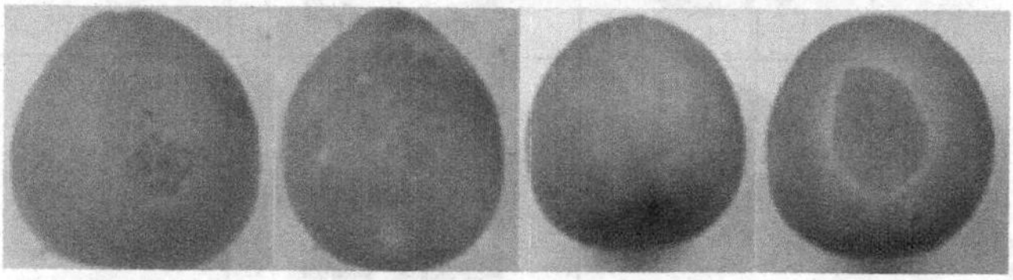

Fig. 7 Some fruits with bruises, scratches, scars on the surface

$\gamma = 0.5$ and $\theta = 60$. In binary (two classes) classification problems, it is helpful to choose k to be an odd number as this avoids tied votes. The experimental results with different k values are shown as the Table 2.

Table 2 Classification results with different k values

k	Training	Testing	True classification	Positive rate (%)
1	100	75	73	97.3
3	100	75	71	94.7
5	100	75	70	93.3
7	100	75	70	93.3
9	100	75	65	86.7

The above table shows that the classification results are highly accurate when the k value is small. So the best k value for our method is 1.

Testing results of some related approaches on the above samples are presented in the Table 3.

Table 3 Testing results with some other approaches

Approach	Testing	True classification	Positive rate (%)
[5]	75	70	93.3
[6]	75	68	90.7
[7]	75	71	94.7

For surface defect counting, the obtained results are similar to manual counting with the eye. The results are shown in the Table 4.

Table 4 Count the surface defects of seven pomelo fruits

Pomelo fruit	Binary a* channel	Region labeling on surface	Number of detected defects
			12
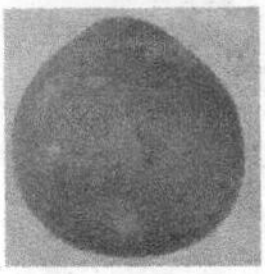			22
			4
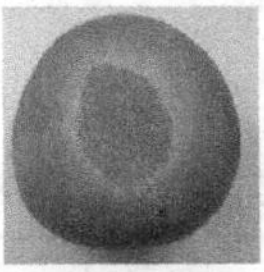			15
			7
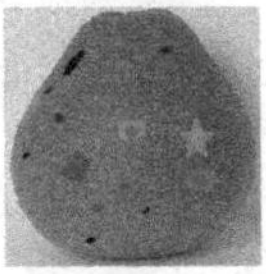			13
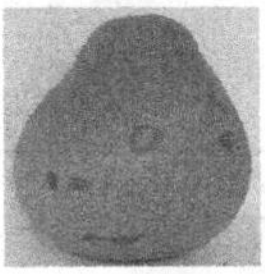			9

5 Conclusion and Discussion

In Vietnam, thousands tons of pomelo are exported each year. The pomelo becomes one of the largest goods brands of Vietnam. The earned profit contributes greatly to the economy of Vietnam. In this paper, a new approach for serving quality pomelo fruit classification is proposed. It helps to improve productivity, reduce costs, save time, and support packing process. This method is based on the compactness and surface texture features extracted from color information in HSV space. The compactness is computed using a size-independent measurement method. With surface texture, five characteristics were used: energy, contrast, correlation, entropy and homogeneity. They are calculated based on gray-level co-occurrence matrix obtained after applying a 2D Gabor filter for the V channel. By combining these features with mean values and standard deviations of three channels (H, S, V), we have the feature vector for classification. The k-nearest neighbor algorithm is used to classify the pomelo fruit into the satisfactory class or unsatisfactory class. In addition, we developed a method to count the number of surface defects using the a* channel in L*a*b* color space. It helps to reduce the classification error rate and improve the accuracy of our approach. Currently, this method is quite effective to apply on fruits, which have surface texture similar to pomelo such as orange, lemon, and tangerine. As further work, our approach will be improved to be applied to other fruits.

Acknowledgments. This work was supported by the DATIC, Department of Computer Science, University of Science and Technology, The University of Danang, Vietnam and the Natural Sciences and Engineering Research Council of Canada (NSERC).

References

1. Good Agricultural Practices, `http://en.wikipedia.org/wiki/Good_Agricultural_Practices`
2. Dacal-Nieto, A., Vzquez-Fernndez, E., Formella, A., Martin, F., Torres-Guijarro, S., Gonzlez-Jorge, H.: A genetic algorithm approach for feature selection in potatoes classification by computer vision. In: IECON 2009. IEEE, Portugal (2009)
3. Deng, Y., Qin, S., Wu, Y.: An Automatic Food Recognition Algorithm with both Shape and Texture Information. In: PIAGENG 2009. SPIE, China (2009)
4. Hetal, N., Patel, R.K., Jain, M.V.: Fruit Detection using Improved Multiple Features based Algorithm. IJCA 13(2) (January 2011)
5. Panli, H.E.: Fruit Surface Defects Detection and Classification Based on Attention Model. JCIS 2012 8(10) (2012)
6. Srivastava, D.K.: Efficient fruit defect detection and glare removal algorithm by anisotropic diffusion and 2d gabor filter. IJESAT 2(2), 352–357 (2012)
7. Kim, D.G., Burks, T.F., Qin, J., Bulanon, D.M.: Classification of grapefruit peel diseases using color texture feature analysis. IJABE 2(3) (September 2009)
8. Alom, M.Z., Lee, H.J.: Gaussian Mean Based Paddy Disease Segmentation. In: ICMV 2010, pp. 522–525 (2010)
9. Sun, D.-W.: Computer vision technology for Food Quality Evalution. Elsevier (2008)
10. Seo, N.: Texture Segmentation using Gabor Filters. In: ENEE 731 Project (November 8, 2006)
11. Jha, S.N.: Nondestructive evaluation of Food Quality Theory and Practice. Springer (2010)

Local Descriptors without Orientation Normalization to Enhance Landmark Regconition

Dai-Duong Truong, Chau-Sang Nguyen Ngoc, Vinh-Tiep Nguyen,
Minh-Triet Tran, and Anh-Duc Duong

Abstract. Derive from practical needs, especially in tourism industry; landmark recognition is an interesting and challenging problem on mobile devices. To obtain the robustness, landmarks are described by local features with many levels of invariance among which rotation invariance is commonly considered an important property. We propose to eliminate orientation normalization for local visual descriptors to enhance the accuracy in landmark recognition problem. Our experiments show that with three different widely used descriptors, including SIFT, SURF, and BRISK, our idea can improve the recognition accuracy from 2.3 to 12.6% while reduce the feature extraction time from 2.5 to 11.1%. This suggests a simple yet efficient method to boost the accuracy with different local descriptors with orientation normalization in landmark recognition applications.

1 Introduction

In context-aware environment, applications or services are expected to wisely recognize and understand current user contexts to generate adaptive behaviors. Context information can be classified into external and internal contexts. External contexts include information can be about location, time, or environmental factors such as light condition, temperature, sound, or air pressure. Internal contexts are related to information that mostly specified by users, e.g. events, plans, or even emotional states. Various methods and systems are proposed to capture and process the wide variety of context information, such as location-based services using network [1] or GPS [2] data, sensor-based [3], audio-based applications [4], or visual based

Dai-Duong Truong · Chau-Sang Nguyen Ngoc · Vinh-Tiep Nguyen · Minh-Triet Tran
Faculty of Information Technology, University of Science, VNU-HCM, Vietnam
e-mail: {nvtiep,tmtriet}@fit.hcmus.edu.vn

Anh-Duc Duong
University of Information Technology, VNU-HCM, Vietnam
e-mail: ducda@uit.edu.vn

V.-N. Huynh et al. (eds.), *Knowledge and Systems Engineering, Volume 1*,
Advances in Intelligent Systems and Computing 244,
DOI: 10.1007/978-3-319-02741-8_34,

systems, e.g. visual search [5], gesture recognition [6], natural scenes categorization [7], template matching [8], etc. With the continuous expansion of capability of personal devices and achievements in research, more and more approaches are proposed to explore contexts to provide better ways to interact with users.

The portability of mobile devices helps people access information immediately as needed. This property makes mobile devices become an essential and promising part of context-aware systems. Derive from social practical demands, especially in tourism industry; landmark recognition is one of the problems with increasing needs. This problem is a particular case of natural scene classification but limits only for places of interest. Landmark recognition applications on mobile devices usually use a general architecture in which the first step is extracting features from captured images. The two most popular approaches for this step are dense sampling [9] and local features extraction. Dense sampling can yield a high accuracy but require a high computational cost which may not be appropriate for mobile devices. Our research focuses on providing a simple method to boost the accuracy of landmark recognition using the local feature approach.

Most of local feature based landmark recognition systems [10, 11, 12] take the traditional approach using local descriptors with orientation normalization. However we show that it is not a good choice for landmark recognition problem. Our idea is inspired by the result of Zhang [13] which shows that descriptors equipped with different levels of invariance may not always outperform the original ones. We take a further step to conduct experiments to prove that in landmark recognition problem, rotation invariance not only is unnecessary but also may decrease the accuracy. This can be explained by the loss of discriminative information of a descriptor during the computation process to make it rotation invariant. It should be noticed that users tend to align their cameras so that images are usually captured in landscape or portrait orientation. This makes rotation invariance become not much efficient in this case.

In this paper, we present our proposed idea then experiments to show that the elimination of the orientation normalization step can enhance the accuracy of common local features. We test three different descriptors, including SIFT, SURF, and BRISK, that require identifying dominant orientation(s) and orientation normalization. We use Bag of Visual Words (BoVWs) [14] which is the basic framework of many state-of-the-art methods in the problem of landmark recognition. Experiments are conducted on two standard datasets: Oxford Buildings [15] and Paris[16] datasets. The remainder of our paper is organized as follows. In section 2, we briefly introduce landmark recognition problem and some state-of-the-art approaches using BoVWs model. In section 3, we describe our proposal. Experimental results are presented and discussed in section 4 while the conclusion and future work are in section 5.

2 Background

One of the major goals of intelligent systems is to provide users natural and simple ways to interact with while still get necessary inputs to generate the response. In order to achieve that, these systems need to be equipped with the capability to recognize external environment and context to generate appropriate behaviors. Visual data is one of the most informative data source for intelligent systems. A popular problem in vision based systems is static planar object recognition with a wide range of applications, especially in augmented reality [17, 18]. The basic solution for this problem is directly matching a query image with existing templates [8]. Different measures can be used to obtain different levels of robustness and distinctiveness. One limitation of template matching is that the relative positions of pixels to be matched should be preserved. A small change in viewpoint may lead to a wrong match. To overcome this disadvantage, proposed methods are mostly focus on efficient ways to detect key points [19, 20], interest points that are stable to brightness, illumination, and transformation variations, and describe them [21]. However, in problems related to scene or object classification with a wide intra-class variation, these methods are not good enough to provide an adequate accuracy. State-of-the-art systems usually use highlevel presentations to eliminate this intra-class variation [9, 14, 22].

To deal with the wide intra-class variation, state-of-the-art approaches in landmark recognition systems obtain a high-level presentation of features in images. The method is called Bag of Features (BoFs) or Bag of Visual Words (BoVWs) [14]. This idea is borrowed from text retrieval problem where each document is described by a vector of occurrence counts of words. In specific, a codebook containing a list of possible visual words corresponding to common patches in scenes is built. Each image will then be described by the histogram of distribution of these visual words. Using a histogram, BoVWs loses the information about spatial distribution of these visual words. Spatial Pyramid Matching (SPM) [22] solves this drawback by introducing a pyramid of histograms in which each histogram captures the distribution of visual words in a specific region at a particular resolution. Locality-constrained Linear Coding (LLC) [9] takes a further step to loosen the constraint that each feature can only belong to a single visual word. In LLC, each feature can belong to several different visual words in its vicinity.

In existing methods of image classification and landmark recognition, uniform dense sampling or local detectors are used to extract key points from images. Despite the advantage of high accuracy, dense sampling requires a high computational cost which is impractical to mobile devices. Therefore, local detector is of preference. The traditional way of using local detectors is taking features with as much invariance as possible. Among the invariant properties of a local feature, rotation invariance is considers an essential property. That is the reason why the orientation normalization step is used in calculating descriptors of popular local features such as SIFT, SURF, and BRISK.

In landmark recognition applications, each landmark appears in users images with minor changes in orientation. Besides, a gyroscope, which is equipped on almost every smart phone, allows estimating the orientation of the image easily. These factors make the rotation invariance of local feature become redundant. Inspired by the result of Zhang [13] which shows that descriptors equipped with different levels of invariance may not always outperform the original ones, combine with two observed properties of landmark recognition problem mentioned above, we propose the idea of eliminating orientation information in extracting local features. In literature, the idea of eliminating orientation information used to be applied in SURF [23]. However, the motivation of the authors is to optimize speed which is different from our motivation. Georges Baatz [24] also conducts an experiment to compare upright-SIFT, SIFT with zero-orientation, with traditional SIFT and concludes that upright-SIFT give the better performance than SIFT. Nevertheless, this experiment is conducted on a dataset containing landmarks whose orientations are normalized to exactly zero. Clearly, this dataset is far different from reality and also cause no rotation difficulty to upright- SIFT. In our experiments, we use standard datasets whose images are collected from real life to prove that in general, eliminating orientation information gives an enhancement in accuracy with variety of local features (SIFT, SURF, and BRISK).

3 Proposed Method

In order to present and illustrate our proposed idea, we need to put the idea in the context of a specific system so that the efficiency of the proposed idea can be evaluated. Through approaches mentioned in section 2, BoVWs is widely used as the core framework for many state-of-the-art systems. Therefore, we also use BoVWs in our proposal of using local features without orientation normalization to recognize landmarks.

In section 3.1, we briefly describe the BoVWs method. We make a small modification in phase 1 local descriptors extraction to integrate the idea. In section 3.2, we describe the orientation normalization processes of common local features and explain why they cause different level of discriminative information loss. In section 3.3, we describe our system in details.

3.1 Bag of Visual Words (BoVWs)

Basically, BoVWs can be divided into six small phases which are described in specific below.

Phase 1: local descriptors extraction. Local features from all images in the training set are extracted. Depending on the problem requirements, various local features can be used to obtain the local descriptors of each image.

Phase 2: codebook building. Every descriptor extracted from the training set is clustered into k clusters. Then, descriptors in one cluster will be represented by the cluster centroid. A centroid, which can be seen as a visual word, describes the most common features of descriptors inside the cluster that are frequently repeated in the images. The set of these visual words forms a codebook. In most systems, k-means is used in the clustering phase. The higher the value of parameter k is, the more discriminative properties of descriptors are preserved. Phase 1 and phase 2 of BoVWs method are illustrated in Fig. 1.

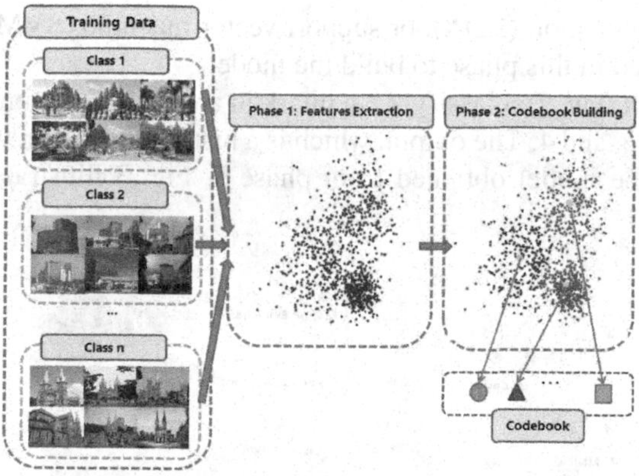

Fig. 1 Features extraction phase and codebook building phase of BoVWs

Phase 3: bag of visual words building. Each local descriptor is characterized by its most similar visual word (nearest in distance). Then, instead of describing an image by all of its local descriptors, the image is presented by a set of visual words. The new representation of an image is conventionally called bag of visual words. This phase is illustrated in Fig. 2.

Phase 4: pooling. We do not stop at representing an image by a list of visual words but keep building a higher-level presentation. A histogram, which has the equal size with the codebook, is taken by counting the number of times each visual word appears in an image. The normalized histogram vectors are then used as the input for the building model phase.

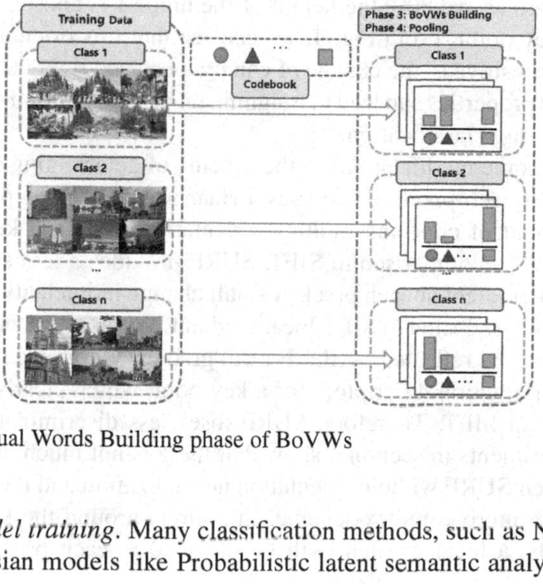

Fig. 2 Bag of Visual Words Building phase of BoVWs

Phase 5: model training. Many classification methods, such as Nave Bayes, Hierarchical Bayesian models like Probabilistic latent semantic analysis (pLSA) and latent Dirichlet allocation (LDA), or support vector machine (SVM) with different kernel, can be used in this phase to build the model.

Phase 6: prediction. To classify a landmark in an image, that image needs to go through phase 1, 3, and 4. The output, which is a histogram vector, will be predicted the label using the model obtained from phase 5. Fig. 3 illustrates this phase in details.

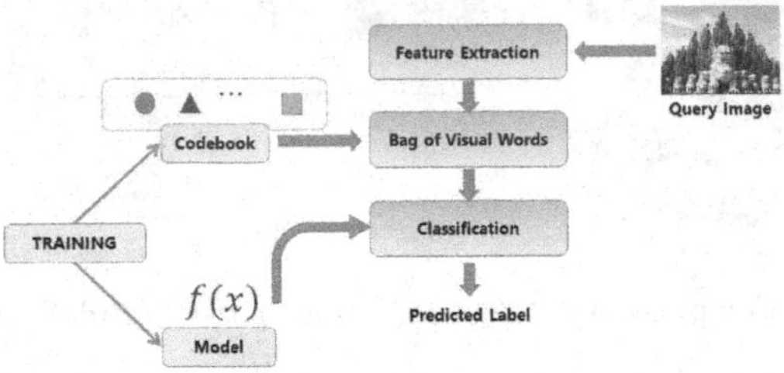

Fig. 3 Predicting a new landmark using BoVWs model

3.2 Orientation Normalization Process

Let us take a further look into the orientation identifying schemes of common local features to understand why it causes the loss of discriminative information. With SIFT, at each pixel in the neighbor region of each key point, the gradient magnitude and orientation are computed using pixel differences. The orientation space is divided into equal bins. Each neighbor pixel votes an amount determined by its gradient magnitude weighted with a Gaussian centered at the key point for the bin to which its gradient orientation belongs. The highest peaks in the histogram along with peaks that are above 80% the height of the highest is chosen to be the orientations of that key point. From Fig. 4, we can see that this orientation identifying scheme is quite sensitive to the change of camera pose and light condition. A small variance of these properties can lead to a significant change of dominant orientations and result in a wrong classification.

SURF makes some modifications in the scheme of determining dominant orientation of SIFT. At each pixel, SURF uses a Haar-wavelet to compute the gradient vector. And instead of choosing multiple orientations for each key point, SURF chooses only one. In comparison to SIFT, SURF provides a less sensitive method to compute local gradient at each pixel. A small change in intensity value of a pixel can be immediately reflected in SIFT local gradient. Whereas, it needs to be a trend of a local region to be reflected in SURF local gradient. Moreover, SURF does not yield different orientation descriptors for a key point which clearly makes it more discriminative than SIFT. Therefore, SURF loses less discriminative information than SIFT. Experiments in section 4 show that there is not much difference in performance between SURF without orientation normalization and the original.

BRISK uses a more complex scheme. In a circle around the key point, which contains n pixels, a local gradient will be obtained at each pair of pixels (from

Fig. 4 Dominant orientations detected by SIFT of the same landmark spots at different viewpoints and light conditions

$n(n+1)/2$ pairs) instead of each pixel like SIFT or SURF. The authors smooth the intensity values of two points in a pair by a Gaussian with σ proportional to the distance between the points. The local gradient is computed using smooth intensity differences. Instead of using bins partition and voting scheme, BRISK directly computes the orientation of the key point by taking the average local gradients of the pair of pixels whose distance over a threshold. Therefore, BRISK prefers long-distance pairs than short-distance pairs. It makes BRISK gradient become less local. This leads to a huge loss of discriminative information which can be seen in section 4.

In conclusion, the rotation invariance is not necessary in the problem of landmark recognition. Moreover, rotation invariant descriptor might reduce the classification performance. Besides, different orientation identification schemes cause different levels of loss of discriminative information. In order to confirm our hypothesis, in the experiments section, we test through SIFT, SURF, and BRISK on the Oxford Buildings and the Paris dataset.

3.3 *Our Specific System*

As mention above, we use the BoVWs model to evaluate the performance of our proposal. In phase 1, we respectively detect key points using detectors of SIFT, SURF, and BRISK. With the set of key points detected by SIFT detector, we describe them by two ways. The first way is using the original SIFT descriptor. The second way is using SIFT descriptor but ignoring the step of rotating the patch to its dominant orientation. We also eliminate key points differed only by their dominant orientation. After this phase, with the set of SIFT key points, we obtain two sets of descriptors: original SIFT descriptors and SIFT without orientation normalization descriptor. In a similar way, with each set of SURF and BRISK key points, we have two sets of descriptors. In phases 2, we use k-means clustering with k-means++ algorithm for choosing initial centroids. We test through different sizes of codebook which range from 512 to 4096 with step 512. We use SVM in phase 5 to train the model.

Multiple binary one-against-one SVM models were obtained to form a multi-class SVM model.

4 Experiments Result

In this session, we report the results of our idea when applied to BoVWs. Experiments are conduct on two standard datasets for landmark recognition problem: Oxford Buildings and Paris. We test through three different local descriptors, which are SIFT, SURF, and BRISK, with varieties of codebook sizes to confirm the efficiency of eliminating orientation information.

4.1 Oxford Buildings

The Oxford Buildings dataset [15] consists of 5062 images from 11 different landmarks. These images range between indoor and outdoor scenes with a great variation in light conditions and viewpoints. Landmarks in some images are difficult to identify because of cluttered backgrounds and occlusions. The similarity of buildings in the dataset even make the classification becomes harder. Some images from the dataset are presented in Fig. 5.

Fig. 5 Some images from the filtered Oxford Buildings dataset

Because images in the Oxford Buildings dataset appear both indoor and outdoor, we filter out the dataset to keep only outside scenes which are suitable for the problem of landmark recognition. All the images in the filtered dataset are resized to be no larger than 640×480 pixels for faster computation. We randomly divide the filtered dataset into 2 subsets: 80% for training purpose and 20% for testing. The codebook and the classification model are built on the training set while samples from test set are used to compute the testing accuracy. We repeatedly run this process 5 times. The average testing accuracy is taking for the final result.

We denote SIFT*, SURF*, BRISK* respectively are SIFT, SURF, BRISK descriptor without orientation information.

Table 1 Accuracy (%) of descriptors with and without orientation information on the Oxford Buildings dataset

	Codebook Size							
Descriptor	512	1024	1536	2048	2560	3072	3584	4096
SIFT	71.80	74.36	75.21	76.92	81.20	76.92	78.63	76.92
SIFT*	**79.83**	**79.83**	**81.54**	**81.37**	**82.05**	**81.37**	**81.37**	**81.20**
SURF	80.34	80.34	81.20	82.91	82.05	82.91	85.47	83.76
SURF*	**81.20**	**82.90**	**83.76**	**85.47**	**88.03**	**87.18**	**88.03**	**87.18**
BRISK	51.45	54.87	55.56	59.83	57.61	57.44	57.44	55.21
BRISK*	**67.01**	**70.43**	**70.43**	**70.09**	**67.69**	**69.74**	**67.18**	**67.86**

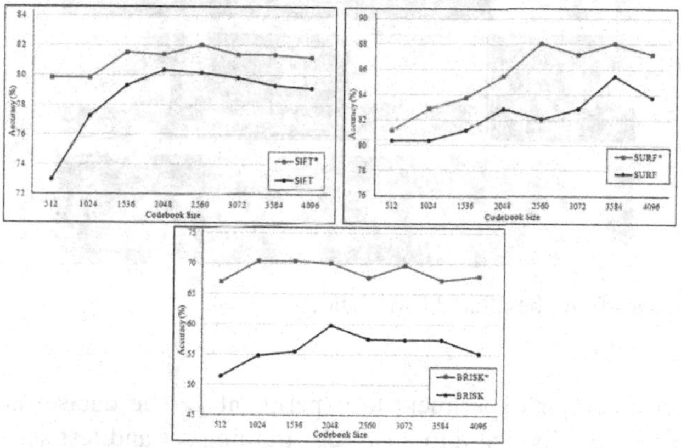

Fig. 6 The experiment result on the Oxford Buildings dataset

The experiment result shows that with or without orientation information, SURF yields the highest precision. On the other hand, BRISK gives the lowest recognition accuracy. Even though landmarks from the dataset do not always appear in the same orientation (Fig. 5), the experiment shows that descriptors without orientation normalization still yield a better performance in comparison to orientation-invariant ones. Table 1 and Fig. 6 show that this elimination gives a remarkable enhancement: on average, it helps boost the accuracy about 4.6% on SIFT, 3.1% on SURF, and 12.6% on BRISK. From the amount of classification performance enhancement of each descriptor, we can conclude that SURF orientation identifying scheme is the one causing least loss of discriminative information. In contrast, BRISK causes a huge loss which means the proposal can give a significant improvement to BRISK. In the best case, it can even raise the accuracy about 15.6%. Another conclusion can be derived from the experiment is that larger codebook size does not always increase the performance while it costs more time for building codebook and training model. In this case, the best codebook size lies between 2048 and 2056.

4.2 Paris

In a similar way to the Oxford buildings dataset, the Paris dataset [16] is collected from Flickr by searching for particular Paris landmarks. The dataset consists of 6412 images from 10 classes. In comparison to the Oxford Buildings dataset, the Paris dataset presents an easier challenge. Landmarks are less obscured. Also, the dissimilarity between classes is quite clear. Some images from the dataset are presented in Fig. 7.

Fig. 7 Some images from the filtered Paris dataset

We conduct a similar experiment to experiment 1. The dataset is filtered and resized and randomly divided into 2 subsets: training set and test set. The classification performances of 3 pairs of descriptors are respectively examined on a variety of codebook sizes.

Table 2 Accuracy (%) of descriptors with and without orientation information on the Paris dataset

	Codebook Size							
Descriptor	512	1024	1536	2048	2560	3072	3584	4096
SIFT	82.19	85.29	85.48	86.84	85.81	85.74	86.32	86.32
SIFT*	**86.45**	**87.36**	**87.81**	**89.23**	**89.03**	**89.03**	**88.90**	**89.68**
SURF	**83.87**	87.74	88.07	87.74	89.68	89.03	**90.32**	**89.68**
SURF*	83.55	**88.07**	**89.03**	**88.71**	**90.00**	**90.00**	89.36	89.03
BRISK	58.97	61.42	61.03	62.77	61.29	61.68	61.55	60.39
BRISK*	**71.03**	**72.71**	**72.65**	**74.58**	**75.10**	**74.97**	**75.42**	**74.84**

Through Table 2 and Fig. 8 we can see that the proposal continue to bring a remarkable enhancement: about 2.9% on SIFT, 0.3% on SURF, and 12.8% on BRISK. It is not surprising that all three descriptors yield better results on the Paris dataset

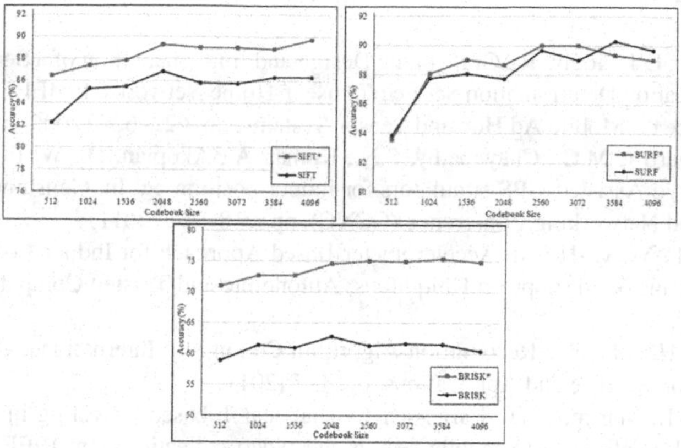

Fig. 8 The experiment result on the Paris dataset

than on the Oxford Buildings dataset. This confirms the hypothesis that the Paris dataset introduces an easier challenge than the Oxford Buildings dataset. Besides, it can be seen form Fig.8 that eliminating orientation information does not always enhance the performance of SURF. However, in most of the cases, it helps boost the accuracy. Moreover, time to extract feature can be save up to 2.5 to 11.1% depend on specific feature.

5 Conclusion

In this paper, we propose the idea of eliminating orientation normalization in calculating visual local descriptors to be applied in landmark recognition problem. The proposal can improve the overall recognition performance of different commonly used local descriptors, such as SIFT, SURF, and BRISK, with a remarkable improvement (12.6% in the best case) while cut down the duration for extracting features. This provides a simple way to boost the efficiency of landmark recognition systems in general.

The results in this paper encourage us to further study and develop landmark recognition systems with more advanced and complex methods, such as LLC or SPM. Besides, we also study to apply results in the field of neuroscience, such as autoencoder, to enhance the accuracy of landmark recognition systems.

Acknowledgment. This research is partially supported by research funding from Advanced Program in Computer Science, University of Science.

References

1. Ahmad, S., Eskicioglu, R., Graham, P.: Design and Implementation of a Sensor Network Based Location Determination Service for use in Home Networks. In: IEEE International Conference on Mobile Ad Hoc and Sensor Systems, pp. 622–626 (2006)
2. Sundaramurthy, M.C., Chayapathy, S.N., Kumar, A., Akopian, D.: Wi-Fi assistance to SUPL-based Assisted-GPS simulators for indoor positioning. In: Consumer Communications and Networking Conference (CCNC), pp. 918–922 (2011)
3. Hsu, C.-H., Yu, C.-H.: An Accelerometer Based Approach for Indoor Localization. In: Symposia and Workshops on Ubiquitous, Autonomic and Trusted Computing, pp. 223–227 (2009)
4. Jarng, S.: HMM Voice Recognition Algorithm Coding. In: International Conference on Information Science and Applications, pp. 1–7 (2011)
5. Adamek, T., Marimon, D.: Large-scale visual search based on voting in reduced pose space with application to mobile search and video collections. In: IEEE International Conference on Multimedia and Expo, pp. 1–4 (2011)
6. Wilhelm, M.: A generic context aware gesture recognition framework for smart environments. In: International Conference on Pervasive Computing and Communications Workshops, pp. 536–537 (2012)
7. Devendran, V., Thiagarajan, H., Wahi, A.: SVM Based Hybrid Moment Features for Natural Scene Categorization. In: International Conference on Computational Science and Engineering, pp. 356–361 (2009)
8. Dai-Duong, T., Chau-Sang, N., Vinh-Tiep, N., Minh-Triet, T., Anh-Duc, D.: Realtime arbitrary-shaped template matching process. In: 12th International Conference on Control Automation, Robotics and Vision, pp. 1407–1412 (2012)
9. Wang, J., Yang, J., Yu, K., Lv, F., Huang, T., Gong, Y.: Locality-constrained linear coding for image classification. In: Conference on Computer Vision and Pattern Recognition (CVPR), San Francisco, CA (2010)
10. Chen, T., Li, Z., Yap, K.-H., Wu, K., Chau, L.-P.: A multi-scale learning approach for landmark recognition using mobile devices. In: 7th International Conference on Information, Communications and Signal Processing, ICICS 2009, Macau (2009)
11. Bandera, A., Marfil, R., Vzquez-Martn, R.: Incremental Learning of Visual Landmarks for Mobile Robotics. In: 20th International Conference on Pattern Recognition (ICPR), Istanbul (2010)
12. Lee, L.-K., An, S.-Y., Oh, S.-Y.: Efficient visual salient object landmark extraction and recognition. In: International Conference on Systems, Man, and Cybernetics (SMC), Anchorage, AK (2011)
13. Zhang, J., Marszalek, M., Lazabnik, S.: Local features and kernels for classification of texture and object categories: A comprehensive study. In: Conference on Computer Vision and Pattern Recognition Workshop, CVPRW 2006 (2006)
14. Fei-Fei, L., Perona, P.: A Bayesian Hierarchical Model for Learning Natural Scene Categories. In: Computer Vision and Pattern Recognition, pp. 524–531 (2005)
15. The Oxford Buildings Dataset, `http://www.robots.ox.ac.uk/~vgg/data/oxbuildings/`
16. The Paris Dataset, `http://www.robots.ox.ac.uk/~vgg/data/parisbuildings/`
17. Shin, C., Kim, H., Kang, C., Jang, Y., Choi, A., Woo, W.: Unified Context-Aware Augmented Reality Application Framework for User-Driven Tour Guides. In: International Symposium on Ubiquitous Virtual Reality (ISUVR), Gwangju (2010)

18. Chen, D., Tsai, S., Hsu, C.-H., Singh, J. P., Girod, B.: Mobile augmented reality for books on a shelf. International Conference on Multimedia and Expo (ICME), Barcelona (2011)
19. Nistér, D., Stewénius, H.: Linear Time Maximally Stable Extremal Regions. In: Forsyth, D., Torr, P., Zisserman, A. (eds.) ECCV 2008, Part II. LNCS, vol. 5303, pp. 183–196. Springer, Heidelberg (2008)
20. Rosten, E., Porter, R., Drummond, T.: Faster and Better: A Machine Learning Approach to Corner Detection. IEEE Transactions on Pattern Analysis and Machine Intelligence 33(1), 105–119 (2010)
21. Chandrasekhar, V., Takacs, G., Chen, D., Tsai, S., Grzeszczuk, R., Girod, B.: CHoG: Compressed histogram of gradients A low bit-rate feature descriptor. In: Conference on Computer Vision and Pattern Recognition, CVPR 2009, Miami, FL (2009)
22. Lazebnik, S., Schmid, C., Ponce, J.: Beyond bags of features: Spatial pyramid matching for recognizing natural scene categories. In: Computer Vision and Pattern Recognition, CVPR 2006 (2006)
23. Bay, H., Ess, A., Tuytelaars, T., Gool, L.: SURF: Speeded Up Robust Features. In: Computer Vision and Image Understanding (CVIU), pp. 346–359 (2008)
24. Baatz, G., Köser, K., Chen, D., Grzeszczuk, R., Pollefeys, M.: Handling Urban Location Recognition as a 2D Homothetic Problem. In: Daniilidis, K., Maragos, P., Paragios, N. (eds.) ECCV 2010, Part VI. LNCS, vol. 6316, pp. 266–279. Springer, Heidelberg (2010)

Finding Round-Off Error Using Symbolic Execution*

Anh-Hoang Truong, Huy-Vu Tran, and Bao-Ngoc Nguyen

Abstract. Overflow and round-off error has been a research problem for decades. With the explosion of mobile and embedded devices, many software programs written for personal computer are now ported to run on embedded systems. The porting often requires changing floating-point numbers and operations to fixed-point ones and here round-off error between the two versions of the program occurs. We propose a novel approach that uses symbolic computation to produce a precise representation of the round-off error for a program. Then we can use solvers to find a global optimum of the round-off error symbolic expression as well as to generate test data that cause the worst round-off error. We implemented a tool based on the idea and our experimental results are better than recent related work.

1 Introduction

Round-off errors are the different between the real result and the approximation result computed using computer. The main reason that causes round-off error is the limitation of number representation in computing devices. Computing devices can be divided in two groups: ones with floating-point unit (FPU) and ones without it and they use floating-point and fixed-point number representation, respectively.

In reality, many software needs to run on both types of devices but because of the different arithmetic precisions between floating-point and fixed-point, a program may produce different results when executing with the two types of numbers. Sometimes the difference is significant and we want to find its threshold, or to make sure the difference is under a given threshold. In history, a big round-off error can make severe consequences such as the Patriot Missile Failure [3].

Anh-Hoang Truong · Huy-Vu Tran · Bao-Ngoc Nguyen
VNU University of Engineering and Technology, Hanoi, Vietnam

* This work was supported by the research Project QG.11.29, Vietnam National University, Hanoi.

V.-N. Huynh et al. (eds.), *Knowledge and Systems Engineering, Volume 1*, Advances in Intelligent Systems and Computing 244,
DOI: 10.1007/978-3-319-02741-8_35, © Springer International Publishing Switzerland 2014

Indeed there are three common types of round-off error: real number versus floating-point number, real number versus fixed-point number, and floating-point number versus fixed-point number. In this paper we address the last type of round-off error for two main reasons. Firstly, with the explosion of mobile and embedded devices, many applications developed for personal computer are now needed to run on these devices. Secondly, even with new applications, it would be impractical and time consuming to develop complex algorithms directly on embedded devices. So many complex algorithms are developed and tested in personal computer using floating-point before they are ported to embedded devices.

Our work was inspired by recent approaches to round-off error analysis [6, 7] that use various kinds of intervals to approximate round-off error. One of the drawbacks of the approaches is the imprecision in the round-off error approximation results. We aim to a precise analysis so we use symbolic execution [4] for the round-off error parts instead of their approximations. The output in our approach is a function of the input parameters that precisely represents the round-off error of the program. Then to find the maximum round-off error, we only need to find the optima of the function in the (floating-point) input domain. We rely on an external tool for this task. In particular we use Mathematica [12] for finding optima and the initial experimental results show that we can find biggest round-off error and at the same time we have the test data that cause the maximal round-off error. Compare to experiments in [6, 7] our maximal round-off errors are smaller.

Our main contributions in this paper are as follows:

- A novel approach using symbolic execution to produce precise functional representation of round-off error on input parameters for the round-off error between floating-point and fixed-point computation.
- A tool that can build the round-off error function automatically and then send to Mathematica to find maximal round-off error.

The rest of the paper is organized as follows. The next section contains some background. In Section 3 we define the symbolic execution that includes round-off error information and so that we can build a precise representation of the round-off error for a program. Then we present our implementation that uses Mathematica to find the maximal round-off error and some experimental results in the Section 4. Section 5 discusses related work other aspects that we do not cover in the previous part. Section 6 concludes and hints to our future work.

2 Background

IEEE 754 [1, 10] defines binary representations for 32-bit single-precision, floating-point numbers with three parts: the sign bit, the exponent, and the mantissa. The sign bit is 0 if the number is positive and 1 if the number is negative. The exponent is an 8-bit number that ranges in value from -126 to 127. The mantissa is the normalized binary representation of the number to be multiplied by 2 raised to the power defined by the exponent.

In fixed-point representation, a specific radix point (decimal point) written "." is chosen so there is a fixed number of bits to the right and a fixed number of bits to the left of the radix point. The later bits are called the integer bits. The former bits are called the fractional bits. For a base b with m integer bits and n fractional bits, we denote the format of the fixed-point by (b,m,n). When we use a base for fixed-point, we also assume the floating-point uses the same base. The default fixed-point format we use in this paper, if not specified, is $(2,11,4)$.

Example 0.1. Assume we use fixed-point format $(2,11,4)$ and we have a floating-point number 1001.010111. Then the corresponding fixed-point number is 1001.0101 and the round-off error is 0.000011.

Note that there are two types of lost bits in fixed-point computation: overflow error and round off error and we only consider the latter in this work as it is more difficult to detect.

3 Symbolic Round-Off Error

First we will present our idea of symbolic execution of round-off error for arithmetic expressions without division, which will be discussed in Section 5.1. Then we will extend to C programs, which will be simplified as a set of arithmetic expressions with constraints.

3.1 Round-Off Error for Arithmetic Operations

Let $\mathbb{L}$ be the set of all floating-point numbers and $\mathbb{I}$ be the set of all fixed-point numbers. $\mathbb{L}$ and $\mathbb{I}$ are finite because of the fixed number of bits are used for their representation. For practicality, we assume that the number of bits in fixed-point format is not more than significant bits in floating-point representation, which means we have $\mathbb{I} \subset \mathbb{L}$.

Let's assume that we are working with an arithmetic expression f of constants, variables, and addition, subtraction, and multiplication. (The division will be discussed later in Section 5.1.) Let $Dom(f)$ be the domain f. We denote f_l and f_i the floating-point and fixed-point versions of f, respectively. Note that not only the domain of of $Dom(f_l)$ and $Dom(f_i)$ are different, their arithmetic operations are also different. So we denote the operations in floating-point by $\{+_l, -_l, \times_l\}$ and in fixed-point by $\{+_i, -_i, \times_i\}$. f_l and f_i can be easily obtained from f by replacing arithmetic operations $+, -, \times$ by the corresponding ones.

Note that here we have several assumptions which based on the fixed-point function is not manually reprogrammed to optimize for fixed-point computation. First, the evaluations of f_l and f_i are the same. The scaling of variables in f_i is uniform so all values and variables use the same fixed-point format.

Because of the differences in floating-point and fixed-point representation, a value $x \subset \mathbb{L}$ usually needs to be rounded to become the corresponding value in $\mathbb{I}$. So we have non-decreasing monotonic function r from $\mathbb{L}$ to $\mathbb{I}$ and for $x \in \mathbb{L}$, $x -_l r(x)$ is called the conversion error. This error is in $\mathbb{L}$ because we assume $\mathbb{I} \subset \mathbb{L}$ (above). Note that we need to track this error as it will be used when we evaluate in floating-point, but not in fixed-point. In other words, the error is accumulated when we evaluating the expression in fixed-point computation.

An input vector $X = (x_1,..,x_n) \in Dom(f_l)$ need be rounded to be the input vector $r(X) \in Dom(f_i)$ of the fixed-point function where $r(X) = (r(x_1),..,r(x_n))$. The round-off error at X is $f_l(X) - f_i(r(X))$. Our research question is to find a global optimum of $f_l(X) - f_i(r(X))$ for all $X \in Dom(f_l)$:

$$\underset{X}{\text{argmax}}\ |f_l(X) - f_i(r(X))|, \text{ for } X \in Dom(f_l)$$

Alternatively, one may want to check if round-off error exceeds a given threshold θ:

$$\exists X \in Dom(f_l) \text{ such that } |f_l(X) - f_i(r(X))| \geq \theta$$

As we want to use the idea of symbolic execution to build a precise representation of round-off error, we need to track all errors when floating-point values (variables and constants) are rounded to the fixed-point ones, their accumulation during the evaluation of the expression, and also new error introduced because of the different between arithmetic operations – $\times_l$ and $\times_i$ in particular.

To track the error, we denote a floating-point x by (x_i,x_e) where $x_i = r(x)$ and $x_e = x -_l r(x)$. Note that x_e can be negative, depending on the rounding methods (example below). The arithmetic operations with symbolic round-off error between floating-point and fixed-point $\boxdot$ = $\{ +_s, -_s, \times_s\}$ are defined as follows. The main idea in all operations is to keep accumulation of error during computation.

Definition 0.1 (Arithmetics with round-off error)

$$(x_i,x_e) +_s (y_i,y_e) = (x_i +_l y_i, x_e +_l y_e)$$

$$(x_i,x_e) -_s (y_i,y_e) = (x_i -_l y_i, x_e -_l y_e)$$

$(x_i,x_e) \times_s (y_i,y_e) = (r(x_i \times_l y_i), x_e \times_l y_i +_l x_i \times_l y_e +_l x_e \times_l y_e +_l re(x_i,y_i))$

where $re(x_i,y_i) = (x_i \times_l y_i) -_l (x_i \times_i y_i)$ is the round-off error when we use the different multiplications.

Note that multiplication of two fixed-point numbers may cause round-off error so the round function r is needed in first part and $re(x_i,y_i)$ in the second part. The addition and subtraction may cause overflow error, but we do not consider it in this work. The accumulated error may not always increase, as in the below example.

Example 0.2 (Addition round-off error). For readability, let the fixed-point format be (10, 11, 4) and let $x = 1.312543$, $y = 2.124567$. With rounding to the nearest, we have $x = (x_i,x_e) = (1.3125, 0.000043)$ and $y = (y_i,y_e) = (2.1246, -0.000033)$. Apply the above definition with addition, we have:

$$(x_i,x_e)+_s(y_i,y_e)=$$
$$(1.3125+_l 2.1246, 0.000043+_l(-0.000033))=$$
$$(3.4371, 0.00001)$$

Example 0.3 (Multiplication round-off error). With x,y in Example 0.2, for multiplication, we have:

$$\begin{aligned}
&(x_i,x_e)\times_s(y_i,y_e)\\
&=(r(1.3125\times_l 2.1246), 0.000043\times_l 2.1246+_l\\
&\quad 1.3125\times_l(-0.000033)+_l 0.000043\times_l(-0.000033)\\
&\quad +_l re(1.3125, 2.1246))\\
&=(r(2.7885375), 0.000048043881+_l re(1.3125, 2.1246))\\
&=(2.7885, 0.000048043881+_l(1.3125\times_l 2.1246)\\
&\quad -_l(1.3125\times_l 2.1246))\\
&=(2.7885, 0.000048043881+_l 2.7885375-_l 2.7885)\\
&=(2.7885, 0.000085543881)
\end{aligned}$$

As we can see in Example 0.3, multiplication of two fixed-point numbers may cause round-off error, so the second part of the pair needs additional value $re()$. This value, like conversion error, is constrained by a range. We will examine this range in the next section.

3.2 Constraints

In Definition 0.1, we represent a number by two components so that we can later build symbolic representation for the round-off error (the second component). In this representation, the two components are constrained by some rules.

Let assume that our fixed-point representation uses m bits for integer part and n bits for fractional part and is in binary base. The first part x_i is constrained by its representation. So there exists $d_1,..,d_{m+n}$ such that

$$\left(x_i=\sum_{i=1}^{m+n} d_{m-i}2^{m-i}\right)\wedge d_j\in\{0,1\}$$

The $x_e = x -_l r(x)$ is constrained by the 'unit' of the fixed-point in base b is b^{-n} where unit is the absolute value between a fixed-point number and its successor. With rounding to the nearest, this constraint is half of the unit:

$$|x_e|\leq b^{-n}/2$$

Like x_e, the $re()$ part also has the same constraint but we do not need it as our implementation covers it automatically.

3.3 Symbolic Round-Off Error for Expressions

In normal symbolic execution, an input parameter will be represented by a symbol. However in our approach, it will be represented by a pair of two symbols with the above corresponding constraints.

As we are working with arithmetic expression, the symbolic execution will be proceeded by first replacing variables in the expression with a pair of symbols, and applying arithmetic expression according to Definition 0.1. The final result will be a pair of symbolic fixed-point result and symbolic round-off error. The later part will be the one we need for the next step - finding its global optima. Before that we discuss about building symbolic round-off error for C programs.

3.4 Symbolic Round-Off Error for Programs

Following [7], we simplify our problem definition as follows. Given a function in C programming language with initial ranges of its input parameters, a fixed-point format and a threshold θ, check if there is an instance of input parameters that causes the difference between the results of the function computed in floating-point and fixed-point above the threshold. Similar to the related work we restrict here the function without unknown loops. That means the program has a finite number of execution paths and all paths terminate.

Fig 1 is the example from [7] that we will use to illustrate our approach for the rest of the paper. In this program we use special comments to specify the fixed-point format, the threshold, and the input ranges of parameters.

```
/*
format: (2, 11, 4); threshold: 0.26; x: [-1, 3]; y: [-10, 10];
*/
typedef float Real;
Real rst;
Real maintest(Real x, Real y) {
   if(x > 0) rst = x*x;
   else rst = 3*x
   rst -= y;
   return rst;
}
```

Fig. 1 An example program

We know that a program in general can be executed in different paths depending on their input values. For each of the possible execution paths we can use normal symbolic execution to build an expression reflecting that path together with constraints (called path condition) of variables for that path. Now we can apply the approach in Section 3.1 combining with the path condition to produce symbolic round-off error for the selected path. For small programs, we can combine all the symbolic expressions for all paths into a single expression.

3.5 Usage of Symbolic Expression of Round-Off Error

Having the symbolic expression of the round-off error for a given program, we can use it for various purposes. To find the largest round-off error or the existence of an error above a given threshold we can use external SMT solvers. We will show this in the next section.

As a function the symbolic expression of the round-off error can also tell us various information about the properties of the round-off error, such as the importance of variables in error contribution. It can be used to compute other round-off error metrics [8] such as the frequency/density of error above some threshold by, for example, testing.

4 Implementation and Experiments

4.1 Implementation

We have implemented the tool in Ocaml [9] and use Mathematica for finding optima. The tool assumes that by symbolic execution the C program for each path in the program we already have an arithmetic expression with constraints (initial input ranges and path conditions) of variables in the expression. The tool takes each of these expressions and its constraints and processes in the following steps:

1. Parse expression and generate an expression tree in Ocaml. We use Aurochs[2] parser generator for this purpose.
2. Perform symbolic execution on the expression tree with $+, -, \times$ to produce round-off error symbolic expression together with constraints of variables. In this step we convert the expression tree from the previous step to a post-order expression.
3. Simplify the symbolic round-off error and constraints of variables using the function `Simplify`[*exp*] of Mathematica. Note that the constants and coefficients in the input expression are also splitted into two parts, the fixed-point part and the round-off error part, both are constants. When the round-off error part is not zero, the simplification step can make the symbolic expression of the round-off error more compact.
4. Use Mathematica to find optimum of the round-off error symbolic expression with constraints of variables. We use function `NMaximize`[{*f*, *constraints*}, {*variables*}] for this purpose. Since Mathematica does not support fixed-point we need to developed a module in Mathematica for converting floating-point numbers to fixed-point numbers and

[2] `http://lambda-diode.com/software/aurochs`

for emulating fixed-point multiplication, see Algorithm 6 and 7 (round to the nearest) in Appendix 6.

4.2 *Experimental Results*

For comparison, we use two examples in [6] as shown in Fig 1 and the polynomial of degree 5 in a 56 bits floating-point computer. Then we experimented with a Taylor series of sine function to see how far we can go with this approach. We tried to make constant coefficients in following examples representable in fixed point so they don't cause any round-off error to simplify error function for testing purpose.

4.2.1 Experiment with Simple Program

For the program in Fig 1, it is easy to compute its symbolic expression for the two possible runs: $(x > 0 \wedge x \times x - y)$ and $(x < 0 \wedge 3 \times x - y)$. We consider the first one. Combine with input range of $x \in [0,3]$ we get $x > 0 \wedge -1 \le x \le 3$ which can be simplified to $0 < x \le 3$. So we need to find the round-off error symbolic expression for $x \times x - y$ where $0 < x \le 3$ and $-10 \le y \le 10$.

Apply the symbolic execution with round-off error part, we get the symbolic expression for round-off error:

$$(x_i \times_l x_i) -_l (x_i \times_i x_i) +_l 2 \times_l x_i \times_l x_e +_l x_e^2 -_l y_e$$

and the constraints of variables (with round to the nearest) are

$$x_i = \Sigma_{j=1}^{15} d_j \, 2^{11-j} \wedge d_j \in \{0,1\} \wedge -1 \le x_i \le 3 \wedge x_i \ge 0 \wedge$$
$$-0.03125 \le x_e, y_e \le 0.03125$$

Next we convert the round-off error symbolic expression and constraints to Mathematica syntax. The result is in Fig 2.

```
boundary = 2^(-5);

NMaximize[{xi^2 + 2*xi*xe + xe^2 - ye - iMul[xi, xi],
Join[{xi == Sum[(Symbol["x"<> ToString[i]])*(2^(11 - i)), {i, 1, 15}]},
Table[0 <= (Symbol["x"<> ToString[i]]) <= 1 && Element[(Symbol["x"<> ToString[i]]),
Integers], {i, 1, 15}], {yi == Sum[(Symbol["y"<> ToString[i]])*(2^(11 - i)), {i, 1, 15}]},
Table[0 <= (Symbol["y"<> ToString[i]]) <= 1 && Element[(Symbol["y"<> ToString[i]]),
Integers], {i, 1, 15}], {-boundary <= ye <= boundary, -boundary <= xe <= boundary,
-10 <= yi+ye <= 10, 0 <= xi+xe <= 3, 0 <= xi <= 3, -10 <= yi <= 10}]},
(Join[{ye,xe,xi,yi}, Table[Symbol["x" <> ToString[i]], {i, 1, 15}],
Table[Symbol["y" <> ToString[i]], {i, 1, 15}]])]
```

Fig. 2 Mathematica problem for example in Fig 1

Mathematica found the following optima for the problem in Fig 2:

- With round to the nearest, the maximal error between floating-point and fixed-point computation of the program is 0.2275390625. The inputs that cause the maximal round-off error are: $x_i = 2.875; x_e = 0.03125$ so $x = x_i +_l x_e = 2.90625$ and $y_i = 4.125; y_e = -0.03125$ so $y = y_i +_l y_e = 4.09375$.

- With round towards -∞ (IEEE 754[1]): $rst = 0.4531245939250891$ with $x_i = 2.8125$; $x_e = \sum_{j=-5}^{-24} 2^j \rightarrow x = x_i +_l x_e = 2.874999940395355225$ and $y_i = 4$; $y_e = -\sum_{j=-5}^{-24} 2^j \rightarrow y = y_i +_l y_e = 3.937500059604644775$.

Compare to [6] using round to the nearest the error is in $[-0.250976, 0.250976]$ so our result is more precise.

To verify our result, we wrote a test program for both rounding methods that generates 100.000.000 random test cases for $-1 \leq x \leq 3$ and $-10 \leq y \leq 10$ and directly compute the round-off error between floating-point and fixed-point results. Some of the largest round-off error results are in Table 1 and Table 2. The tests were ran many times but we did not find any inputs that cause larger round-off error than our approach.

Table 1 Top round-off errors in 100.000.000 tests with round to the nearest

No.	x	y	err
1	2.9061846595979763	-6.530830752525674	0.22674002820827965
2	2.9061341245904635	-4.4061385404330045	0.2267540905421832
3	2.9062223934725107	-3.2184902596952947	0.22711886001638248

Table 2 Top round-off errors in 100.000.000 tests with round towards $-\infty$

No.	x	y	err
1	2.8749694304357103	-0.874361299827422	0.4523105257672544
2	2.874964795521085	-5.4371633555888055	0.45258593107439893
3	2.8749437460462444	-1.249785289663956	0.4525868325943687

4.2.2 Experiment with a Polynomial of Degree 5

Our second experiment is a polynomial of degree 5 from [6]:

$$P5(x) = 1 - x + 3x^2 - 2x^3 + x^4 - 5x^5$$

where fixed-point format is $(2, 11, 8)$ and $x \in [0, 0.2]$.

After symbolic execution, the symbolic round-off error is:

$$0. +_l 3 \times_l x_i^2 -_l 2 \times_l x_i^3 -_l 5 \times_l x_i^5 -_l x_e +_l 6 \times_l x_i^2 \times_l x_e +_l 4 \times_l x_i^3 \times_l x_e -_l 25 \times_l x_i^4 \times_l x_e +_l 3 \times_l x_e^2 -_l 6 \times_l x_i \times_l x_e^2 +_l 6 \times_l x_i^2 \times_l x_e^2 -_l 50 \times_l x_i^3 \times_l x_e^2 -_l 2 \times_l x_e^3 +_l 4 \times_l x_i \times_l x_e^3 -_l 50 \times_l x_i^2 \times_l x_e^3 +_l x_e^4 -_l 25 \times_l x_i \times_l x_e^4 -_l 5 \times_l x_e^5 -_l 3 \times_l iMul[x_i, x_i] +_l 2 \times_l iMul[iMul[x_i, x_l], x_i] -_l iMul[iMul[iMul[x_i, x_i], x_i], x_l] +_l 5 \times iMul[iMul[iMul[iMul[x_i, x_i], x_i], x_l], x_i]$$

and the constraints of variables with round to the nearest are

$$x_i = \Sigma_{j=1}^{19} d_j\, 2^{11-j} \bigwedge d_j \in \{0,1\} \bigwedge 0 \le x_i \le 0.2$$
$$\bigwedge -0.001953125 \le x_e \le 0.001953125 \bigwedge 0 \le x_i +_l x_e \le 0.2$$

For this problem, Mathematica found the maximum 0.007244555 with $x_i = 0.12890625; x_e = -0.001953125$ so $x = x_i +_l x_e = 0.126953125$. In [6], their error is in $[-0.01909,\ 0.01909]$ when using round to nearest, so our result is better.

We verify our results with round to the nearest by directly compute the difference between the fixed-point and floating-point with 100.000.000 random test cases for $0 \le x \le 0.2$. The lagest error we found is 0.00715773548755 which is very close, but still under our bound.

4.2.3 Experiment with Taylor Series of Sine Function

The last experiment we want to see how far we can go with our approach, so we use Taylor series of sine function.

$$P7(x) = x - 0.166667x^3 + 0.00833333x^5 - 0.000198413x^7$$

where fixed-point format is $(2,11,8)$ and $x \in [0,0.2]$.

The largest round-off error is 0.00195312 with $x_i = 0; x_e = 0.00195313$ so $x = x_i +_l x_e = 0.00195313$. Compare to [6] using round to the nearest the error is in $[-0.00647,\ 0.00647]$ so our result is much better. We tried with longer Taylor series but Mathematica cannot solve the generated problem.

5 Discussion and Related Work

5.1 Discussion

We have shown our approach using symbolic execution to build a functional representation that precisely reflects the round-off error between floating-point and fixed-point versions of a C program. Finding global optima of the representation allows us to find the inputs that cause the maximum round-off error. Some advantages and disadvantages of this approach are discussed below.

First, our approach assumes that Mathematica does not over approximate the optima. However, even if the optima is over approximated, the point that cause the optima is still likely the test case we need to find and we can recompute the actual round-off error at this point.

Second, it is easy to see that our approach may not be scalable for more complex programs. The round-off error representation will grow very big with complex programs. We need to do more experiments to see how far can we go. Some simplification strategy may be needed such as sorting the contribution to the round-off error of components in the round-off error expression and removing components that is complex but contributes insignificant to the error. Or we can divide the expression into multiple independent parts to send smaller problems to Mathematica.

Third, if a threshold is given, we can combine testing to find a solution for the satisfiability problem. This approach has been useful

Finally, we can combine our approach with interval analyses so the interval analyses will be used for complex parts of the program while in other parts we can find their precise round-off errors. We will investigate this possibility in our future work.

Note that the largest round-off error is only one of the metrics for the preciseness of the fixed-point function versus its floating-point one. In our previous work [8], we proposed several metrics. Our approach in this work can be used to calculate other metrics since the round-off error function contains very rich information about the nature of round-off error.

5.2 *Division*

We have ignored the division in the above sections to ease the illustration of our idea. This $\div_s$ operator can be defined similar to multiplication as follows:

Definition 0.2. Division with round-off error

$$(x_i, x_e) \div_s (y_i, y_e) = (r(x_i, y_i),\ (x_i +_l x_e) \div_l (y_i +_l y_e) -_l (x_i \div_l y_i) +_l de(x_i, y_i))$$

where $de(x_i, y_i) = (x_i \div_l y_i) -_l (x_i \div_i y_i)$ is the round-off error when we use the different divisions.

The division can be implemented similar to multiplication. We leave this evaluation for future work.

5.3 *Related Works*

Overflow and round-off error analysis has been studied from the early days of computer science because we all know not only fixed-point but also floating-point number representation and computation have the problem. Most work address both overflow and round-off error, for example [10, 1] but since round-off error are more subtle and sophisticated, we focus on it in this work but our idea can be extended to the overflow error.

In [5] the authors list three kinds of overflow and round-off errors: real numbers versus floating-point, real numbers versus fixed-point, and floating-point numbers versus fixed-point numbers. We focus on the last one in this work so we will only discuss the related work of the last one.

The most recent work that we are aware of is of Ngoc and Ogawa [7, 6]. The authors develop a tool called CANA for analyzing overflows and round off errors. They propose a new interval called extended affine interval (EAI) to estimate round-off error ranges instead of classical interval [2] and affine interval [11]. EAI avoids the problem of introducing new noise symbols of AI but it is still imprecise as our approach.

6 Conclusions

We have proposed a method that can find the largest round-off error between the floating-point function and its fixed-point version. Our approach is based on symbolic execution extended for the round-off error so that we can build a precise representation of the round-off error. It allows us to get a precise maximal round-off error together with the test case for the worst error. We also built a tool that uses Mathematica to find the optimum of the round-off error function. The initial experimental results are very promising.

We plan to do more experiments with complex programs to see how far we can go with this approach, because like other systems that are based on symbolic execution we depend on external solvers. We also investigate possibilities to reduce the complexity of round-off error function before sending to the solver by approximating the symbolic round-off error. For real world programs, especially the ones with loops, we believe that combining interval analysis with our approach may allow us to balance between preciseness and scalability. For example, the parts of the program without loops and a lot of multiplication/division will be estimated by our approach, and we range approximation techniques for the remaining parts.

References

1. Goldberg, D.: What Every Computer Scientist Should Know About Floating-Point Arithmetic. ACM Computing Surveys, 5–48 (1991)
2. Goldsztejn, A., Daney, D., Rueher, M., Taillibert, P.: Modal intervals revisited: a mean-value extension to generalized intervals. In: Proc. of the 1st International Workshop on Quantification in Constraint Programming (2005)
3. Higham, N.J.: Accuracy and Stability of Numerical Algorithms. SIAM: Society for Industrial and Applied Mathematics (2002)
4. King, J.C., Watson, J.: Symbolic Execution and Program Testing. Communications of the ACM, 385–394 (1976)
5. Martel, M.: Semantics of roundoff error propagation in finite precision calculations. In: Higher-Order and Symbolic Computation, pp. 7–30 (2006)
6. Ngoc, D.T.B., Ogawa, M.: Overflow and Roundoff Error Analysis via Model Checking. In: Conference on Software Engineering and Formal Methods, pp. 105–114 (2009)
7. Ngoc, D.T.B., Ogawa, M.: Combining Testing and Static Analysis to Overflow and Roundoff Error Detection. In: JAIST Research Reports, pp. 105–114 (2010)
8. Pham, T.-H., Truong, A.-H., Chin, W.-N., Aoshima, T.: Test Case Generation for Adequacy of Floating-point to Fixed-point Conversion. In: Proceedings of the 3rd International Workshop on Harnessing Theories for Tool Support in Software (TTSS). ENTCS, vol. 266, pp. 49–61 (2010)
9. Smith, J.B.: Practical OCaml (Practical). Apress, Berkely (2006)
10. Stallings, W.: Computer Organization and Architecture. Macmillan Publishing Company (2000)
11. Stolfi, J., de Figueiredo, L.H.: An introduction to affine arithmetic. In: Tendencias em Matematica Aplicada e Computacional (2005)
12. Wolfram, S.: Mathematica: A System for Doing Mathematics by Computer. Addison-Wesley (1991)

Appendix

Multiply in fixed-point on Mathematica with round to the nearest is defined in Algorithm 7. Assume that the fixed-point number has fp bits to represent fractional part. The inputs of multiplication are two floating-point numbers a and b and we need to produce the result in fixed-point.

Firstly, we shift left each number by fp bits to get two integer numbers. Then we take their product and shift right $2 * fp$ bits to receive the raw result without rounding. With round to the nearest, we shift right the product fp bits, store in *i_mul_shr_fp*, and take the integer and fractional part of *i_mul_shr_fp*. If the fractional part of *i_mul_shr_fp* is larger than 0.5 then integer part of *i_mul_shr_fp* is increased by 1. We store the result after rounding in *i_mul_rounded*. Shifting left *i_mul_rounded* fp bits produce the result of the multiplication in fixed-point.

Algorithm 6: Converting floating-point to fixed-point

Input : x: floating number
Output: fixed-point number of x
Data : bin_x: binary of x
Data : fp: the width of fractional part
Data : x_1: result of shifting bits left bin_x
Data : *ip_number*: integer part of number x_1
Data : *digit* : the value of nth *bit*
Data : *fixed_number*: the result of converting

Procedure `convertToFixed` (x, fp);
begin
    Convert a floating-point x to binary numbers bin_x;
    x_1 = Shift left bin_x by fp bits;
    ip_number = Integer part of x_1;
    Take the $(fp+1)^{th}$ bit of bin_x as *digit*;
    if *digit equals 1* **then**
        if $x > 0$ **then**
            ip_number = *ip_number* $+_l$ 1;
        else
            ip_number = *ip_number* $-_l$ 1;
    else
        ip_number = *ip_number*;
    fixed_number = Shift right *ip_number* by fp bits;
    return *fixed_number*

Algorithm 7: Fixed-point multiplication emulation in Mathematica

Input : a: floating number
Input : b: floating number
Output: multiplication of $a \times_i b$
Data : *a_shl_fp*: result after shifting left *fp* bits of a
Data : *i_a_shl_fp*: integer part of *a_shl_fp*
Data : *b_shl_fp*: result after shifting left *fp* bits of b
Data : $i\_b_shl\_fp$: integer part of *b_shl_fp*
Data : *i_mul*: multiplication of two integer *i_a_shl_fp* and *i_b_shl_fp*
Data : *i_mul_shr_fp*: result after shifting right *fp* bits of *i_mul*
Data : *ipart_i_mul* : integer part of *i_mul_shr_fp*
Data : *fpart_i_mul* : fraction part of *i_mul_shr_fp*
Data : *truncate_part*: result after shifting left 1 bit of *fpart_i_mul*
Data : *round_part*: integer part of *truncate_part*
Data : *i_mul_rounded*: result after rounding
Data : *result*: product in fixed-point

`Procedure iMul` (a, b);
begin
Convert a floating-point x to binary numbers bin_x;
a_shl_fp = Shift left a by *fp* bits;
i_a_shl_fp = Integer part of *a_shl_fp*;
b_shl_fp = Shift left b by *fp* bits;
i_b_shl_fp = Integer part of *b_shl_fp*;
i_mul = multiply two integer *i_a_shl_fp* and *i_b_shl_fp*;
i_mul_shr_fp = Shift right *i_mul* by *fp* bits
and then take integer and fraction part of i_mul is that *ipart_i_mul* and *fpart_i_mul*;
truncate_part = Shift left 1 bit *fpart_i_mul*;
round_part = Take Integer part of *truncate_part*;
i_mul_rounded = *ipart_i_mul* + *round_part*; *with rounding to the nearest*
result = Shift right *i_mul_rounded* by *fp* bits;
return *result*

Author Index